Nuclear Safety: A Human Factors Perspective

Nuclear Safety: A Human Factors Perspective

EDITORS:

JYUJI MISUMI
Institute of Social Research
Institute of Nuclear Safety System Inc.

BERNHARD WILPERT
Institute of Psychology – Research Center Systems Safety
Berlin University of Technology

RAINER MILLER
Institute of Psychology – Research Center Systems Safety
Berlin University of Technology

TAYLOR & FRANCIS
ALERE FLAMMAM
1798 – 1998

UK Taylor & Francis Ltd, 1 Gunpowder Square, London, EC4A 3DE
USA Taylor & Francis Inc., 325 Chestnut Street, 8th Floor, Philadelphia, PA 19106

British Library Cataloguing-in-Publication Data
A catalogue record for this book is available from the British Library
ISBN 0–7484–0818–5 (cased)

Library of Congress Cataloging in Publication Data are available

Cover design by Jim Wilkie
Typeset in Times 10/12pt by Graphicraft Limited, Hong Kong
Printed by T.J. International Ltd, Padstow, UK
Cover printed by Flexiprint, Lancing, West Sussex

Contents

List of contributors

Miklós Antalovits is Professor of Work Psychology in the Department of Psychology and Ergonomics at the Technical University of Budapest. He holds a degree in engineering from the Technical University of Budapest and in psychology from Eötvös Loránd University in Budapest. He received his PhD in 1986. His major research interests are the impacts of information technology on work, skill development through the use of simulators, and the development of computer-aided teaching aids and assessment methods especially for operators' training in the process industry.

Address: Technical University of Budapest, Department of Ergonomics and Psychology H-1111 Budapest, Egry J. u. 1. E. III, Hungary
e-mail: antalovits@erg.bme.hu

George Apostolakis, Ph.D. is a professor of nuclear engineering at the Massachusetts Institute of Technology. He is also affiliated with the Institute's Operations Research Center and the MIT Program for Environmental Engineering Education and Research. He is Editor-in-Chief of the international journal *Reliability Engineering and System Safety* and founder of the International Conferences on Probabilistic Safety Assessment on Reactor Safeguards of the US Nuclear Regulatory Commission. His current research areas include risk-informed and performance-based regulation, environmental risk management involving multiple stakeholders, the influence of organisational factors on safety, and software dependability.

Address: Dept. of Nuclear Engineering, Rm. 24–221, Massachusetts Institute of Technology, 77 Massachusetts Avenue, Cambridge, MA 02139–4307, USA
e-mail: apostola@mit.edu

Robert Baggen has a degree in psychology and started his scientific career with research on mental workload, video display technology, and software ergonomics at the University of Wuppertal. Since 1995 he has been a member of the Research

Center Systems Safety and teaches courses in work and organisational psychology at the Berlin University of Technology. Current research interests are the development of computer support tools for event analysis and software ergonomic evaluation methods.
Address: Institute of Psychology, Research Center Systems Safety (FSS), Berlin University of Technology, FR 3–8, Franklinstr. 28, D-10587 Berlin, Germany
e-mail: robert.baggen@tu-berlin.de

Gerhard Becker has a degree in mechanical engineering and gained his experience in safety analysis by studying nuclear accident scenarios for various nuclear power plants in Germany and by participating in the international reactor safety experiments in Marviken, Sweden. The influence of human behaviour on the safety of complex technical systems has been the main focus of his work over the last 20 years. Based on comprehensive experience gained as head of an interdisciplinary team concerned with the ergonomic design of technical systems, which is seen as the precondition for the safe performance of human operators, his recent main interest is directed to the relevance of organisation and management for the safety of industrial technology. He is currently project manager at the Institute for Nuclear Technology and Radiation Protection (IKS) of TÜV Rheinland. The results of his work have been documented in several research reports, studies, and publications.
Address: TÜV Rheinland e.V., Institute for Nuclear Technology and Radiation Protection (IKS), Am Grauen Stein, D-51105 Köln, Germany
e-mail: nc-tuevrik@mail.netcologne.de

Dr. Mathilde Bourrier earned her PhD in sociology at the Institut d'Etudes Politiques de Paris in December 1996. During her dissertation she worked with the Centre de Sociologie des Organisations in Paris, founded by French sociologist Michel Crozier, and was a visiting fellow at the High Reliability Organisations Group (University of California at Berkeley). She was recently appointed assistant Professor at the University of Technology at Compiègne and works as a consultant for the French Atomic Energy Commission. Her current research interests focus on organisational reliability.
Address: University of Technology at Compiègne, Department of Technology and Human Sciences, Center Pierre Guillaumat, B.P. 649, 60206 Compiègne, Cédex, France
e-mail: mathilde.bourrier@utc.fr

Louis C. Buffardi, PhD, is a professor at George Mason University, where he has been Director of the Applied Experimental Psychology Program and Director of the doctoral program in psychology. He has published his research in journals such as *Science*, *Quarterly Journal of Experimental Psychology*, and *American Psychologist*. He was co-director (with Dr. Fleishman) of the Project on Human Error Probability Prediction sponsored by the US Nuclear Regulatory Commission.
Address: George Mason University, Fairfax, VA 22030, USA
Fax: +1 (703) 993–1359

Babette Fahlbruch holds a degree in psychology from the Berlin University of Technology. She has been a member of the Institute of Psychology's Research Center Systems Safety since 1992 and an assistant lecturer since 1995. She is working on her doctoral thesis with the provisional title of 'Causal cognition and accident analysis'. Her interest is in safety research.
Address: Institute of Psychology, Research Center Systems Safety (FSS), Berlin University of Technology, FR 3–8, Franklinstr. 28, D-10587 Berlin, Germany
e-mail: babette.fahlbruch@tu-berlin.de

Edwin A. Fleishman, PhD is Distinguished University Professor and Director of the Center for Behavioral and Cognitive Studies at George Mason University. He is former Editor of the *Journal of Applied Psychology* and was the recipient of the Distinguished Scientific Award for the Applications of Psychology from the American Psychological Association (APA). He was President of APA's Division of Engineering Psychology, the Division of Evaluation, Measurement, and Statistics, and the Society for Industrial and Organizational Psychology. He was also the President of the International Association of Applied Psychology.
Address: 11304 Spur Wheel Lane, Potomac, MD 20854, USA
e-mail: mrieaf@aol.com

Andreas Gans holds a degree in economics. He has been a member of the Research Center Systems Safety for several years.
Address: Institute of Psychology, Research Center Systems Safety (FSS), Berlin University of Technology, FR 3–8, Franklinstr. 28, D-10587 Berlin, Germany

Frank Guldenmund has a degree from the University of Leiden in cognitive sciences with a specialisation in methodology and has worked at the Safety Science group in Delft since 1991. His research centres around safety management and safety culture in the process industry and the health service. He acts as adviser on research methodology to the group.
Address: Frank Guldenmund, Lecturer in Safety Science, Delft University of Technology, Kanaalweg 2b, 2628EB Delft, Netherlands.
e-mail: frank.guldenmund@wtm.tudelft.nl

Andrew Hale is a British occupational psychologist by background, but has worked in the Netherlands since 1984. His interests have developed from human factors, accident analysis and safety training to his current research concerns of safety management, regulation and certification. He has set up and/or run courses for safety professionals at undergraduate, post-graduate and post-experience levels in the UK, Algeria and the Netherlands. He chairs, or is a member of a number of scientific advisory groups nationally and internationally in Europe and chairs the Dutch Expert Committee on Certification of Health and Safety Management Systems.
Address: Andrew Hale, Professor of Safety Science, Delft University of Technology, Kanaalweg 2b, 2628EB Delft, Netherlands.
e-mail: a.r.hale@wtm.tudelft.nl

Dr. Marin Ignatov is senior research fellow at the Bulgarian Academy of Sciences in Sofia. He majored in work and organisational psychology at the University of Technology, Dresden, Germany. He was organisational consultant in several international East-European enterprise-restructuring projects (PHARE, Carl-Duisberg-Gesellschaft). He also published on cross-cultural communication. He was visiting professor at the universities of Bamberg, Frankfurt, and Berlin, Germany; Salzburg University, Austria; Wesleyan University, Connecticut, USA; as well as in the Cranfield School of Management, England. At present he is working within the research group of Professor Wilpert in Berlin on an international project concerning implicit behavioural safety rules.
Address: TU Berlin, Institute of Psychology, Franklinstr. 28, FR 3–8, D-10587 Berlin, Germany
e-mail: marin.ignatov@tu-berlin.de

Katsuji Isobe holds a bachelor's degree in cognitive psychology from CHIBA University. He worked at the Hokuso Electrical Substation of the Tokyo Electric Power Company (TEPCO) from where he was transferred to the Human Factors Dept. in the Nuclear Power R&D Center of TEPCO. In 1992 he was a visiting researcher at the Center for the Study of Language and Information (CSLI) at Stanford University, USA, for one year. He currently works as a staff researcher at the Human Factors Dept. in the Nuclear Power R&D Center of TEPCO.
Address: Human Factors Department, Nuclear Power R&D Center, Tokyo Electric Power Co., Inc., Egasaki 4–1, Tsurumiku, Yokohama, 230 Japan.
e-mail: isobe@rd.tepco.co.jp

Lajos Izsó is an associate professor in the Department of Psychology and Ergonomics at the Technical University of Budapest. He graduated first in chemistry and later received a teacher's degree and a degree in psychology. His doctoral dissertation dealt with factors influencing human reliability in human–computer systems. Dr. Izsó's major professional interests are safety and human reliability and the ergonomic design and evaluation of human–computer interfaces.
Address: Technical University of Budapest, Department of Ergonomics and Psychology, H-1111 Budapest, Egry J. u. 1. E. III, Hungary
e-mail: izsolajos@erg.bme.hu

When this paper was written Barry Kirwan was a lecturer in ergonomics at the University of Birmingham, UK, and on a sabbatical at the Delft University of Technology, Netherlands. He has worked extensively both in universities, in consulting organisations and in the nuclear industry (at British Nuclear Fuels) as a researcher and practitioner in human reliability, human factors and safety, with a concentration more recently on safety management modelling. He has written and edited books on task analysis and human reliability techniques. He took up his current post in 1996 running a human factors research and development group for air traffic control.

Address: Barry Kirwan, Head of Human Factors, ATMDC, National Air Traffic Service, Bournemouth Airport, Christchurch, Dorset BH23 6DF, UK.
e-mail: bkirwan@atmdc.nats.co.uk

Mitsuhiro Kojima, PhD, is a senior researcher at the Human Factors Research Center of the Central Research Institute for Electric Power Industries (CRIEPI). He studied physical and cultural anthropology in graduate school and is interested in the ethnology of the electric communication such as the Internet.
Address: Central Research Institute of the Electric Power Industry, Human Factors Research Center, 2–11–1, Iwado Kita, Komae-shi, Tokyo 201, Japan
e-mail: kojima@criepi.denken.or.jp

Fumio Kotani holds a degree in marine engineering and joined the Kansai Electric Power Company in 1969. After working mainly in the maintenance section of nuclear power plants and at the Institute of Human Factors at the Nuclear Power Engineering Corporation, he became the general manager of the human factors project at the Institute of Nuclear Safety System.
Address: Institute of Nuclear Safety System, 64 sata, Mihama-cho, Mikata-gun, Fukui, 919–1205, Japan
e-mail: kotani@inss001.inss.co.jp

Dr. Meshkati is an Associate Professor of Civil/Environmental Engineering and an Associate Professor of Industrial and Systems Engineering at the University of Southern California (USC). He is also the Director of School of Engineering Continuing Education programmes. He is a Certified Professional Ergonomist (CPE #650) and an Adjunct Scholar at the Department of Human Work Sciences and the Center for Ergonomics of Developing Countries, Lulea University of Technology in Sweden. Prior to joining USC in 1985, he was Project Director for Decision Dynamics Corporation, Research Engineer at SKC Research Inc., and lectured at the Art Center College of Design in Pasadena, California.
Address: Dept. of Civil/Environmental Engineering, Dept. of Industrial & Systems Engineering, University of Southern California, Los Angeles, CA 90090–0021, USA
e-mail: meshkati@usc.edu

Rainer Miller holds a degree in psychology from the Berlin University of Technology. He is a member of the Research Center Systems Safety and has worked as a research assistant in several projects on the improvement of human factors, incident analysis and incident reporting. In 1996 Mr. Miller was a visiting researcher at the Institute of Social Research of the Institute of Nuclear Safety System, Inc., Kyoto, Japan.
Address: Institute of Psychology, Research Center Systems Safety (FSS), Berlin University of Technology, FR 3-8, Franklinstr. 28, D-10587 Berlin, Germany
e-mail: rainer.miller@tu-berlin.de

Juyji Misumi, Doctor of Literature, and recipient of the Kurt Lewin Memorial Award in 1994 has held numerous professorships since 1963. At present he is Director of the Japan Institute for Group Dynamics and Director of the Institute of Social Research at the Institute of Nuclear Safety System, Inc.

Address: Director, Japan Institute for Group Dynamics, 14F Nishinippon Shimbun Kaikan, 1–4–1, Tenjin, Chuo-ku, Fukuoka 810, Japan
Fax: +81 96–713–1309

Neville Moray took a degree in philosophy, psychology and physiology at Oxford University, and has worked in universities in the UK, Canada, the USA and France, both in departments of psychology and in departments of engineering. He has worked for two human factors and ergonomics consulting companies. He is currently the DERA Professor of Applied Cognitive Psychology at the University of Surrey at Guildford, UK. He has worked on reports on the human factors of nuclear safety, on ergonomics needs in the coming decade, on the design of complex systems and on human error. His current interests are the role of humans in highly automated systems, and in the contribution of ergonomics to global problems of the 21st century.

Address: Department of Psychology, University of Surrey, Guildford, Surrey GU2 5XH, UK
e-mail: N.Moray@surrey.ac.uk

Yoshimasa Nishijima is Deputy Director of the Human Factors Research Center, CRIEPI. He holds a bachelor's degree and a master's degree in mechanical engineering from Keio University.

Address: Human Factors Research Center, CRIEPI, 2–11–1, Iwadokita, Komae-shi, Tokyo 201, Japan
e-mail: nisijima@criepi.denken.or.jp

James Reason has been Professor of Psychology at the University of Manchester since 1977, from where he graduated in 1962. He obtained his PhD from the University of Leicester. He is the author of *Human Error* (1990), and *Managing the Risks of Organisational Accidents* (1997).

Address: Department of Psychology, University of Manchester, Manchester M13 9PL
e-mail: reason@hera.psy.man.ac.uk

Alex Regenass is a research assistant. His special interests are computer training, software ergonomics, errors, and accident prevention.

Address: University of Bern, Department of Psychology, Muesmattstr. 45, 3000 Bern 9, Switzerland
e-mail: regenass@psy.unibe.ch

Gene Rochlin is Professor of Energy and Resources at the University of California, Berkeley, where he teaches the politics and sociology of energy and environmental issues. His research focuses on social, political, and organisational dimensions of scientific and technical decision-making and socio-technical systems, including arms control, organisational behaviour, and social and societal interactions. His recent book *Trapped in the net: The unanticipated consequences of computerisation* (Princeton: Princeton University Press, 1997) deals with the long-term and indirect effects of advanced computers and networks.

Address: Energy and Resources Group, 310 Barrows Hall, University of California, Berkeley, CA 94720–3050, USA
e-mail: armsis@socrates.berkeley.edu

Kunihide Sasou graduated from Keio University with a Masters in engineering in 1989 and joined the Central Research Institute of Electric Power Industry (CRIEPI) the following year. In 1996 he was at the University of Manchester as a visiting research associate. He is interested in team behaviour under abnormal operating conditions at nuclear power plants, particularly methods for analysis of team behaviour, the simulation of team behaviour, the study of team errors, and the education of operators.

Address: Human Factors Research Center, Central Research Institute of Electric Power Industry, 2–11–1, Iwato-kita, Komae-shi, Tokyo 201, Japan
e-mail: sasou@criepi.denken.or.jp

Norbert Semmer, PhD, is Professor of Work and Organisational Psychology. His special interests are stress and health, errors, accident prevention, training, group processes, and the regulation of self-esteem at work.

Address: University of Bern, Department of Psychology, Muesmattstr. 45, 3000 Bern 9, Switzerland
e-mail: semmer@psy.unibe.ch

Shinya Shibuya holds a bachelor's degree in electrical engineering from Kitami Institute of Technology. He started work with TEPCO in 1989 at the maintenance department of the Fukushima nuclear power station. Since 1994 he has been a staff researcher at the Human Factors Dept in the Nuclear Power R&D Center of TEPCO.

Address: Human Factors Department, Nuclear Power R&D Center, Tokyo Electric Power Co., Inc., Egasaki 4–1, Tsurumiku, Yokohama, 230 Japan

Hirofumi Shinohara, PhD, received his early degrees in the field of education, concentrating on educational psychology and group dynamics. From 1970 to 1972 he worked at the Institute for Group Dynamics in Fukuoka City. In 1973 he joined the Faculty of Education at Kumamoto University, where he is now Professor of Psychology.

Address: Kumamoto University, Faculty of Education, 2–40–1 Kurokami, Kumamoto City 860, Japan
e-mail: sinohara@gpo.kumamoto-u.ac.jp

Tomohiro Suzuki has been temporarily transferred from the Chubu Electric Power Company to the Central Research Institute of the Electric Power Industry (CRIEPI).

He used to be an operator of the Hamaoka Nuclear Power Station. At CRIEPI, he is mainly in charge of human errors analysis for the Japanese version of the Human Performance Enhancement System.
Address: Human Factors Research Center, Central Research Institute of Electric Power Industry, 2–11–1, Iwato-kita, Komae-shi, Tokyo 201, Japan
e-mail: sasou@criepi.denken.or.jp

Nobuyuki Tabata holds a bachelor's degree in nuclear power engineering from the University of Tokyo. He entered the Tokyo Electric Power Company in 1970 where he has been working in the Nuclear Power Division. He has work experience in test operations of NPP, nuclear fuel, NPP safety, and development of a demonstration Fast Breeder Reaction. Currently, he is the manager of the Human Factors Dept. in the Nuclear Power R&D Center.
Address: Human Factors Department, Nuclear Power R&D Center, Tokyo Electric Power Co., Inc., Egasaki 4–1, Tsurumiku, Yokohama, 230 Japan

Kenichi Takano, Dr. Eng., is a research fellow at the Human Factors Research Center, CRIEPI. His research area covers the analysis methodology of human-related incidents and physiological measurements of humans at work, and his most recent interest is in cognitive modelling of nuclear power operators.
Address: Central Research Institute of the Electric Power Industry, Human Factors Research Center, 2–11–1, Iwado Kita, Komae-shi, Tokyo 201, Japan
e-mail: kojima@criepi.denken.or.jp

Isao Tanaka is Senior Associate Vice President and Director of the Human Factors Research Center, CRIEPI. He holds a bachelor's degree in electric engineering from Shizuoka University.
Address: Human Factors Research Center, CRIEPI, 2–11–1, Iwadokita, Komae-shi, Tokyo 201, Japan

Teruo Tokuine has been working as an engineer for 20 years in the field of nuclear power generation. He is currently in charge of quality assurance and human factors in the nuclear generation management division.
Address: General Office of Nuclear & Fossil Power Production, Kansai Electric Power Company, 3–22. Nakanoshima 3-Chome, Kita-ku, Osaka 530–70, Japan
e-mail: K524500@kepco.co.jp

Tetsuya Tsukada has a degree in electrical engineering from Yonago Technical College. From 1972 to 1996 he was at the Kansai Electric Power Company, working in the maintenance sections of the thermal power plant and then the nuclear power plant. Since 1996 he has been Senior Researcher in the Human Factors Project at the Institute of Nuclear Safety System.
Address: Institute of Nuclear Safety System, 64 sata, Mihama-cho, Mikata-gun, Fukui, 919–1205, Japan
e-mail: tukada@inss001.inss.co.jp

Björn Wahlström is presently a research professor in systems engineering at the Technical Research Centre of Finland (VTT). His interests include nuclear safety, risk analysis, human factors, decision-making, organisation and management, electricity markets, and technology management. Professor Wahlström was the director of the Electrical and Automation Engineering of VTT laboratory from 1983 to 1993. From 1989 to 1991 he was on leave from VTT and worked with the International Institute for Applied Systems Analysis (IIASA) in Laxenburg, Austria.
Address: VTT Automation, POB 13002, FIN-02044 VT, Finland
e-mail: Bjorn.Wahlstrom@vtt.fi

Bernhard Wilpert has been Professor of Work and Organisational Psychology since 1978 and Director of the Research Center Systems Safety. He is member of the Reactor Safety Committee of the German Federal Ministry of Environment, Nature Protection, and Reactor Safety. In 1989 he received an honorary doctorate from the State University Gent, Belgium.
Address: Institute of Psychology, Research Center Systems Safety (FSS), Berlin University of Technology, FR 3–8, Franklinstr. 28, D-10587 Berlin, Germany
e-mail: bernhard.wilpert@tu-berlin.de

Osamu Yamaguchi is Senior General Manager and Deputy Executive General Manager of the Plant Management Headquarters, The Japan Atomic Power Company. He holds a bachelor's degree in electric engineering from Kyoto University.
Address: The Japan Atomic Power Company, 1–6–1 Ohtemachi, Chiyoda-ku, Tokyo 100, JAPAN

Michio Yoshida, Master of Education, graduated from Kyushu University in 1973 where he became a research assistant in 1976. Since 1986 he has been an associate Professor at Kumamoto University.
Address: Center for Educational Research and Training, Kumamoto University, 5–12, Kyomachi Honcho, Kumamoto 860, Japan
e-mail: yoshida@gpo.kumamoto-u.ac.jp

Dr. Seiichi Yoshimura joined CRIEPI in 1976. He is engaged in studying the automation of maintenance works in nuclear power plants and a support system for plant diagnosis. His interest is in the modelling of human emotions and learning. He received his doctorate from Hokkaido University in 1995.
Address: Human Factors Research Center, Central Research Institute of Electric Power Industry, 2–11–1, Iwato-kita, Komae-shi, Tokyo 201, Japan
e-mail: sasou@criepi.denken.or.jp

Preface

This volume is the result of long-standing cooperation between the Institute of Social Research within the Institute of Nuclear Safety System Inc. (ISR/INSS, Mihama, Japan) and the Forschungsstelle Systemsicherheit (FSS – Research Center System Safety) of the Berlin University of Technology. ISR/INSS and FSS decided in 1993 to organise a series of international conferences on human factor research in nuclear power operations (ICNPO) under joint financing, tutelage, and organisation. The stated objective of this conference series is 'to initiate and facilitate contacts and interaction among the dispersed ongoing social science and human factor research in Nuclear Operations of various countries in order to improve further safety and reliability of nuclear power operations' (mimeographed Conference Report of ICNPO I, Berlin, 31 October–2 November, 1994). ICNPO I brought together 40 participants, representing as many as a dozen disciplines. That first review of ongoing research on human factors in nuclear power operations resulted in a lively, interactive and worldwide network of information exchange and co-operation among researchers engaged in the field of nuclear operations. The contacts established in ICNPO I laid the foundations for a more focused conference, ICNPO II (Berlin 28–30 November, 1996), the substance of which we document in this publication for a wider audience.

Three people stand out among the many without whose efforts we could not have achieved our aspirations to unite the ICNPO II contributions in a comprehensive volume: Ulrike Wiedensohler, who managed all the administrative details, including the retyping of many manuscript revisions; Hans Maimer, who redesigned most of the graphic presentations; and David Antal, whose scrutiny and precision drastically improved and unified the style of presentation and our use of the English language. We gratefully acknowledge their critical contributions. Last, but not least, we express our sincere thanks to the ISR/INSS, whose financial support made the editorial work on this volume possible.

THE EDITORS
Berlin/Mihama, February 1998

xix

Introduction

BERNHARD WILPERT AND RAINER MILLER

HUMAN FACTORS RESEARCH IN NUCLEAR POWER OPERATIONS: TOWARDS AN INTERDISCIPLINARY FRAMEWORK

In recent years both engineering and the human sciences have reflected a growing interest in the often intricate interaction of technical, social, psychological, managerial, and political influences on the safety of industrial systems. Systems safety, as distinct from occupational safety, is becoming the catchword. The scope of thinking in the area of occupational safety has continuously expanded from a concern with purely personal characteristics to the impact of the workplace and of organisational and managerial factors. Nevertheless, the focus of occupational and work safety still remains intra-organisational. Systems safety, by contrast, may be defined in line with Roland and Moriarty (1990, 7) as the 'quality of a system that allows the system to function without major breakdowns under predetermined conditions with an acceptable minimum of accidental loss and unintended harm to the organisation and its environment' (Fahlbruch and Wilpert, in press).

This broadening of interest has been stimulated by the variety and speed of social change:

- Technology is accelerating in many areas of social activity such as land traffic and transport, shipping, oil-drilling, and manufacturing, and in such process industries as nuclear energy production and the chemical industry.

- The ever-growing number of large-scale technical systems carries the threat of serious accidents and the potential for considerable loss and damage to human life and the environment.

- Growing international competition brings the risk of 'short-termism' rather than of long-term investment in the safety and reliability of technical installations.

These developments represent a challenge to management structures and strategies that often surpasses their capacity to change and adapt (Baram, 1998). Industrial systems safety is emerging, therefore, as an interdisciplinary problem, requiring researchers and practitioners to cooperate closely and to go beyond settled national boundaries and established scientific traditions. Hence the need for comprehensive analytic frameworks. To integrate the contributions presented in this volume, we propose to combine the open sociotechnical systems approach (Emery and Trist, 1960) with a broad understanding of human factors.

The open sociotechnical systems approach provides a context within which to examine work systems as a social and a technical subsystem. The social subsystem comprises components such as organisational culture and managerial practices, inter-organisational relations, and the division of labour and roles. Only the best quality and combination of technical equipment and the components of the social system will optimise system outputs such as productivity, reliability, and safety.

Analysing the functioning of systems from the perspective of open sociotechnical systems allows the complex interaction of relevant factors to be considered on different systems levels (Reason, 1990). However, since any given system may be seen as a component of a larger system, the question is: where do the boundaries of a given system lie? The answer to this question depends on the ultimate goal of the analysis. In our case it is to ensure systems safety. In other words, the system and its boundaries must be conceptualised as the totality of all factors, elements, and components that contribute to, and otherwise influence, safety.

For our purposes, the social system is divided into four parts (Wilpert *et al.*, 1994):

- The individual subsystem (the acting person (operator) and those functions relating to his or her workplace).

- The *work* group or team (several persons with a common work task and group-specific characteristics – e.g. competencies, norms, and social relations).

- The organisation (comprising managerial and organisational structures, rules, and regulations).

- The extra-organisational environment (all groups or organisations lying outside the focal organisation but contributing to the goal of safety).

This conceptualisation enables us to offer an overall 'landscape' within which to locate the respective foci of the chapters in this volume (see Figure 1). The main topic that they address may be either within one of the five subsystems or at the interface of two or more subsystems.

What should be regarded as the 'human factors'? The issue is important because the everyday understanding of the expression differs from its scientific use. For many practitioners human factors denote human error. In narrow ergonomic conceptualisations only immediate human–machine interface elements are considered as part of human factors. Our view of human factors is, however, broader than that

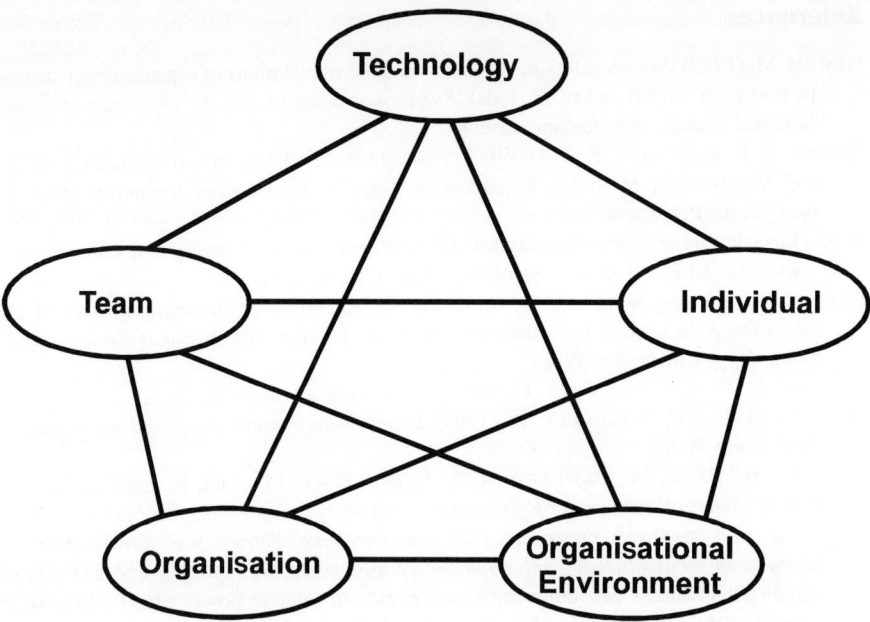

Figure 1   The five subsystems of a nuclear power plant as a sociotechnical system.

in common parlance or in narrow scientific definitions. It was best defined by the Electric Power Research Institute (EPRI) of the US nuclear power industry. Human factors are 'all the variables in a work situation that interactively shape personnel performance' (EPRI, 1988, p. 2). Such factors may include extra-organisational influences such as overregulation and tight control from regulatory bodies. These variables can ultimately reduce the creativity and responsibility of individual operators in their work setting.

The book is divided into four main sections corresponding to the four social subsystems described above. Each section begins with introductory notes that relate each chapter to the five subsystems shown in Figure 1. Contributions to Part One focus on the environmental aspects of nuclear operations: general cultural issues and inter-organisational relationships. Covering the whole gamut of this subject, the authors present findings of comparisons between nuclear power plants in different countries, demonstrate cultural impacts on safety behaviour and technical design, examine characteristics and approaches to measurement of safety culture, and discuss environmentally induced problems and interventions for change. Organisational features and their impact on nuclear safety are the primary focus of Part Two, where the authors explore such topics as organisational structure and leadership, and consider various points of views on optimisation measures. Part Three relates to group and individual performance. Part Four offers a cross-sectional perspective: learning from experience through different strategies for studying incidents and accidents.

References

BARAM, M. (1998) Process safety management and the implications of organisational change. In HALE, A. and BARAM, M. (eds), *Safety management and the challenge of organisational change*. Amsterdam: Elsevier.

EMERY, F. E. and TRIST, E. L. (1960) Socio-technical systems. In CHURCHMAN, C. W. and VERHULST, M. (eds), *Management, science, models and techniques* (vol. 2). New York: Pergamon.

EPRI (Electric Power Research Institute) (1988) *Human factors primer for nuclear utilities managers* (EPRI NP-5714). San Diego: Essex Corporation.

FAHLBRUCH, B. and WILPERT, B. (in press) System safety – An emerging field of I/O psychology. In COOPER, C. and ROBERTSON, I. (eds), *International Review of I/O psychology*. Chichester: Wiley.

REASON, J. (1990) *Human error*. Cambridge: Cambridge University Press.

ROLAND, H. E. and MORIARTY, B. (1990) *System safety engineering and management*. New York: Wiley.

WILPERT, B., FANK, M., FAHLBRUCH, B., GIESA, H.-G., MILLER, R. and BECKER, G. (1994) *Weiterentwicklung der Erfassung und Auswertung von meldepflichtigen Vorkommnissen und sonstigen registrierten Ereignissen beim Betrieb von Kernkraftwerken hinsichtlich menschlichen Fehlverhaltens* [Improvement of reporting and analysis of significant incidents and other registered events in nuclear power plants in terms of human errors] (BMU-1996–457). Bonn: Bundesminister für Umwelt, Naturschutz und Reaktorsicherheit.

Nuclear power operations and their environment: culture and inter-organisational relations

Introduction

The contributions in Part One present different viewpoints about the way nuclear safety is embedded in an organisational or cultural context. In terms of the socio-technical systems approach, these chapters deal with the subsystems called 'organisation' and 'organisational environment' and with their manifold interactions.

Touching on basic epistemological issues, Rochlin takes the position that nuclear safety cannot be understood in terms of objective rules and procedures, but must instead be regarded as a socially constructed element of culture.

The influence of national culture on safety is investigated in different ways. Wahlström identifies differences in the safety practice of two countries, whereas Moray shows how the design of control rooms, including ergonomic recommendations, are influenced by culture. Bourrier, in her comprehensively researched contribution identifies different strategies of resource allocation in French and American nuclear power plants.

The increasing importance of safety culture, which may be placed at the interface of organisation and environment, is taken into consideration in the contributions by Meshkati, Semmer and Regenass, and Ignatov. Meshkati shows how dimensions of national culture interact with components of safety culture. The contributions by Semmer and Regenass and by Ignatov address the problem of measuring social norms as an important element of safety culture.

A holistic viewpoint of safety is taken in the contribution by Becker and that by Wilpert, Fahlbruch, Miller, Baggen, and Gans. Both chapters are an examination of the interaction between organisational factors and the organisational environment. In discussing obstacles to open communication about safety aspects, Becker points out intra-organisational barriers (e.g. hierarchical constraints) as well as environmental factors (e.g. the political situation in Germany). Wilpert *et al.* describe the 'interorganisational field' of safety in Germany, meaning all organisations and institutions – their interactions – contributing to the safety of German nuclear power plants.

CHAPTER ONE

The social construction of safety

GENE I. ROCHLIN

University of California, Berkeley, Cal.

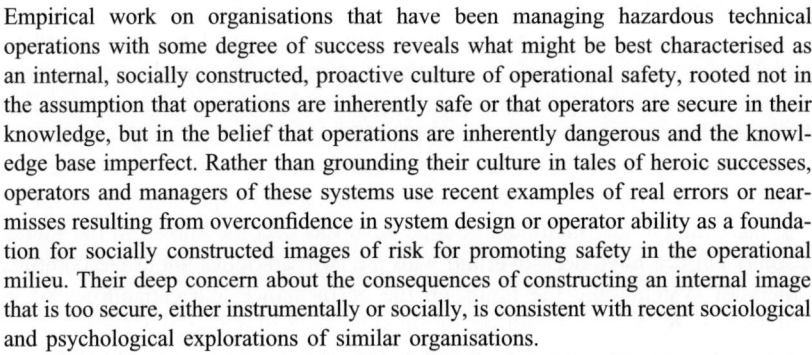

Empirical work on organisations that have been managing hazardous technical operations with some degree of success reveals what might be best characterised as an internal, socially constructed, proactive culture of operational safety, rooted not in the assumption that operations are inherently safe or that operators are secure in their knowledge, but in the belief that operations are inherently dangerous and the knowledge base imperfect. Rather than grounding their culture in tales of heroic successes, operators and managers of these systems use recent examples of real errors or near-misses resulting from overconfidence in system design or operator ability as a foundation for socially constructed images of risk for promoting safety in the operational milieu. Their deep concern about the consequences of constructing an internal image that is too secure, either instrumentally or socially, is consistent with recent sociological and psychological explorations of similar organisations.

The difficulty for analysts and system designers is understanding how to identify such cultures and the means by which they are created and maintained. Because these cultures are socially constructed, they arise as much from interpersonal and intergroup interaction as from the more commonly studied interactions with external designers and regulators. Because the interaction is anchored in cultural dynamics, there are wide variations in its manifestation even for similar plants in roughly similar settings. The analytical problem is to identify the means to correlate the social construct of operational safety with empirical observations concerning the nature and role of human operators in maintaining reliability and controlling risk.

There are old pilots, and there are bold pilots, but there are no old, bold pilots.

Anon.

When the Berkeley 'high-reliability organisation' (HRO) group first set out to examine the role of operators in maintaining safety in a variety of complex, high-technology operations, the intention was to produce a reasonably ordinary,

5

organisational study, essentially positivist in orientation and based on a combination of behavioural analysis, survey research instruments, and empirical observation (La Porte, 1996; Rochlin, 1993). My HRO colleagues and I sought to link what operators believe, or say they believe, about their performance and the performance of their systems with observations about operational 'reliability' that involve measurable indicators. In some cases, such as aircraft launches from an aircraft carrier, indicators of this sort could be derived if the data were made available. In others, such as nuclear power plant operations or air traffic control, it soon became clear that there was no baseline for measuring failures. Moreover, and more surprisingly, it became clear that, even if such data were available, they would not provide indicators of operational safety. Paradoxically, operators' perceptions of potential risk were an essential element in their construction of an environment of safe operations (Roberts, 1993; Rochlin, 1996).

Operational safety is more than a culture. Even within the organisation itself safety is a social construct. It is therefore difficult not only to measure but to analyse and characterise with the usual analytic tools of organisational theory. It involves *élan* and modesty, pride in past accomplishments and concern about future ones, cooperation between operators and technical cadres and also tensions between them, as well as an acknowledgment of the importance of close oversight by managers and regulators, and a demand for independence of action. The sense of accomplishment is defined as a blend of autonomy, safety, and production goals (Roberts, Rousseau, and La Porte, 1993). But what sort of blend? And to what extent can a constructed belief in operational safety be measured against some set of observations or empirical measures, say, of reliability or of the control of risk?

Most of the literature on risk and reliability in potentially high-consequence, sociotechnical operations lacks not only an adequate, 'operationalisable' definition of operational safety, but even a means for defining and expressing it. As La Porte (1996) has pointed out, attempts to define what has become known in the literature as a *safety culture* have been overlaid with the twin anxieties about the complexity of emergent systems and the limits on the performance and capabilities of individual human beings. The culture of reliable operation that the Berkeley HRO group has reported on is not what most people mean by a culture of safety (International Nuclear Safety Advisory Group [INSAG], 1991): it does not correspond to searches for excellence (Deal and Kennedy, 1982; Peters and Waterman, 1982), and it is not brought about at the institutional level by the assumption of collective regulatory responsibility (Rees, 1994). Weick (1987) has perhaps come closest in characterising it as bringing empirical research on operational norms and procedures to bear on more familiar organisational studies of safety culture (Weick and Roberts, 1993).

Scarry (1985) pointed out that most human languages have only a very limited and indirect vocabulary for expressing the character or nature of pain. Safety seems to have that same elusive quality, both as a term and as a construct. It is much talked about in general, but little examined in detail. Even professional analysts of risk and reliability tend to resemble the popular press in paying quite a lot of attention to examining what goes wrong and why, very little attention to situations where something could have gone wrong but didn't (near-misses), and even less attention to those cases where, in principle, quite a lot could go wrong, but very little has. In the remainder of this chapter, I first mirror the argument to discuss not risk, but safety, as a social construct, and then use the empirical work of the Berkeley HRO group to carry that discussion over into the more circumscribed universe of the operation of nuclear power plants and other sociotechnical systems that have, as an inherent property, the potential to cause considerable and dramatic public harm and, as an observable property, their failure to have done so.

1.1 THE SOCIAL CONSTRUCTION OF RISK

A fundamental premise of much of the literature on risks, crises, and accidents, including a great deal of the organisational literature, takes a default definition of safety as the antithesis of error, or as a residual set defined in terms of the negation or absence of risk (Bonaca, 1991; INSAG, 1991; Pidgeon, 1991; Ralph, 1988). This definition is an instrumentalist orientation that rests upon a set of philosophical, epistemological, ontological, and normative assumptions that can be summarised in terms of a formalist, functionalist perspective of control that emphasises positive action and predictive models ('t Hart, 1993). The professional literature on risk in complex, technical systems has long been dominated by formal, engineering-oriented definitions and approaches that largely eliminate the social dimension, or at the very least externalise it, severely constraining the range and scope of social and institutional analysis that can be applied (Clarke and Short, 1993). That per-spective may be sufficient for automatic systems, but is woefully inadequate for sociotechnical ones.

This paradigm is most deeply embedded in the classic engineering definition of risk as the product of hazard (potential consequence) and probability of occurrence. After the risk has been calculated according to some method (such as fault-tree analysis combined with probabilistic failure rates), certain numbers are generated (for example, x events of a certain type per million operations, or per year of operation). Where human operators and operations are concerned, the result may be further augmented by a variety of empirical or psychological means for estimating the probability of single-mode or collective human error. This procedure yields a more sophisticated approach to estimating how often a system will experience an error or failure with a certain range of consequences, but what then? It may be risk in a formal sense, but it does not map well on to the world of human action. It does not deal with the issue of whether such a risk is accepted, or acceptable, to the operators, the operating company, the corporation, regulators and other watchers, or the general public.

It was always clear that neither individual nor public response to risk was in accordance with the rank ordering and analytic findings of objective risk analysis. In its most classic expression, many people who are afraid to fly are hardly concerned at all with the risk of driving to the airport. The positivist response to this paradox has been to cast the subjective dimension in terms of a formal scientific-economic model, in which contingent valuation is used in the form of revealed preferences. The fundamental premise, more widely elaborated since, is that there are different categories of risk perception and risk acceptance that depend on the interaction between the individual and the source of potential harm (Hiskes, 1996).

The classic analysis by Starr (1969) incorporated the social and behavioural dimension by dividing risk exposure into voluntary and involuntary. This approach was intended to preserve the positivist paradigm of rational calculation and planning. However, the work of psychologists such as Slovic and Fischhoff showed that no purely economic-rational model, including that of revealed preferences, could adequately encompass the non-rational ordering and cycling preferences displayed in actual practice (Slovic, Fischhoff, and Lichtenstein, 1976). Instead, it was necessary to extend the notion of preferences to include not only the nature of the risk (i.e. what the consequence is), or its familiarity, but also such behavioural characteristics as whether the individual believes that he or she is in control of the risk-producing activity, or can only react to it.

In recent years, social and political aspects of the divisions in the interpretation and meaning of risk have been emphasised in views based rather explicitly on both individual and collective dimensions of risk perception (Clarke, 1989). This perspective was most notably framed in cultural terms in the work of social theorists such as Beck (1992) and Luhmann (1993), who sought to define risk in social rather than technical terms. Taking a slightly different path, theorists such as Thompson, Douglas, Ellis, and Wildavsky have developed a methodology for policy analysis that is conceptually similar to the economic-psychological division of voluntary and involuntary risks. They seek to comprehend variations of risk acceptance in terms of cultural variations in political and social groupings, each of which constructs socially a different interpretation and meaning for the nature and importance of different categories and types of risk (Douglas and Wildavsky, 1983; Thompson, Ellis, and Wildavsky, 1990).

Hiskes (1996) and others working on the social construction of risk from a more postmodern perspective have gone further, arguing that risk is neither entirely instrumental nor wholly external, but that it is an emergent property of social (or, in this case, sociotechnical) systems. A product of people's common lives, it emerges from the interactions of individuals, and is manifested in the social construction of common narratives of the perils and progress of collective interaction. From this perspective, risk is a group phenomenon with a dual character, as a set of more or less objectively determined properties of the external world and as a collection of subjective responses to that world. In such constructivist views the perception and acceptance of risk is an intersubjective social act, a 'speech act' if you will (Searle, 1969).

The cultural-theory school has, however, been strongly criticised not only for being fundamentally normative but also for dealing with risk as if it consisted almost entirely of subjectivity and perceptions (Shrader-Frechette, 1991). In contrast, Turner (1978) and others have sought to find a balance by adopting a quasi-instrumental perspective, defining collective errors as a negentropic property of social systems based on the human ability to self-organise damaging events and circumstances that are not decomposable at the individual level. Over time, other schools of thought on risk have emerged and have sought better coupling between the social sciences and the technical realm, yet these have remained framed within essentially social rather than technical definitions, such as 'the definition of risk is essentially a political act' (Kasperson, 1992, p. 155) or a sociopolitical act (Turner, 1989). Even those who have chosen to argue that risk hardly exists at all as an unchangeable feature of an objective physical world, but is a construct that the human imagination overlays on the surrounding world, acknowledge the world as a tangible, and not just a social, reality (Jasanoff, 1993).

I do not intend to delve extensively here into the interesting and provocative debate about the two bodies of risk theory, the objective and the subjective, or into the question of whether they are ontologically complementary or epistemologically incompatible. Rather, my purpose is to point out that, whatever the merits and strengths of objective risk analysis, it does not provide either a definition or a description of the social interpretation of being at risk. Positivist calculation of risk may be considered to have a certain degree of objective reality, particularly when it expresses purely technical probabilities, but it is given meaning by the filtering of empirical data and formal models through subjective interpretations and inter-subjective constructions of representation and meaning. Constructivist-positivist inter-pretations such as that by Turner (1989) have gone much further in exploring the notions of collective interaction and social construction, but attempts to encompass intersubjectivity, representation, and interpretation more fully have been difficult at best. Although abstract and constructivist, workable models of risk from tech-nical systems that present physical dangers must still acknowledge that repres-entations of risk are not entirely separable from the objective reality represented; it is interpretation and meaning, narrative and discourse, if you will, that shape responses, beliefs, and actions.

1.2 THE SOCIAL CONSTRUCTION OF SAFETY

Three recent sets of news stories may be used to frame the growing importance of studying safety as a positive attribute of sociotechnical systems. One, having to do with an unprecedented rash of crashes by high-performance F-14 naval fighter aircraft, questioned whether the pilots were becoming overconfident both in their abilities and in the airframes, thereby rejecting the familiar homily quoted as the epigraph of this chapter (Schmitt, 1996). The second set centred on Vaughan's (1996) admirable analysis of the *Challenger* accident, in which she examines the consequences of constantly reconstructing perceptions of operational safety to in-corporate and normalise survived errors (Broad, 1996). The third set, with opposite implications, arose from the discovery that the US Federal Aviation Administration (FAA), anticipating the introduction of a new and more advanced air traffic control automation suite, had already cut back on the number of maintainers familiar with the present, quite antiquated equipment, and that controllers were not only nervous about the consequences, but manifestly concerned that operational safety was being compromised (Wald, 1996).

The perspective adopted in all three cases was not the usual, normal-accident argument – because complex systems are inherently incapable of being operated safely, accidents are only to be expected (Perrow, 1984; Sagan, 1993). In contrast to these traditional approaches, which take safe operation to be a period between accidents, all three sets of news stories accepted the premise that accidents are untoward events in these systems, and that operational success, defined as avoiding both public harm and operational failure, was, or should have been, the normal and appropriate circumstance, even in a set of very demanding operational conditions with very high potential consequences. That risk is increasingly discussed as a social construct is no longer surprising; what is remarkable is the growing perception that safety is socially constructed as well.

It is entirely possible to define safety as the negation of risk, just as it is possible to define health as the condition of not being noticeably ill, or life as the state of not being noticeably dead. Although such framing is not uncommon when one has recently been exposed personally or vicariously to a direct risk or the threat of illness or death, a statement that one is 'alive and well' is rarely deconstructed in such objective, positivist, medical terms in the wider and more general social and cultural context. Playing against such narrow meanings, expressions of good health or good life are most often social constructions that involve meanings and inter-pretations that are more complex, more general, and more difficult to quantify.

Referents to safety in common usage are similarly dichotomised. *Webster's Seventh New Collegiate Dictionary*, for example, first defines the word *safe* pas-sively and positivistically: 'free from harm or risk,' the opposite of hazardous or dangerous. But the gates of ambiguity are then opened by *Webster's* second defini-tion: 'Secure from threat of danger, harm, or loss,' which is not quite the opposite of danger, harm, or loss, respectively. Although it is possible to interpret the latter statement narrowly and objectively, it would be broader to consider both *threat* and *security* as perceptions, in which case safety is largely subjective, or even to go further and consider *threat* to be an expression of intentionality, and *secure* to be a perceptual frame, in which case both are inherently and irrevocably intersubjective.

Divergences between individuals with regard to their perceptions of being safe may be as wide as the divergences between individual and collective perceptions of safety and the results of objective analysis. *Being* safe and *feeling* safe can be quite different things. In parallel with the literature on risk, the question of whether an individual is in a situation over which he or she has control, or one in which that person does not, may strongly affect perceptions. But there remains a fundamental difference. Risk is subject to logical parsing, because it is an affirmation, an expres-sion that something that could happen might happen. Safety, defined as the state of being 'not at risk,' is not. It is a conditional negation, an expression that something that will cause harm or loss might not happen, or that something that will happen might turn out not to be harmful. Moreover, one must distinguish between the active construction of safety and the feeling that one is safe: the latter may be based more on suppression or repression of the knowledge of risk than on its calculation.

The analogy with health can be and has been extended further. Reason (1995), for example, pointed out that health can be thought of as 'an emergent property inferred from a selection of physiological signs and lifestyle indicators' (Reason, 1995, p. 1716). He then constructed an analogy with complex, hazardous systems, commenting that assessing an organisation's safety health is like assessing the physical health of a patient, in that it requires sampling from a potentially large number of indicators. However, to argue further that these systems can be parsed to generate separable dimensions along which organisational safety might be assessed is to frame the notion of emergence in an overly positivistic way.

To take a real world example, I feel safe at the moment in my study in my home in Berkeley, California. This feeling encompasses many things. I am probably accurate in not being concerned about hurricanes, floods, tigers, or stampedes. I do not imagine (except by an effort of perverse will) that I might be at risk from falling aeroplanes, meteorites, terrorist bombs, assassination, or the explosion of a gas main. But my perceived risk of the potential and very real possibility of an earthquake is kept low partially by denial, partially by a number of measures taken to reinforce the house and provide for medical and other contingencies, and partially by a belief that there is a collective response system that will help. I am not calculating the risk of heart attack or cancer, over which I have very little control, or of repetitive strain injury, over which I imagine I have a lot, although I might calculate those risks if I were all alone or in a remote area where medical and emergency help would be difficult to find. When I went out this morning, I crossed the street, drove to the store, came in proximity to or touched people who might well have incurable com-municative diseases, purchased several food products that could (conceivably) prove harmful when ingested, and bought a package of cigars (no comment).

In short, my own construction of safety (and risk) shows an absolutely typical mix of evaluations of individual and collective behaviour, personal and social judg-ment, subjectivity and intersubjectivity, and representation and interpretation. It involves not only voluntary actions to control risks in the environment, and often implicit assessments of the low risk presented by other potential sources of harm over which I have no control, but also the acceptance and definition of my indi-vidual, personal safety as a series of interactive social contracts and constructions – for example, the quite different perceptions and reactions evoked when I am out on the streets of a completely strange city at night.

This effort at analysis is more socially grounded than most, yet it still omits almost completely another important set of considerations arising from the active nature of the social and political dimensions of individual or organisational environ-ments. Safety, like risk, can be created and manipulated by others, whose purposes, objectives, and interpretations in its shape and direction are either unknown or masked. This interpretation goes beyond the social constructivist perspective that social reality cannot be objectively assessed, introducing in addition the extent to which safety is, or may be, contingent upon different subjective constructions given to it by different actors ('t Hart, 1993).

As with the business of trying to generate the perceptual and cognitive dimen-sions of the social construction and definition of risk, there may be a wide gap between the subjective evaluation of what is safe and any set of comparative empirical indicators that might be constructed from data sets or observations. But that discrepancy is even more difficult to determine for safety, because the objective dimensions of risk are more easily defined than the subjective, particularly in those cases where the source of the risk is entirely external and largely probabilistic. Moreover, the two notions of subjective and objective risk are not entirely unconnected. Where safety is an interactive social property, a collective social property, or both, where it depends upon subjective indicators such as beliefs and behaviour, as well as upon intersubjective ones such as signalling or behavioural assumptions, there are many situations where safety is a collective and self-reinforcing property of the system. However, it must also be admitted that there are cases where actions predicated on the perception of safety may also, objectively, decrease it. As with risk, the most important, and least commonly addressed, analytic problem is to separate, identify, and characterise these situations in a way that provides guidance for identifying which of the cases is which, and why.

1.3   OPERATIONAL SAFETY AS A SOCIAL CONSTRUCT

The epigraph of this chapter has been familiar to pilots for so long that its origins are lost. What is important, however, is not how it originated but how commonly and widely it is known, and how often it is invoked as a token within the community. It has also been expressed in another significant way by both civil and military pilots, all of whom will repeat, in one way or another, the maxim that pilots are most at risk when they have gained enough experience to be overconfident – that is, to begin believing that flying is a safe environment rather than an inherently dangerous one. What increases safety over time is not just experience, but the recognition that they work in an environment that is always subject to unexpected events or unfamiliar conditions.

To recreate the paradox raised at the beginning of this chapter in particular circumstances, pilots' continuing belief that flying is inherently risky and that the environment in which they operate is potentially hostile is acknowledged by them to be a major part of their culture of safe operation. Purely exogenous or uncontrollable dramatic technical failure aside (e.g. bombs, fuel-tank explosions), what has created the present low level of serious operational errors and accidents is a dialectic of safety: the result is to foster high confidence in the inherent quality and design of the equipment and the training and expertise of oneself and other members of the crew, while at the same time remaining alert and attentive for signs or signals that a circumstance exists, or is developing, in which confidence is erroneous or misplaced. One of the most remarkable findings of the HRO team's extensive semistructured interviews with pilots, air-traffic controllers, and nuclear plant operators and maintenance people is the ease and unselfconsciousness with which they appear to maintain this duality of representation (Bourrier, 1996; Rochlin and von Meier, 1994). They clearly see no contradictions in the dichotomous responses my colleagues and I can elicit by varying the framing and wording of specific questions put to them.

This finding is perhaps clearest in an analysis of US operators' responses to the incredible web of detailed and formal rules and regulations with which they must comply (La Porte and Thomas, 1995). Almost every operator who was interviewed felt that regulation at the required level was intrusive, at times distracting, and often far more detailed than required for the task. Nonetheless, almost every operator also argued that the rules and regulations they lived with were, by and large, necessary and important. There was little evidence that this seeming contradiction created in them a sense of cognitive dissonance or permanent tension between seemingly incompatible beliefs. Instead, the lack of resolution of the dialectic is almost always explained away with an often implicit hypothesis that a harmonious synthesis is achievable, but that it is blocked by exogenous factors such as bureaucratic difficulties, public sensitivities, or political ignorance. What they generally could not tell us was the extent to which potential stresses were relieved by incorporation rather than resolution, by designing into the operators' regular procedures regulation-compliant behaviour that was specified to be functional whether it was or not. And incorporation is a very accurate description of the process: there were many aspects of operator task design and performance for which operators could no longer tell whether the design and specification had originally arisen from internal evaluation and review, or had been imposed to satisfy the expectations or demands of outside watchers and regulators.

In an article on 'hero stories,' another member of the HRO research team, Paul Schulman, provides the groundwork for beginning to understand such behaviour and parse it according to industry, activity, and function (Schulman, 1996). What he noted first was a sharp contrast between organisations in which hero stories were prevalent and welcome (hero-embracing) and those in which such stories were rare and not encouraged (hero-avoiding). Hero-embracing strategies emphasise extraordinary performance and rapid response, the need for individual actions to accomplish goals and maintain performance. This need is most evident in organisations such as the Marine Corps, fire departments, electric utility linemen, and, in fact, naval pilots. These stories serve, among other things, as an effective means for organisational learning and the maintenance of cumulative knowledge.

In contrast, not only were hero stories almost completely absent among nuclear power plant operators, their presence would have been disturbing. In this context 'hero' is a term of derision or criticism, describing an operator who will act on his or her own personal achievements and judgment, with little regard for the collectivity or for rules and regulations (in the United States, the contempt is often summed up in the derogatory term 'cowboy,' in itself worthy of an entire deconstructivist essay). Safety is sought through collective action and interaction, through shared knowledge and responsibility. Organisational learning is formalised, and cumulative knowledge is transmitted not through legends or narratives but through the creation and modification of formal procedures. The hero-embracing organisation sees its task environment as decomposable into a series of unstable and unpredictable events, and defines safety in terms of the capacity to cope with a constant stream of emergencies. The hero-avoiding organisation sees its task environment as a holistic one, in which exigencies will arise in context and must be dealt with collectively within the scope of integrated organisational design.

The two frames for constructing and defining operational safety intersubjectively and collectively lead to quite sharply contrasting sets of social and individual behaviour. The first establishes a context in which responsibility for performance of those tasks that are not routine (particularly those that put the organisation most at risk) rests primarily with individuals, heroes, or at least potential heroes, to be sought out and nurtured. What lends confidence in collective performance is the belief that any contingency will be met by one or the other. The second rejects such an approach as dangerous even (or perhaps especially) for those tasks that present the greatest potential for harm. Instead, threats are to be identified, assessed, and routinised as much as possible (for example, through drills and simulations). And even when unexpected events do occur, they must be responded to through collective action. Not only does safety not depend on heroic individual initiative, it may well be threatened by it.

The dangers of overadherence to the first perspective are well known (particularly from military histories). Operators, teams, and organisations may overestimate their knowledge, or their capacity, or may simply try to do more than they are capable of, because the organisation expects it of them. The F-14 crashes referred to previously are an example. The potential risks arising from the second perspective are more indirect and can affect the organisation as a whole as well as individual operators or groups. At the operational level the operator or operating team may underestimate uncertainty, otherwise misread their operating environment, or become so confident in the equipment (or their performance) that they ignore procedures or become inattentive even to manifestly displayed signals. There are even reported cases where operating teams assumed that a clearly displayed indication of a malfunction signalled a problem with the instrument and not the plant. Potential consequences arising at the organisational level are well illustrated by the case of NASA, where the *Challenger* accident was almost certainly a result of near-accidents or other warning signals becoming internalised as expressions of confidence in operational robustness rather than as indicating the risk of impending failures.

What the Berkeley HRO group found in its own work was an additional, third type of potential consequence common to both perspectives, which is in some way exemplified by the third set of introductory stories cited in the preceding section of the chapter, concerning the maintenance of air-traffic control computers. Reducing the instrument maintenance staff was seen by the FAA as a reasonably cost-effective strategy that, at worst, threatened short-term reliability in the interests of promoting long-term improvement. But, because it affected the operators' construction of safety in operations, it was seen by them as threatening to cause actual harm to the public as well. Moreover, because operators understood implicitly or explicitly that operational safety is a social as well as a technical construct, there was positive feedback: they believed that operations were becoming objectively less safe, because of the increased concern that their construction of operational safety was not being properly taken into account by the FAA.

This kind of response is neither entirely unusual nor unique to operations that manage technical systems that are capable of causing considerable harm. It has long been observed that the more the structure of an organisation or other social 'system' is derived from the institutionalisation of myth and ritual, the more elaborate are the displays of confidence, satisfaction, and good faith (Meyer and Rowan, 1977). Moreover, these instances are in most cases not simply vacuous affirmations of institutionalised myths, but form the basis for a commitment to support that facade by collective action ('t Hart, 1993). Operational safety, it seems, may well be a version of that chimerical notion, the social construction of reality, in the sense that it is not the source but rather an outgrowth of the organisation's institutional enactment of its commitment to safety as an institutionalised (socially constructed), collective (intersubjective) property.

It is sometimes difficult for people raised in positivistic disciplines and trained to shave their observations regularly with Occam's razor to appreciate fully the importance or robustness of these richly textured, intersubjective constructs. Many people are not entirely comfortable either with the language or with the terminology. Nevertheless, and despite the seeming lack of precision in the constructivist discourse, the complaints of many operators (and not a few of their managers) that no one outside their operational environment can fully understand it or evaluate what is or is not valuable to the culture of reliable and low-error operation are often in accord with constructivist propositions.

1.4 PROTECTING THE OPERATIONAL ENVIRONMENT

One of the striking findings of the research by the Berkeley HRO group is the depth and extent of concern among operators and operational managers that regulators, 'experts,' or other outsiders seeking to extirpate residual operational errors (and there will always be a residual where human beings are involved) may impose upon them changes that erode or interfere with their confidence in the safety of operations. However well-meaning they may be, external parties who have not immersed themselves completely in the details and milieu of a particular operational context are unlikely to be sufficiently informed about or sensitive to the processes and interactions from which operational cultures are constantly constructed and reconstructed. The literature already contains examples of cases where positivist, objective efforts to improve the operating environment have been antithetical to constructed, representational norms and routines (Roth, Mumaw and Stubler, 1992). As the level of plant automation increases, the role of the operator is more and more that of a supervisory controller of the equipment rather than of the plant, a limitation of scope that can interfere with operators' sensitivity to warning signs or ability to respond to unexpected events (Zuboff, 1984, Bindon and John, 1995, Rochlin, 1997). If strategies that modify equipment, task environment, or organisation in the name of reducing risk are not based on a full understanding of social and cultural factors, they may well result in increasing operational error, by interfering with the less explicitly observable processes through which safety is created and maintained (Dougherty, 1993).

This point became quite clear in the work my colleagues and I have done on intercommunication in aircraft carrier flight operations, where the constant stream of on-ship communications traffic during operations was directed specifically at maintaining the integration of the collectivity, and reassuring all participants of its status, even when the level of operational activity was low. In a related example, an analysis of extensive video and audio recordings of conversations between controllers and pilots reveals many rules and interactions that are not formalised, but are meant to nurture communication and cooperation (Sanne, 1995). Their structure and content were shaped to a great extent by the recognition by all parties concerned that safety was indeed a collective property that emerged from the inter-action between them.[1]

Similar findings for the training of nuclear plant operators have been reported in the literature. Preserving intercommunication is the fundamental premise of the work of the Westinghouse group (Roth *et al.*, 1992) and of others. Janssens, Grotenhus, Michels and Verhaegen (1989) described what they called 'social organisational safety' as depending deeply on the 'spontaneous and continuous exchange of information relevant to *normal* functioning of the system in the team, in the social organisation' and made such intercommunication a primary goal of safety training programmes for operators. This view exactly parallels the observations and findings that my colleagues and I recorded in similar work across the whole range of organisations we investigated. Although no operational team that we inter-viewed had any formal language to describe safety, each one considered it some-thing that in their organisation 'arose from' or 'depended upon' the free flow of information at *all* times, because that flow kept the team integrated, and because experienced operators monitoring the flow of what often seems to be trivial or irrelevant information can often discern in it warning signs of impending trouble.

Although the HRO organisations that my colleagues and I have observed differ greatly in the way such behaviour and sociobehavioural constructs are displayed, there remain many other things that are also held in common. Among them are: *élan* and strongly held expectations about the value of skilled performance; peer group pressures to excel as a team; a high degree of cooperation; willingness to move in and 'own' emerging problems rather than dodge them; and reward for reporting rather than disguising errors (La Porte, 1996). In addition, where time is a critical factor in operations, collective as well as individual situational awareness is complemented by a high degree of discretion and responsibility (Roberts *et al.*, 1993; Rochlin, La Porte, and Roberts, 1987). On the other hand, autonomy and discretion are not without cost. Because most of the systems we observed require both a high degree of technical skill and considerable operational knowledge and experience, these two realms are often divided formally into different occupational domains, operational domains, or both. For the same reason, the people who install and maintain the machinery itself are often divided formally from those who design and operate it.

Furthermore, there are in many cases serious limits to the extent to which regu-lators and other so-called watchers can be neglected as part of the 'text' from which operational cultures (and safety) are constructed. In the case of FAA air-traffic control, for example, the regulatory presence is so integral even to task design and performance that it is an essential and primary element even for such relatively simple constructs as the framing of particular tasks (La Porte and Consolini, 1991; Sanne, 1995). For nuclear power plants the number of regulations and the amount of regulatory paperwork alone constitute a major element of working life (La Porte and Thomas, 1995), complemented in many cases by the physical presence of permanently assigned regulatory personnel.

Accordingly, at a typical nuclear power plant there are (or may be) differently socialised cadres of managers, operators, engineers, maintenance personnel, and regulatory compliance personnel, in addition to financial officers, administrators, and others in the front office. These findings are similar to those of organisational ethnographers in other domains (see Zuboff, 1984; Barley, 1992; Barley and Bechky, 1994). Because each of the operational groups bases its collective identity on a different operational characteristic and defines its role differently, and because social relations and interdependencies are stronger within groups than between them, one result is an intergroup tension between units organised around different professional skills, approaches, or training (von Meier, 1995). In US plants the dynamics are exemplified most strongly by the tensions between operational and technical experts (Rochlin and von Meier, 1994) and between operators and maintenance people (Schulman, 1993). Similar tensions are manifest in the work of Bourrier and others on French plants, most characteristically those between managers and workers (Bourrier, 1996).

Such tensions are difficult to read, let alone interpret, without complete familiarity with the social and cultural dimensions of the plant and its operating environment. In some cases intergroup rivalries may be a warning sign of difficulties or tensions within the plant. In others (including several my colleagues and I have observed) they are part of the dynamic process of making sure that the right group owns an impending error or difficult task. In still others, the conflict may be formally or culturally ritualised, and therefore not just intersubjective but also structural.[2] The temptation to view such tensions as interfering with operational safety, or as being largely exogenous to it, is great, but the pattern we have observed in plants that are operating well and safely is one of a balance between collaboration and conflict, and between mutual respect and occupational chauvinism, a balance that serves the dual purpose of maintaining group self-identity and cohesion and providing a necessary channel for redundancy and backup. In one classic nuclear plant observation operators, faced with a problem that taxed their competencies (but posed no immediate operational threat or regulatory intervention), kept that information away from the engineering staff until they were certain that it could not be solved in the control room. At that point, in an interesting duality that maps the above discussion well, they turned the problem over completely to engineering, while at the same time devaluing it (in their own context) by calling it 'only' an engineering problem.

The challenge for the group's work – indeed, for all safety analysis – is to take such examples, together with the work of others who have come to similar conclusions, and to try to construct from them a definition of operational safety that is analytical, robust, predictive, and not just contingent on past performance (or luck); and to be able to decide what is and what is not valuable and important for the social construction of safety, without marginalising those who speak for experience, history, culture, social interactions, or other difficult-to-quantify sociocultural and socioanthropological variables. It is possible, as some postmodernist critics have suggested, that to invoke intersubjectivity, or emergence, is necessarily to frustrate analysis by invoking a holistic perspective in which everything must be learned and nothing can be generalised (Harvey, 1990). But it should be remembered that constructivist and interpretive approaches were, in turn, developed because of the obvious shortcomings of reductionist, empirical analysis (see, for example, Crozier, 1967; Czarniawska-Joerges, 1992).

The problem was, and is, that many analysts of, and consultants to, the nuclear industry, including, in many cases, on-site as well as off-site regulatory personnel, are grounded in approaches that are fundamentally rationalistic. This limitation is true not only of technical model-building and organisational analysis, but also of human-technical interactions, psychological theories of framing and scenario-building, and even cognitive analysis. Ritualisation, seemingly non-functional displays, and expressions of confidence that are not grounded in empirically determinable observables are all too often disregarded or, worse, trampled on or over-written in an effort to impose theoretically argued strategies that are claimed to be functional solely in terms of abstractly constructed models and rationalised and formalised observables.

1.5   CONCLUSION

It has long been recognised by sociologists and other students of the sociocultural and interactive dimensions of organisational culture that organisations whose formal structures become increasingly elaborated in response to increased complexity, technical considerations, and external social demands, but whose institutional structures also incorporate the myths and rituals of the operational and institutional environments, tend to respond to the strain of trying to resolve this double image by decoupling structures from each other and from ongoing activities (Meyer and Rowan, 1977). Formal coordination, inspection, and evaluation may remain, and be formally complied with in great detail, but may be less important for ensuring safety than for maintaining the logic of confidence, good faith, and mutual respect.

This view is easily extended to encompass the behaviour that my colleagues and I have seen in nuclear power plant control rooms and other places where operational safety is a property that emerges from the dynamic social interaction of operators with their equipment, their environment, and each other. Moreover, it goes beyond even the more sophisticated, roughly contemporaneous rationalist approach according to which open systems respond to uncertain environments (Emery and Trist, 1960). The strategy of creating informal networks in formal organisations in order to facilitate inter-unit coordination that is not specified in the organisational hierarchy has been well analysed as an adaptive closed-system approach to contingencies that often arises consciously and rationally in response to an increasingly poor match between the interdependencies of tasks and the separations enshrined in the formal organisational design, especially in those cases where both the organisational environment and the tasks to be managed grow increasingly complex (Scott, 1981). Intersubjective, social constructivist approaches, on the other hand, extend the characterisation of institutional behaviour to encompass enactment of drama and ritualisation of myth. Although I would not go quite that far for nuclear plant operations in general, there were certainly many elements of observed behaviour that could, and perhaps should, be characterised this way.

In different contexts and in different manifestations encountered throughout our research, my colleagues and I observed that operational safety cannot be captured simply as a set of rules or procedures, or of simple, empirically observable properties, externally imposed training or management skills, or decomposable cognitive and behavioural scripts. Although much of what the operators do can be empirically described and framed in positivistic terms, a large part of how they operate and, more importantly, a large part of how they operate safely, is holistic, in the sense that it is a property of the interactions, rituals, and myths of the social structure and beliefs of the entire organisation, or at least of a large segment of it. Moreover, at least in the case of nuclear plants, air-traffic control, and aircraft carrier flight operations, it is intersubjective and emergent not just at the individual but also at the intergroup level. The more technically complex the system (and nuclear plants probably do occupy that end of the spectrum), the less possible it is to decompose the interlocking set of organisationally and technically framed social constructs and representations of the plant in search of the locus of operational safety (La Porte and Consolini, 1991).

This blending of operational, technical, social, professional, and task-oriented frames also draws heavily on local as well as regional and national cultures, beliefs, rituals, and myths. Though there are many aspects of operational reliability that apply generally across all the plants we have observed, the specifics of behaviour and adaptation in US plants are quite different from those in French, German, or Swedish ones (Rochlin and von Meier, 1994). Indeed, some fairly wide dissimilarities have been observed even between US plants in different regions of the country (Bourrier, 1996). To some extent, the divergence may be explained by differences in theoretical orientation, analytical tools, or both, as is the case between the contingency theory that underlies the Berkeley HRO group's approach, and the method of strategic analysis that is more commonly used in the European context (Terssac, 1992). But it may also be grounded in the different cultural and social milieux within which each plant constructs and reconstructs its own representations and meanings (Carnino, Nicolet and Wanner, 1990). The construction of operational culture is known to be vastly different in Russian plants, just as it is in other eastern European plants, and from the many conversations with our colleagues from abroad, we researchers from the Berkeley HRO group are certain that new information and new interpretations will be needed for the study of operational safety in Japan, China, or Korea.

What I have argued in this chapter is very much in the direction of encouraging further research on the cultural, interpretive, intersubjective, and socially constructed dimensions of nuclear plant operation. However, just as in the previous discussion of the social construction of risk, I do not mean that one can discard or neglect either the constraints and demands of the technical systems by which nuclear plants are operated, or the very real consequences that may ensue inside or outside the plant from accidents or errors. In his wonderful work *Aramis*, Latour (1996) argued that the Aramis project failed ultimately because it was unloved rather than technically flawed. But it should also be noted that the real-world constraints on the operation of the sociotechnical system that would have had to be in place to make Aramis work were so stringent and so far removed from current reality as to make it unlovable. Postmodern analytic tools, contentious enough for purely social constructs, need also to be augmented not only by the constraints of technical systems when they are involved, but also by the forms of adaptation and compliance required to put them into operation. Perhaps an entirely new term is needed, although descriptively accurate constructs such as inter-(subject/object)-ivity are to be avoided, not only because they are ugly neobarbarisms, but because they are entirely too simple to capture the variety of what is out there to study.

Representations of operational safety, like those of risk, are not entirely separable from the objective reality that they depict, but they cannot be entirely framed in objective terms either. Between the Scylla of social construction of reality and the Charybdis of positivist empiricism lie troubled and roiled theoretical waters. The consequence for the analyst, or at least for this analyst, is to be trapped perpetually in a crossfire directed, on one side, by schools of epistemological rationalism and positivistic method (largely engineers and psychologists) and, on the other, by the non-models that postmodernists (largely sociologists and political scientists) have deconstructed and abstracted from the real world. It is a very interesting place to work. And there is quite a lot of interesting work that still remains to be done.

Notes

1   A similar, if more one-sided example is the response of emergency-room physicians in Los Angeles to the frequent near-drownings of children to which they respond. They deliberately avoid chastising or questioning parents about fencing or otherwise protecting pools to avoid imposing feelings of guilt or negligence that might cause them to hesitate to call for help, or to bring their children in immediately (Karlene Roberts, private communication).

2   It has been a fundament of organisational strategic analysis applied to French techno-industrial systems that managers know little of technical details, and workers little of management politics (Crozier, 1967). The culturally elaborated *détente* is for managers to write detailed procedures that may not be workable, and workers do what they have to to get the job done without attempting to modify them, and each expresses some contempt for the other (Crozier and Friedberg, 1993). Efforts to apply 'American' standards of reducing conflict by inter-group meetings, and getting better correspondence between formal and informal procedures would almost certainly be counter-productive, particularly if applied without sensitivity to the long sociocultural history by which these arrangements were arrived at (Terssac, 1992).

References

BARLEY, S. R. (1992) Design and devotion: Surges of rational and normative ideologies of control in managerial discourse. *Administrative Science Quarterly*, **37**, pp. 363–99.

BARLEY, S. R. and BECHKY, B. A. (1994, February) In the backrooms of science: The work of technicians in science labs. *Work and Occupations*, **21**, pp. 85–126.

BECK, U. (1992) *Risk Society*. London: Sage.

BINDON, F. and JOHN, L. (1995, December) The role of the operator in the safety of the nuclear industry. *Power Engineering Journal*, **9**, pp. 267–71.

BONACA, M. V. (ed.) (1991) *Living probabilistic safety assessment for nuclear power plant safety management*. Paris: OECD/Nuclear Energy Agency.

BOURRIER, M. (1996) Organising maintenance work at two American nuclear power plants. *Journal of Contingencies and Crisis Management*, **4**, pp. 104–12.

BROAD, W. J. (1996, January 28) Risks remain despite NASA's rebuilding. *The New York Times*, p. 1.

CARNINO, A., NICOLET, J. L. and WANNER, J. C. (1990) Man and risks: Technological and human risk prevention (Norman, S. tr.). In KLING, A. L. (ser. ed.), *Occupational Safety and Health* (Vol. 21). New York: Marcel Dekker.

CLARKE, L. (1989) *Acceptable risk? Making decisions in a toxic environment*. Berkeley: University of California Press.

CLARKE, L. and SHORT, J. F. (1993) Social organisation and risk – some current controversies. *Annual Review of Sociology*, **19**, pp. 375–99.

CROZIER, M. (1967) *The bureaucratic phenomenon*. Chicago: University of Chicago Press.

CROZIER, M. and FRIEDBERG, E. (1993) *Le pouvoir et la règle*. Paris: Editions du Seuil.

CZARNIAWSKA-JOERGES, B. (1992) *Exploring complex organisations: A cultural perspective*. Newbury Park, CA: Sage.

DEAL, T. E. and KENNEDY, A. A. (1982) *Corporate culture: The rites and rituals of corporate life*. Reading, MA: Addison-Wesley.

DOUGHERTY, E. M. (1993) Context and human reliability analysis. *Reliability Engineering & System Safety*, **41**, pp. 25–47.

DOUGLAS, M. T. and WILDAVSKY, A. B. (1983) *Risk and culture: An essay on the selection of technical and environmental dangers*. Berkeley: University of California Press.

EMERY, F. E. and TRIST, E. L. (1960) Socio-technical systems. In EMERY, F. E. (ed.), *Systems Thinking*, (pp. 281–96). London: Penguin.

HARVEY, D. (1990) *The condition of postmodernity*. Oxford: Basil Blackwell.

HISKES, R. P. (1996, September) Hazardous liaisons: Risk, power, and politics in the liberal state. Paper presented at the Annual Meeting of the American Political Science Association, San Francisco.

INSAG (International Nuclear Safety Advisory Group) (1991) Safety Culture. (Safety Series No. 75-INSAG-4). Vienna: International Atomic Energy Agency.

JANSSENS, L., GROTENHUS, H., MICHELS, H. and VERHAEGEN, P. (1989, November) Social organisational determinants of safety in nuclear power plants: Operator training in the management of unforeseen events. *Journal of Organisational Accidents*, **11**, pp. 121–9.

JASANOFF, S. (1993) Bridging the two cultures of risk analysis. *Risk Analysis*, **13**, pp. 123–9.

KASPERSON, R. E. (1992) The social amplification of risk: Progress in developing an integrative framework. In KRIMSKY, S. and GOLDING, D. (eds), *Social Theories of Risk*, (pp. 153–78). Westport, CT: Praeger.

LA PORTE, T. R. (1996) High reliability organisations: Unlikely, demanding, and at risk. *Journal of Contingencies and Crisis Management*, **4**, pp. 60–71.

LA PORTE, T. R. and CONSOLINI, P. M. (1991) Working in practice but not in theory: Theoretical challenges of High-Reliability organisations. *Journal of Public Administration Research and Theory*, **1**, pp. 19–47.

LA PORTE, T. R. and THOMAS, C. W. (1995) Regulatory compliance and the ethos of quality enhancement: Surprises in nuclear power plant operations. *Journal of Public Administration Research and Theory*, **5**, pp. 109–37.

LATOUR, B. (1996) *Aramis, or the love of technology* (C. Porter, tr.). Cambridge, MA: Harvard University Press.

LUHMANN, N. (1993) *Risk: A Sociological Theory* (Barrett, R. tr.). New York: de Gruyter (Original work published 1991).

MEYER, J. W. and ROWAN, B. (1977, September) Institutionalised organisations: Formal structure as myth and ceremony. *American Journal of Sociology*, **83**, pp. 340–63.

PERROW, C. (1984) *Normal accidents: Living with high risk technologies*. New York: Basic Books.

PETERS, T. J. and WATERMAN, R. H. (1982) *In search of excellence: Lessons from America's best-run companies*. New York: Harper and Row.

PIDGEON, N. F. (1991, March) Safety culture and risk management in organisations. *Journal of Cross-Cultural Psychology*, **22**, pp. 129–40.

RALPH, R. (ed.) (1988) *Probabilistic risk assessment in the nuclear power industry: Fundamentals and applications*. New York: Pergamon Press.

REASON, J. (1995, August) A systems approach to organisational error. *Ergonomics*, **39**, pp. 1708–21.

REES, J. V. (1994) *Hostages of each other: The transformation of safety since Three-Mile Island*. Chicago: University of Chicago Press.

ROBERTS, K. H. (ed.) (1993) *New challenges to understanding organisations*. New York: Macmillan.

ROBERTS, K. H., ROUSSEAU, D. and LA PORTE, T. R. (1993) The culture of high reliability: Quantitative and qualitative assessment aboard nuclear power aircraft carriers. *High Technology Management Research*, **5**, pp. 141–61.

ROCHLIN, G. I. (1993) Defining high reliability organisations in practice: A taxonomic prolegomenon. In ROBERTS, K. H. (ed.), *New challenges to understanding organisations*, (pp. 11–32). New York: Macmillan.

ROCHLIN, G. I. (1996) Reliable organisations: Present research and future directions. *Journal of Contingencies and Crisis Management*, **4**, pp. 55–9.

ROCHLIN, G. I. (1997) *Trapped in the net: The unanticipated consequences of computerisation*. Princeton: Princeton University Press.

ROCHLIN, G. I., LA PORTE, T. R. and ROBERTS, K. H. (1987, Autumn) The self-designing high reliability organisation: Aircraft carrier flight operations at sea. *Naval War College Review*, **40** (4), pp. 76–90.

ROCHLIN, G. I. and VON MEIER, A. (1994) Nuclear power operations: A cross-cultural perspective. *Annual Review of Energy and the Environment*, **19**, pp. 153–87.

ROTH, E. M., MUMAW, R. J. and STUBLER, W. F. (1992) Human factors evaluation issues for advanced control rooms: A research agenda. In HAGEN, E. W. (ed.), *IEEE Fifth Conference on Human Factors and Power Plants* (pp. 254–60). New York: Institute of Electrical and Electronics Engineers.

SAGAN, S. D. (1993) *The limits of safety: Organisations, accidents, and nuclear weapons*. Princeton: Princeton University Press.

SANNE, J. M. (1995, October) Air traffic control as situated action: A preliminary study. Conference on Social Studies of Science and Technology, Charlottesville, VA.

SCARRY, E. (1985) *The body in pain: The making and unmaking of the world*. New York: Oxford University Press.

SCHMITT, E. (1996, January 31) Fighter pilot killed in Nashville crashed in sea in '95 Navy says. *The New York Times*, p. 1.

SCHULMAN, P. R. (1993) Negotiated order or organisational reliability. *Administration and Society*, **52** (3), pp. 356–72.

SCHULMAN, P. R. (1996) Heroes, organisations, and high reliability. *Journal of Contingencies and Crisis Management*, **4**, pp. 72–82.

SCOTT, W. R. (1981) *Organisations: Rational, natural, and open systems*. Englewood Cliffs, NJ: Prentice-Hall.

SEARLE, J. (1969) *Speech acts: An essay in the philosophy of language*. London: Cambridge University Press.

SHRADER-FRECHETTE, K. S. (1991) *Risk and rationality: Philosophical foundations for populist reforms*. Berkeley: University of California Press.

SLOVIC, P., FISCHHOFF, B. and LICHTENSTEIN, S. (1976) Cognitive processes and societal risk-taking. In CARROLL, J. S. and PAYNE, J. V. (eds), *Cognition and social behaviour*, (pp. 165–84). Potomac, MD: Lawrence Elbaum.

STARR, C. (1969) Social benefits versus technological risks. *Science*, **165**, pp. 1232–8.

'T HART, P. (1993) Symbols, rituals and power: The lost dimensions of crisis management. *Journal of Contingencies and Crisis Management*, **1**, pp. 36–50.

TERSSAC, G. DE (1992) *Autonomie dans le travail*. Paris: Presses Universitaires de France.

THOMPSON, M., ELLIS, R. and WILDAVSKY, A. (1990) *Cultural theory*. Boulder, CO: Westview Press.

TURNER, B. A. (1978) *Man-Made disasters*. London: Wykeham Publications.

TURNER, B. A. (ed.) (1989) *Organisational symbolism*. Berlin: de Gruyter.

VAUGHAN, D. (1996) *The* Challenger *launch decision: Risky technology, culture, and deviance at NASA*. Chicago: University of Chicago Press.

VON MEIER, A. (1995) Cultural factors in technology adoption: A case study of electric utilities and distribution automation. Unpublished doctoral dissertation, University of California, Energy and Resources Group.

WALD, M. L. (1996, January 26) Ambitious update of air navigation becomes a fiasco. *The New York Times*, p. 1.

WEBSTER'S SEVENTH NEW COLLEGIATE DICTIONARY (1969). Springfield, MA: G. C. Merriam.

WEICK, K. E. (1987) Organisational culture as a source of high reliability. *California Management Review*, **29**, pp. 12–127.

WEICK, K. E. and ROBERTS, K. H. (1993, September) Collective mind in organisations: Heedful interrelating on flight decks. *Administrative Science Quarterly*, **38**, pp. 357–81.

ZUBOFF, S. (1984) *In the age of the smart machine: The future of work and power*. New York: Basic Books.

CHAPTER TWO

Constructing organisational reliability: the problem of embeddedness and duality

MATHILDE BOURRIER

University of Technology at Compiègne, France
Department of Technology and Human Sciences

This chapter attempts to make a contribution to the study of organisational reliability in high-risk industries through a sociological analysis of four scheduled outages in France and the United States. The field research took place between 1991 and 1995. I maintain that organisational reliability is neither an attribute, nor an attitude that can be imposed solely from outside. Rather, it is a result of a complex compromise between resource allocation and actors' strategies *within* the organisation. I also maintain that the working out of responsibilities, the allocation of resources, and the delegation of power greatly influence the ability to build and sustain organisational reliability.

In this chapter I also point out the intrinsic fragility of organisational reliability, which is largely dependent upon essentially dual actors' strategies. Consequently, I advocate a systematic assessment of social interactions and their motives in every high-risk organisation, so that the fundamentals of organisational reliability can be understood. Sociological and anthropological methodologies can greatly facilitate this sorely needed task.

It is too often believed that one can successfully achieve organisational reliability by implementing safety management systems (e.g. systematic assessments or sophisticated audits). Safety management systems may play their part, but I believe that they are not sufficient to ensure organisational reliability (for presentation of the newest safety management system see Hale and Baram, in press). I suggest that organisational reliability should be investigated and seen as a property of the social systems embedded in what Rochlin (1993) calls 'reliability-seeking organisations' (p. 12). In my view, the social construction of organisational reliability can best be analysed through a *systemic analysis* (Chisholm, 1989; Crozier and Friedberg, 1977; Crozier and Thoenig, 1975; Landau, 1991) thus helping to focus on systemic effects.

In the first section, I review the literature that has contributed to shaping the concept of organisational reliability. I then discuss the limitations of current theories, arguing that too little attention is paid to the embeddedness of reliability (Granovetter, 1985). In the second section I describe the organisational setting at each of the four plants studied (scheduling and planning, creating and updating of maintenance procedures, coordination between operations and maintenance, and subcontracting practices), arguing that each plant has its own organisational strategy for coping with the complexity of scheduled outages. In the third section, I analyse how this diversity of organisational features influences the strategies of actors. Finally, I attempt to characterise the way each plant develops its organisational reliability.

2.1 THE ORIGINS OF ORGANISATIONAL RELIABILITY

Striving for organisational reliability is like searching for the Holy Grail. No one can actually prove that a given organisation is safer, more reliable, more resilient, or more robust than the next one. No single best way ever exists, and high-risk industries are no exception to this. But the fact that an increasing number of organisations conduct activities that require a superior level of performance and an equally high level of safety only makes the issue more intriguing (on nuclear power plants, air-traffic control, airlines, chemical plants, see Gras, Moricot, Poirot-Delpech and Scardigli, 1990; La Porte and Consolini, 1991; La Porte, Roberts and Rochlin, 1987; Roberts, 1993). As a result, organisational approaches have attracted increasing interest (sometimes at the expense of ergonomic or psychological perspectives).

Scholars nevertheless intuitively suggest that there is an organic link, somewhat still obscure, between the division of labour, the organisation of activities and the safety and reliability of an organisation (Ackermann and Bastard, 1991; Clarke, 1989). Moreover, organisational failures such as Chernobyl (Reason, 1987, 1990b), the French blood scandal (Setbon, 1993), or the *Challenger* accident (Vaughan, 1996) have indeed emphasised the contribution of organisational factors.

2.1.1 Organisational failure

Research on organisational reliability was initially framed mostly in terms of *organisational failure* (Perrow, 1984; Setbon, 1993; Shrivastava, 1987; Vaughan, 1996). High-risk industries have been studied chiefly when they have failed rather than when they have succeeded, perhaps because their failure can hardly go unnoticed.

Perrow (1984) claimed that, for technological and organisational reasons, high-risk industries cannot escape a normal accident (i.e. an endogenously produced failure). In this view, there is no point in speaking about organisational reliability. As a topic it is thus neglected in favour of organisational failure, despite the fact that both concepts could be interpreted as two sides of the same coin. (See Amalberti, 1996, and Reason, 1990b, for example, who argue that error and performance are closely linked; or Faverge, 1970, 1980; Guillermain and Mazet, 1993; and Poyet, 1990, who suggest that human activity is a source of both reliability and failure.)

2.1.2 Organisational reliability

A different, albeit in many respects complementary, perspective was adopted by the High-Reliability Organisations Group (HRO group), founded in Berkeley, California, by T. R. La Porte with G. Rochlin, K. H. Roberts and P. Schulman and, more generally, by 'reliability theorists', as Sagan (1993, p. 14) calls them (La Porte, Roberts, Rochlin, Schulman, Wildavsky, Weick).[1] HRO scholars are not so much interested in the fact that these organisations are doomed to fail sooner or later, but rather in the conditions that allow them to fail so rarely. Their agenda was thus to examine the foundations of organisational reliability (La Porte and Consolini, 1991; La Porte *et al.*, 1987). HRO authors pointed out a set of selective criteria that, according to them, make HROs special thus requiring specific methodologies, as opposed to classic organisations, which can rely on common trial-and-error learning processes.

According to HRO theorists an HRO displays four characteristics (see Rochlin, 1988):

1 Members of the organisation totally agree on its goals.

2 One can observe the use of redundant decision channels and the use of redundant controls and supervision between staff.

3 Comprehensive training programmes help employees develop their expertise in new domains while at the same time refreshing their memories.

4 The power to make decisions is both highly centralised and highly decentralised, meaning that anybody (even from the lowest rank of the hierarchy) can stop any kind of activity if he or she judges that installations or employees are at risk.

For HRO authors these four criteria identify the capacity of an organisation to sustain a high level of reliable functioning.

2.1.3 Organisational culture and reliability

Weick (1987, 1993) and Rochlin and von Meier (1994) have attempted to extend the HRO research agenda. In different ways they have emphasised the role of cultural factors in the development of organisational reliability, and have developed a series of analyses based on international comparisons or on the cultures of firms and small groups. Weick went further in studying the mechanisms leading to the loss of *sensemaking* in an organisation – that is, to the destruction of roles, structures, and the ability to cooperate – a process that greatly compromises coherence, resilience, and, finally, organisational reliability (Weick, 1993, 1995).

2.1.4   The embeddedness of organisational reliability

In many respects all these approaches are ground-breaking: they all emphasise critical organisational features that deserve closer scrutiny. As scholars have shown, organisational reliability (as well as its twin notion, organisational failure) is embedded in day-to-day organisational intricacies, so one has to investigate the social fabric of cooperative behaviour in order to trace its sources. Unfortunately, the advocates of these approaches tend to overlook (or rather take for granted) that reliability is a *social construct*, which one may approach better by adopting a framework which seeks to disentangle the variety of employees' motivations, strategies and interactions about issues that include, *but are not limited to*, safety. This is wholly different from Rochlin's *Social Construction of Safety* which narrowly focuses on employees' opinions regarding safety. I would be inclined to call this perspective *cultural* rather than *social*.

Perrow (1984, p. 67), for example, overlooked the ability of the people on the job to intervene to counteract adverse consequences of high-risk technologies. He saw in-house personnel as mainly passive or, at any rate, totally overawed by the complexity and unpredictability of the installations which they run. In reality, however, they are the most important contributors to the quest for reliability.

HRO theorists, too, have flaws of their own. First, I am unconvinced that simply adopting the four criteria cited above will automatically ensure the reliability of any organisation. These criteria may well be crucial, but what if organisation X, which has been known for years as highly reliable, does not meet these criteria? Conversely, what could be said of organisation Y if it meets exactly the same criteria as organisation X, but has a poor safety record? Secondly, HRO theorists (just like Perrow) do not study the construction of reliability from the actors' perspective. If they did, they would find it difficult to claim that goal-sharing can ever be achieved: dissent, compartmentalisation, power struggles, and goal displacement are the features of most organisational life (Crozier and Friedberg, 1977; Gouldner, 1954a, 1954b; Selznick, 1949; Whyte, 1948).

I suggest that because reliability-seeking organisations are also normal organisations, organisational reliability should be researched through the study of social interactions and professional relations. I maintain that organisational reliability depends upon the quality and the nature of social relations, and hence upon the strategies adopted by the work force. This leads me to investigate organisational reliability through the prism of work relations.

Table 2.1 Ratio of number of employees on site to outage duration

Nuclear power plant	Number of employees	Make, power, and number of unit	Outage duration	Ratio (employees: outage days)
Bugey	1400 permanent 1500 contractors	Framatome (4 PWR, 900 MW 5 1 Graphite-1994)	Decennial outage 145 days	20
Nogent	600 permanent 1000 contractors	Framatome (PWR, 1300 MW) 2	Annual outage 47 days	34.04
Diablo Canyon	1900 permanent 1900 contractors	Westinghouse (PWR, 1100, MW) 2	Annual outage 63 days	60.31
North Anna	1000 permanent 1500 contractors	Westinghouse (PWR, 900 MW) 2	Annual outage 57 days	43.85

• Figures for Bugey cannot be strictly compared with those of the three others because they refer to a decennial outage, a much longer period (5 months).

2.2 CONFRONTING ORGANISATIONAL DIVERSITY

The examples in this chapter are drawn from fieldwork carried out in four nuclear power plants – Bugey and Nogent in France and Diablo Canyon and North Anna in the United States – between 1991 and 1994 (Bourrier, 1994, 1996a, 1996b). The most striking feature that emerged from my extensive studies was the diversity of organisational strategies and responses for coping with scheduled outages. Although the four plants were comparable in terms of design (Westinghouse, Framatome) and power (900 MW, 1100 MW, 1300 MW) (see Table 2.1), organisational choices were found to be totally different from one plant to the next.

2.3 OUTAGE PROBLEMS AND CHALLENGES

2.3.1 Outages

Scheduled outages are required by law; they are part of a licensee's obligations. During a scheduled outage, hundreds of tests, repairs, and revisions are performed in order to ensure plant safety and component reliability. During that period, part of the fuel is also changed and reloaded. When the core is partly unloaded, favourable radiological conditions allow workers to perform most of the maintenance activities.

Since the early 1990s the variance of outage duration between companies, plants, and countries has decreased. Nowadays, a 'normal' annual outage lasts between 30 and 50 days. There is strong incentive for the electrical companies to reduce this duration further in order to foster productivity, since for a reactor not to produce means an interrupted supply. All efforts are geared to minimising the outage. Therefore, scheduling and planning options as well as management strategy are intensely debated and carefully thought through.

2.3.2   Organisational challenges

A scheduled outage is interesting in many respects. First, it is a highly complex technical task requiring the dedication of hundreds of people with very different backgrounds, skills, and crafts (mechanics, instrumentation and control [I&C] technicians, electricians, welders, pipe-fitters, engineers, planners, and schedulers, to name a few) and involving a wide variety of activities. Because the outage must proceed according to a rigorous plan, issues of coordination and cooperation are crucial.

Secondly, due to the radiological risk, the activities have to be performed in compliance with certain rules in order to minimise workers' exposure. As a result, each job has to be formally authorised at each sequence according to the type of work, the geographical area, and the exact state of the core. A procedure that authorises a job is called a *clearance*. Unlike other industries, such as chemical plants, it is impossible to perform all tasks at once. Hence, each task is given a priority and a specific work order, which is attached to a work package.

Thirdly, because the operation of the installations does not cease completely, operation crews stay on board. This arrangement means that maintenance crews do not have full control over the outage. They must negotiate constantly with operations personnel, who are formally in charge of authorising work. More fundamentally, operations and maintenance have diametrically opposed attitudes toward outages. On the one hand, operations personnel dislike disturbances and new situations because it is their primary job to keep the plant running safely, or as they like to say, to 'fly the plane'. Unfortunately, during an outage they will have to deal daily with new and unpredictable situations, especially because they will be under pressure from the maintenance workers, who like to see all materials available at once. In addition, it has been shown that maintenance jobs cause problems and sometimes errors, a finding that contradicts what plant personnel had believed for years: that outages were less risky than full-power mode. As a result, operations personnel are more cautious than they used to be. Maintenance workers, on the other hand, truly enjoy the outage period. It is rich and exciting for them, because they can open, test, and repair a variety of materials and see how components have fared.

The number of people involved requires a great deal of coordination. Most of the people carrying out the operations are subcontractors. (At Bugey and North Anna, 70 per cent of the maintenance activities are performed by outside contractors; at Diablo Canyon, 80 per cent; and at Nogent, 90 per cent.) Monitoring and mixing outside contractors with in-house personnel is an additional challenge.

2.3.3   Achieving organisational reliability: the problems

As mentioned above, work can be performed only if it is explicitly authorised by operation personnel. Maintenance crews are therefore highly dependent upon operational decisions. Yet, maintenance crews are responsible for carrying out the work correctly and on time. Organisation theory holds that this close interdependence might give rise to conflicts, which will have to be mitigated one way or another. Otherwise, the risk of jeopardising the outage might increase. Creating the conditions for efficient cooperation is also a matter of integrating contractors on site, especially when it comes to feedback from the field and information exchange between contractors and in-house employees.

A second set of problems is the difficulty of working under time constraints with a vast number of detailed procedures – one of the major challenges faced daily by maintenance crews. Scholars who have studied work inside nuclear power plants have all been impressed by the number of regulations in use (see Hirschhorn, 1993; Osty and Uhalde, 1993; Schulman, 1993; Terssac, 1992). It is thus crucial to investigate the question of the conditions under which it is possible for maintenance people to follow these procedures without jeopardising the outage planning and, hence, the outage duration.

This subject has been treated in many different studies in ergonomics and sociology (Chisholm, 1989; Crozier, 1963; Downs, 1967; Girin and Grosjean, 1996; Gouldner, 1954a, 1954b; Linhart, 1978; Leplat and Terssac, 1990; Terssac, 1992), according to which no procedure, formal script, or formal structure can ever be strictly followed. The reasons have been well documented: unpredictable events can make a rule inapplicable, because the exact conditions of an activity cannot be totally determined in advance. As a result, employees are forced to adjust and adapt in order to perform the required tasks (Chabaud and Terssac, 1987; Coninck, 1995; de Keyser and Nyssen, 1993). One of the least controversial results in social science appears to be that prescribed work never matches real work. However, I will demonstrate that this view is not fully supported by the empirical evidence which I have gathered. The whole issue of the natural strategy of deviation should be thoroughly revised.

2.4 DIVERSITY OF RESPONSES

The challenges described above are diversely taken into account at Bugey, Nogent, Diablo Canyon, and North Anna. In this section I will summarise the various organisational devices provided at each plant in order to cope with the four identified challenges.

In a nuclear power plant operations and maintenance are the biggest employers (see Table 2.2). Journeymen, foremen, general foremen, and section or department managers constitute the main players of maintenance services. They are supported by schedulers and planners (especially in the United States), engineers (*components*

Table 2.2 Maintenance personnel versus operations personnel

	Bugey	Nogent	Diablo Canyon	North Anna
Maintenance	400	180	459	250
Operations	130 (unit. 4/5)	142	380	125

engineers or sometimes *systems engineers*), and various technical staff. On the operations side the hierarchy goes from technicians, through operators (both licensed and unlicensed), to shift supervisors and their assistants. They are supported by operations engineers and technical staff. How does each site provide for the challenges they face during an outage?

2.4.1 Scheduling and planning an outage

At Bugey (studied in 1991), planning and scheduling were not top priorities. Only two months were devoted to the planning and preparation of a scheduled outage. A real planning section did not exist. Headquarters in Paris provided the site with the main planning frame, and only limited adjustments could be made on site by two planners and a small team of operations personnel. The planning strategy that was adopted allowed a limited number of contingencies to be built into the planning. *Préparateurs* were in charge of preparing for the outage. They created new procedures upon request and made sure that all work packages were available to foremen. When Bugey was under study, the *préparateurs* were behind schedule, and the outage started with a third of the work packages missing.

Compared to Bugey, Nogent (studied in 1994) displayed increasing attention to planning issues, with six months being devoted to preparation of the outage. This prolonging of the preparation period could be attributed to management's new priorities and concerns at France's state-run electricity supply monopoly, the *Electricité de France* (EDF). It was perceived that a successful outage depended upon extensive preparation.

Two months prior to the outage all secondary plannings had to be frozen in order to eliminate the potential risk of disorganisation and unplanned rush (something Bugey's *préparateurs* were constantly arguing about). A planning section worked closely with a small operations team much like the one at Bugey. EDF headquarters was still providing the main frame, but the site became increasingly responsible for the detailed preparation and planning of each phase.

This new attention to planning and scheduling issues did not change the situation as far as normal operations were concerned. EDF plant managers believed a specific scheduling function was unnecessary because it was understood that every employee should be able to schedule his or her own work.

At Diablo Canyon employees devoted obsessive attention to planning and scheduling. Planners and schedulers were key actors. Their explicit role was to coordinate the constraints placed on maintenance and operations and on maintenance activities themselves. Sophisticated tools existed for planning and anticipating problems at various stages. These planning devices were in use not only during an outage, but all year long. Such continuous planning was a key difference between Bugey's and Nogent's approaches to outages.

Diablo Canyon was quite remarkable for its extensive planning philosophy. A work planning centre, consisting of planners and schedulers (100 people in all) worked out a daily plan, for which activities from all services were reported. Daily schedulers were in charge of creating a plan that served both maintainers' and operations' needs. The key issue was to provide the operations team with materials. A similar document (called the *matrix*) reported both the systematic revisions and tests by the maintenance crew and the periodic testing requirements of the operations people. In this way each department could see the other's constraints and act accordingly. Schedulers worked closely with general and maintenance foremen to collect information ranging from problems with parts delivery and illness to training constraints and plant conditions. During the outage the same organisational devices were in place. Daily schedulers were replaced by outage schedulers, and daily planning became outage planning. The coordination of maintenance and operations was designed in advance and not worked out on the job once the outage had started.

In addition to these considerable, but still normal, efforts, 18 months were deemed necessary to plan all outage activities. More than 30 special groups, called *high-impact teams*, were in charge of preparing the outage. These groups consisted of individuals from various service areas such as access badging, turbine generators, valves, steam generator blowdown, containment fan cooler, and spare parts, and represented all ranks (journeymen, foremen, schedulers, and engineers). Their main job was to find the best way to perform the tasks related to a single activity in a corresponding time frame, or window. They had no say in determining the windows themselves, whose length was set by the outage manager. They were, however, responsible for allocating resources within the window. As soon as the outage started, the *high-impact teams* were disbanded. It was up to maintenance foremen and general foremen to execute the agreed plans: on Diablo Canyon's high-impact teams, see Haber, Barriere, and Roberts (1992); and Haber, Shurberg, Barriere, and Hall (1992). Of the four nuclear power plants studied, Diablo Canyon was the only one that allocated so many resources to the preparation of the outage.

At North Anna, preparing an outage was 'business as usual' (as the plant manager declared). A group of 37 employees (including schedulers and planners) prepared future activities all year long. Maintenance staff, operators, and schedulers worked together in order to design the plan of the day, the plan of the week, and the plan of the month. At North Anna outage preparation lasted one year. Unlike Diablo Canyon, however, no specific organisation was in charge of planning the outage. Moreover, unlike the three other cases, there was no outage control centre (*structure d'arrêt*) at North Anna.

Planning and preparing an outage used different resources, depending on the plant in question. Bugey and Diablo Canyon were the two opposite types; North Anna and Nogent lay somewhere between those two poles. This diversity was clearly reflected in the comparison of the outages at these four nuclear power plants.

2.4.2 Cooperation between maintenance and operations

At Bugey, work relations between maintenance and operations were improvised. The many tensions between the two areas were particularly noticeable during outages, but they were also present all year long. They could be regarded as a key characteristic of organisational life at Bugey. Operations people claimed that maintenance personnel did not pay attention to the materials (especially small and fragile materials like sensors). They claimed that maintenance workers did not strictly follow the rules, and that they constantly put pressure on them to get the job done faster. On the other hand, maintainance staff believed that operations people deliberately ignored planning requirements, and that they failed to understand the implications of working under a very tight schedule, especially when this implied working with subcontractors who were faced with an incentive scheme to discourage delays.

During the outage a small operations team located at the outage control centre was in charge of 'lubricating' relationships between operations and maintenance. The centre enabled clearances and processed authorisations in order to support the shift supervisor, who ran both units at the same time. More important, this compact team had the ability to grant maintenance workers the flexibility they needed to complete their work. Unfortunately, the negotiations between the two departments remained difficult, despite this valuable assistance.

At Nogent, cooperation was difficult and a bad atmosphere prevailed. Although the outage was prepared in greater detail than at Bugey, it started with a vast backlog of clearances and work authorisations. This backlog was due to a serious mismatch between the number of activities and the number of clearances that could be issued in one night by a regular operations crew. There was a lack of coordination between the maintenance department, which prepared the work packages and their related clearance requests, and the operations department, which could not keep up with the paperwork. Since the early 1990s, maintenance and operations departments in French nuclear power plants have undergone massive reorganisation with employees being asked to modify their routines and adjust to new work practices. At Nogent these changes caused maintenance and operations personnel to neglect part of their outage preparation and planning, notably that of clearances. Despite six months of scheduling and preparation, clearances could not be obtained in time even on the first day of the outage, a problem that angered contractors. The clearance issue has revealed an important organisational flaw. Priorities among maintenance jobs had not been ranked. The EDF technicians who were in charge of these logistics declared that they had lacked time to perfect their planning as they would have liked, so they requested all clearances at once. This disorder generated a great deal of resentment against reforms. Undoubtedly, employees of the Nogent plant adopted various strategies to slow down or impede the process, and the preparation of the outage became the focus of dissatisfaction.

At Diablo Canyon peaceful relationships between maintenance and operations reflected carefully managed cooperation strategy. Conflicts were avoided. Contrary to the practice in the other plants we have studied, clearances were prepared by maintenance workers themselves. Although all the members of the clearance team came from the operations department, they all reported to maintenance. As former licensed operators, they were fully skilled in the management of clearances. Their role was explicitly to ensure that maintenance workers could work on the materials according to operations requirements. In this organisation shift supervisors remained the formal authorisers, because they had to sign off clearances electronically. More-over, maintenance foremen at Diablo Canyon were permitted to print their own clearances each morning.

At North Anna the atmosphere between operations personnel and maintenance workers was highly conflictual and seemed even to be deteriorating at the time of the study (1993). Complaints from both sides were non-stop. Shift supervisors had become extremely cautious about work authorisations, an attitude denounced by maintenance foremen and their subordinates, who believed that operations employees were purposely slowing down the process. For example, shift supervisors would not hesitate to ask questions in order to double-check that a requested clear-ance was appropriate. They also frequently stopped the process if they were not satisfied with the answer. It was noticed that more than half of the shift supervisors were genuinely trying to obstruct work clearances as much as possible, in order to minimise hazardous maintenance activity on their shifts.

This difference in attitudes towards maintenance work remained unexplained until it was observed that the management at North Anna had recently adopted a new audit system that traced the numbers and types of problems and errors back to each shift, thereby allowing an assessment of a shift's relative performance. This new system, which had been implemented after operations personnel at North Anna committed two violations of Nuclear Regulatory Commission regulation during a maintenance period, was not welcomed in the operations department. Some shift supervisors became so antagonistic that they refused to allow the execution of any maintenance job during their shift that involved risk (as most of them do). This attitude led to their pushing work on to those shift supervisors who were more willing to take risks. And, as I discovered, the supervisors who were prepared to take on extra maintenance were the more experienced ones.

Instead of increasing nuclear safety, the new audit system may have thus pro-gressively undermined the overall level of safety. A vicious circle was formed. First, maintenance foremen had to adjust to this new system and, in order to protect themselves from last-minute changes or obstacles introduced by operations person-nel, they began to pick and choose. The so-called easy ones (such as preventive maintenance, standard changes, cleaning-up, and oil-checks) were put forward and usually accepted by operations staff without any difficulty, whereas the tricky ones (protracted repair jobs or troubleshooting, which were new to the maintenance workers and which were difficult to time accurately in advance) were put on hold and sometimes even rejected, depending on the attitude of the shift supervisor. As a result the outage planning, or the plan of the day, began to suffer unofficial and unacknowledged adjustments.

It is conceivable that a difficult job might in fact have got less attention than it should, simply because it ended up being left to the last (obliging) shift supervisor. Indeed, the risk of error had increased because the tricky jobs were disguised and not distributed fairly among shift supervisors. Ultimately, the actual planning of the outage came to diverge more and more from what had been intended. Management appeared to ignore the number of changes and delays that were concealed by these subtle manoeuvres and countermoves. Clearly, the new audit system did not lead to enhanced safety. Because of the atmosphere of competition between shifts, it encouraged shift supervisors to work strictly by the book. It can also be predicted that shift supervisors will not be willing to continue accommodating non-routine maintenance requests. The natural trend will inevitably lead all operations employees to become extremely suspicious of maintenance work and to block the outage altogether.

2.4.3   Integrating subcontractors

The French context

In France, contracting companies have to be chosen from an established list, which includes only those who have been audited and accepted as nuclear contractors by EDF. They are in charge of selecting employees who have the proper training and qualifications. EDF foremen used to have informal influence over the choice of contractors and could veto the choice of employees in some cases. They preferred giving the same job to the same firm in the belief that such continuity reduced the risk of errors. Current contracting practice, which is based on competitive job application, reduces the maintenance foremen's ability to interfere with market procedures. Resistance to this new trend, however, is already perceptible among maintenance workers.

More important, labour legislation has completely transformed professional relations during maintenance activities. As amended by the regulation of 12 July 1990 (No. 90613) – known as Title IV, Article 29 – which governs contracting practices and the hiring of licit labour (*soustraitance et prêt de main d'oeuvre licite*), Article L.125-2 of the French labour code stipulates that contractors have to be directly supervised by their own foremen and not by EDFs. The era when EDF maintenance foremen were all-powerful and regarded as the real masters is over. They have lost a considerable amount of power, whereas contractors, who are faced with new responsibilities, especially with regard to quality-control measures, have gained in status. At first contractors were somewhat daunted by this new responsibility, since it meant a commitment and dedication they had never had the opportunity to demonstrate fully. They adjusted to the challenge, however, and have become extremely knowledgeable (see the section on *Rule-following at Nogent*, page 40).

The resentment among EDF employees still rankles, however. Unions have taken up the issue, since they see subcontracting as a managerial strategy to get rid of the skilled craftsmen – their main on-site supporters. EDF employees claim they are being deskilled and believe that they will shortly lose all status through being deprived of access to the necessary information.

The US context

In the United States contractors are recruited either through job shops or, when special crafts people such as pipe-fitters, welders, or mechanics are needed, through Union placement officers. Some contractors (often I&C technicians) are freelance and negotiate their contracts directly with the plant.

At North Anna maintenance foremen were in charge of recruitment. They acknowledged that they try to get the same contractors each year because they had already established a degree of mutual recognition and trust with these workers. By contrast, employees usually disliked working with contractors, whom they did not fully trust. They preferred to work on separate work-sites and to avoid mixed teams. Thus, maintenance foremen had to decide carefully how to allocate work between contractors and North Anna employees. Know-how and skill were clearly at stake: jobs that obviously required skills and/or materials that were not supplied in-house were left entirely to contractors. Similarly, undesirable tasks, such as preventive maintenance, would also be given to contractors under the supervision of North Anna foremen. Maintenance workers retained those in-house jobs that they felt were vital to the plant and jobs that protected the skills of North Anna employees (e.g. the checking, repairing and testing of valves).

At Diablo Canyon, too, maintenance foremen were in charge of recruiting contractors. They claimed that they had more control over the list provided by the job shop than over the one provided by the Union, from which it was usually difficult to delete names. And just as at North Anna, Diablo Canyon was divided over the issue of employing contractors. However, suspicions about contractors were not distributed equally among the hierarchy. As demonstrated by the high percentage of contractors whose engagement was renewed, managers trusted the contractors they recruited, whereas maintenance foremen – even though many of them did not voice their misgivings freely – by and large felt that too many inexperienced contractors were still being given responsibility during outages. Nevertheless, at Diablo Canyon contractors were always mixed with permanent employees, and some in-house journeymen became outage foremen in order to supervise outsiders on site.

2.4.4 Creating and updating maintenance procedures

At Bugey, a handful of individuals were in charge of the crucial task of creating and modifying procedures: only 17 *préparateurs* created and updated maintenance procedures, work packages and work orders. Each *préparateur* had an area of expertise (mechanics, electricity, I&C) and was in charge of a certain type of equipment (pumps, valves). He was an expert in measuring, technical drawings, and procedures, but his expertise was not fully recognised by maintenance foremen, who believed they should be the experts because they had hands-on experience. It appeared that *préparateurs* were systematically overworked. By the time they finished their work packages and drafting new procedures, they were already focusing on the next outage. *Préparateurs* were not part of the outage control centre. A temporary organisation was set up to manage and coordinate the outage. During an outage, *préparateurs* were seldom seen on work sites.

Despite the limited availability of the *préparateurs*, workers were not authorised to change a procedure in any way without prior approval.

At Nogent, a handful of individuals were in charge of the crucial task of creating and modifying procedures. Twenty *préparateurs* created work packages, and wrote work orders and maintenance procedures. Unlike their Bugey counterparts, they belonged to the outage control centre, although they did not get around much during the outage. As in Bugey, they were formally responsible for modifying procedures or work orders. They were also in charge of tracing all reports of instances of non-compliance and deviation. The unit could not be started up until the *préparateurs* had cleared up all deviations, a requirement that put a heavy load on them.

Corrections and updates were easy at Diablo Canyon, where section engineers were responsible for creating and modifying maintenance procedures, among other things. There were as many section engineers as maintenance foremen. In addition, a group of 20 procedure writers were specifically in charge of I&C procedures. Section engineers and foremen had comparable status in their respective hierarchy. (Foremen are in practice slightly better paid, since they qualify for overtime pay.)

During the outage, section engineers left their usual offices for the foremen's quarters in order to be closer to the crews and to be ready to adjust, modify, or wholly or partly rewrite the procedures. They insisted that during an outage it was their primary job to remain available to the crews at all time. During an outage, they were described by workers as 'acting foremen'. Moreover, both foremen and section engineers at Diablo Canyon had to sign for procedural updates. This requirement enhanced cooperation between the two groups.

At North Anna foremen and crews updated their procedures. A separate service was responsible for creating new maintenance procedures. By contrast with arrangements at the three other plants, the one at North Anna placed maintenance foremen in charge of updating procedures whenever their crews found them impossible to work with. Hence, foremen had both the authority and the responsibility to initiate changes. They decided with their subordinates whether a procedural change was required. If the procedural change involved something that was important for safety equipment (as defined in the quality control manual), the request had to be reviewed and assessed by a managerial committee called the *Security and Nuclear Safety and Operations Committee*, which met daily. After a full briefing delivered by the foreman, the committee members could ask additional questions to improve their understanding of the reasons behind the request for a procedural change. The workers (who by then had stopped their work) received rapid feedback on the course of action to be adopted. If the procedural change was accepted, they would help the foreman draft a new procedure.

I now turn to the effects that these organisational differences had on the workers' behaviour with regard to following or breaking rules. My hypothesis is that the more difficult it is to update and change procedures, the more likely it is that rule-breaking will occur.

2.5 WORKFORCE STRATEGIES AT THE SOURCE OF ORGANISATIONAL RELIABILITY

2.5.1 Arrangements and informal networks at Bugey: a necessity

At Bugey many informalities, violations and improvisations[2] were observed. Most employees confessed that they were rarely able to follow work procedures strictly. It was common practice to repair a pump or a valve without following the procedure. For workers, procedures were not a support but merely a statement of what ought to be done. For instance, prescribed procedures did not take into consideration problems with wear and tear on components, the availability or non-availability of tools, delays or broken parts. Systematic deviant behaviour was indeed observed. Maintenance staff were constantly improvising small repairs without a specific work order, inventing new sequences when their procedures were too ill-defined, or cutting corners by skipping procedures that they considered irrelevant. For all the discretion the Bugey staff exercised, however, they did observe limits. Workers wanted these arrangements to remain secret, otherwise, if something went wrong, workers and foremen would get the blame. Employees shared among themselves a set of unspoken, or private, rules. Adjustments were routinely made, but they did not exceed certain self-imposed limits. These boundaries were learned through a long socialisation process, akin to *compagnonnage* (a craftsman's time of service after apprenticeship).

Most maintenance staff and their foremen had extensive experience with components. Foremen recorded in small notebooks crucial figures or important information about past repairs and tests. In principle, these observations should have been incorporated into maintenance procedures and brought to the attention of *préparateurs*, but they rarely were. Staff members recalled that they used to inform the preparation team but, when they realised that modifications were never made or rarely taken on, they decided to keep those observations to themselves in their notebooks.

By regularly touring the work sites, foremen made sure that everybody understood that some necessary, yet secret adjustments, were accepted practice but that others were not. Foremen had a great deal of power on work sites. They were the ultimate decision-makers on breaking the rules. Much autonomy thus existed on site, largely reflecting the difficulties workers had when performing their tasks. What were the reasons for this strategy?

As noted previously, there was a lack of field assistance for maintenance employees. Bugey journeymen had no official power to change anything in a procedure. Nor could they expect quick support from the overworked *préparateurs*. As a result, they were *bound* to adjust procedures informally and secretly. Had they not performed these daily adjustments, they would never have completed their tasks. The maintenance workers took full responsibility for these tacit micro-adjustments. They were part of their expertise and were also the best way to accumulate private know-how and to develop power and autonomy inside the plant. Rule-breaking emerged both as a necessity and as a power strategy. (For similar descriptions, see Terssac, 1992, pp. 79–211.)

Similar deviant patterns have often been documented in other settings and are conventionally interpreted as resulting from the incompleteness of prescribed procedures (Crozier and Friedberg, 1977). Nevertheless, violations and secret adjustments at Bugey are largely brought about by structural flaws inside the organisation. The important point here is not that rules are structurally incomplete, but rather that a flawed organisational design provides the incentive for violations.

2.5.2   Rule-following at Nogent: a solid protection for contractors

At Nogent the balance of power, which favoured contractors, seemed paradoxical. Because contractors could not afford to be found breaking the rules, they carefully complied with them and therefore held a strong position in the plant. This strategy of compliance was disturbing to EDF plant personnel, who were used to breaking the rules in order to get the job done. Contractors, on the contrary, developed a compliance strategy, because they could not afford to be found breaking the rules.

This strategy raises the question of how it was possible that EDF journeymen at Bugey in 1991 could not perform their tasks without adjusting them, whereas a few years later contractors at Nogent in 1994 were able to comply. Maintenance procedures were written by the same *préparateurs*, whose availability on work sites did not appear to have improved dramatically. The fact is that contractors succeeded in forcing *préparateurs* to change maintenance procedures upon request.

In order to explain this unexpected behaviour, one must realise that the responsibilities of contractors on work sites increased greatly between 1991 and 1994. Contractors are no longer entirely under the control of EDF foremen. They must now demonstrate that they can sustain a very high level of performance while complying with procedures. Knowing that a contracting firm found unable to comply with regulations would probably be dropped from the established list, contractors responded to this challenge by carefully drafting their contracts to spell out their duties *vis-a-vis* the EDF in greater detail than in the past. Specifically, contractors are not obliged to make up for any delay recognised as resulting from faulty organisation by the EDF, unless the contract is revised and either additional money or extra time provided. Contractors were consistently observed to be extremely tough during the negotiations conducted before the outage. They were clearly intent on minutely defining what could and could not be expected or required of them.

2.5.3   Compliance at Diablo Canyon: a feasible strategy

At Diablo Canyon an astonishing level of compliance was discernible. I had evidence that the workforce, especially those on work sites, followed the rules to the letter. Management developed an operating philosophy called *verbatim compliance*. According to other scholars who had been studying the same plant (La Porte and Thomas, 1995; Schulman, 1993), verbatim compliance resulted from the specific regulatory environment within which Diablo Canyon was (and still is) operating.

They reported that regulatory pressure was so important that it was cost-effective for Diablo Canyon to anticipate future requirements and comply with them before they became mandatory. As a result, staff at all levels would be persuaded to adopt this remarkably compliant attitude.

Although I on the whole agree with this argument, the question remains, how and why do employees, at their respective levels, manage to comply with numerous regulations? Management's general insistence on promoting compliance does not necessarily mean that such behaviour is observable on the shop floor. After all, this was not the case at Bugey.

To my great surprise, however, the philosophy of *verbatim compliance* seemed to be embedded in workers' practices. In situations where workers from Bugey would just go ahead and solve the problem by themselves, Diablo Canyon's workers invariably summoned their foremen and asked for guidance. Diablo Canyon's workers never extended the scope of their work without proper authorisation from their superiors (except in the case of emergency) and never took the initiative in their tasks.

This attitude is at odds with the conventional wisdom that such behaviour cannot be maintained by workers since, sooner or later, they will be confronted by the dilemma of having to cope with an unplanned situation requiring adjustments, while under pressure to perform the job with no delays. Under these circumstances, workers will not choose to comply. However, according to the same conventional wisdom, the other possible strategy is the compliant one of 'doing it by the book'. In other words, employees decide to risk delays. Compliance is then considered a highly subversive violation of the rules.

At Diablo Canyon some elements of going by the book were observable. Foremen and general foremen, who were somewhat remote from the field, frequently complained about the difficulty of motivating their crews, particularly when it came to taking responsibility for a problem or a solution. But they were also the first to acknowledge that the regular calls for assistance by the workers allowed them to remain constantly in touch. As a result, the position taken by the foremen was somewhat ambivalent: they would have liked to see their subordinates more prepared to take the initiative, yet they welcomed the implications of what was sometimes called *malicious compliance.*

Nevertheless, I believe that compliance at Diablo Canyon cannot be solely interpreted as malicious. The reason that workers took no interest at all in breaking the rules was that they were able to comply with them. Indeed, workers had easy access to section engineers, whose major job during an outage was to modify procedures and work orders upon request. By contrast with Bugey's situation, access to procedure writers was not 'rationed' structurally, an arrangement that allowed Diablo Canyon's maintenance staff to interrupt their work as much as they wanted without jeopardising their planning, as numerous section engineers were available. Accordingly, it was not only in the staff's interest to comply with the rules, it was also feasible.

This degree of compliance cannot be fully explained, however, without taking into consideration the amount of preparation required by the outage. Diablo

Canyon's personnel devoted 18 months to the planning and scheduling of an out-age. All forces were geared to minimising the gap between what was written and what actually happened. High-impact teams certainly made for the advance clear-ing of problems and potential conflicts in the field. It is almost as if the scripts had been rehearsed, tested, and coordinated beforehand. As a result, little improvisation was left to Diablo Canyon employees. In this regard, the question of causality remains open. The astonishing amount of advance planning could be interpreted as a means of counteracting or coping with malicious compliance. In other words, such a devotion to planning and scheduling issues may have reflected a necessary caution, given the attitude of the work force. Conversely, such attention to planning and preparation steadily worked against maintenance staff ever developing their own sense of initiative.

2.5.4 Compliance and violation at North Anna: the two sides of empowerment

At North Anna, journeymen complied with maintenance rules. If confronted with unplanned difficulties, they informed their foremen and sought their advice. They would ask for a new work order whenever the scope of the work turned out to be either inappropriate or unclear. But this type of compliance was not comparable to that at Diablo Canyon. North Anna journeymen complied, not because they bene-fited from unlimited assistance, but because they were involved both in creating and updating procedures. The regulations lay within the scope of their work.

Maintenance foremen had the explicit authority to initiate changes whenever their crews needed an adjustment or an update. Journeymen had no interest in adjusting procedures; they knew they could obtain the change formally. After all, the managers' committee met every day and was on call all night in case of emergency. Procedural changes were routinely made, and managers devoted time to this recurrent task.

North Anna's situation contradicted the conventional wisdom that maintenance workers in nuclear power plants have no other choice than to break the rules if they cannot change them when they need to. When rules and regulations are designed to remain feasible for those who have to follow them, workers have far less interest in breaking them, since their concerns are taken into account when the rules are being formulated. One can find further justification for this conclusion by consider-ing the most important example of procedures being violated – the exception that proves the rule. As previously mentioned, there were tensions between shift super-visors and maintenance foremen, with the supervisors stopping the greatest possible number of tasks they deemed to be 'risky', and the foremen retaliating by postpon-ing 'tricky' jobs until a more accommodating shift supervisor was on duty. Thus, complex jobs tended to fall more and more often to the same shift supervisors, a result which may cast doubt on the level of safety achieved thereby. Upon closer examination it appeared that shift supervisors were not only angry about the new auditing system (which increased competition among them), but also resentful that they had no say in decisions on planning issues. Their incentive to comply with planning was therefore diminished. Conversely, on work sites maintenance personnel took part in the various stages of the entire process. The workers gave their advice when procedures were drafted, they tested those procedures and they could modify them when necessary. This greater degree of involvement increased their commitment and, hence, their compliance.

North Anna, therefore, provided a sort of laboratory experiment to test conditions under which both compliance and violations develop. This study's main conclusion is that the independent variable is the degree to which the staff are allowed to participate in the design of a prescribed set of procedures (whether planning or maintenance).

2.6 ACHIEVING ORGANISATIONAL RELIABILITY: THE PROBLEM OF DUALITY

I have attempted to demonstrate that there are many different factors influencing social interactions in nuclear power plants, and that there are grounds for believing that these social interactions are closely interwoven with organisational reliability. The trouble is, however, that social interactions cut both ways. They are essential for establishing organisational reliability, but can also be a major obstacle to it. Hence, I believe that a systematic assessment of social interactions is what is required.

2.6.1 Organisational reliability through autonomy and opacity at Bugey

Bugey clearly illustrates the problem of the intrinsic duality of every social adjustment. Organisational reliability at Bugey was supported by the interaction between maintenance foremen and their staff on the one side, and management on the other. The members of the group working on the job were ready to dedicate themselves to their tasks as long as they had the benefit of autonomy (Terssac, 1992). This relative freedom required their commitment to, and skill at, monitoring deviations. This gave journeymen and foremen a clear advantage, because it implied that they had developed and continued to demonstrate a great deal of expertise to do with components. It also meant that Bugey's workers were used to responding to unplanned situations. Does this conclusion mean that 'unofficial' methods are at the core of organisational reliability? Undoubtedly, they play a key role, but it can also be a problem for an organisation to rely almost exclusively on tacit knowledge and tacit arrangements, which do not transfer particularly well outside the boundaries of specific groups. At Bugey it was found that informal adjustments did in fact lie at the core of the power enjoyed by those who executed the maintenance procedures. The more successfully they carried out such 'unofficial' maintenance, the more they appeared as key players inside the organisation. Repairs have then to be analysed both as a means of bargaining and as vehicles of reliability. Organisational reliability at Bugey was achieved through tacit and secret adjustments. They ensured and guaranteed the quality of the work and of performance, but they also posed great challenges to an organisation undergoing massive reform. Most of the foundations of organisational reliability were difficult to uncover without disrupting power relations. Unfortunately, new contracting practices are progressively changing the rules of the game, and will inevitably 'dethrone' EDF foremen. As a result, one should expect a period of upheaval until a new equilibrium is established.

2.6.2   Organisational reliability through compliance and apathy at Diablo Canyon

At Diablo Canyon it is certain that compliance on work sites was part of organisational reliability. But at what cost? While workers' ability and willingness to comply could surely be viewed as a sound basis for a reliability-seeking organisation, it may also have its downside. Staff at Diablo Canyon showed little or no initiative, partly because they exercised self-restraint and partly because initiative was not invited anyway. This strategy of non-involvement has cast doubt on their ability to respond to unplanned situations, where a quick reaction might make a difference. Such potential apathy is often cited by critics of organisations that rely too much on regulations (Wildavsky, 1989). Diablo Canyon's compulsive planning and scheduling did provide a partial solution to this potential risk. Organisational reliability at Diablo Canyon was thus rooted in the complex interaction between the work force's strategy of compliance and non-involvement, on the one hand, and, on the other the adoption of many mutually reinforcing organisational and planning devices and constant assistance in the field.

2.6.3   Organisational reliability through self-correction at North Anna

At North Anna organisational reliability was achieved through other means, specifically through generalisation and facilitation of feedback and updates in real time. In a sense, North Anna was close to the definition of a 'self-correcting organisation' given by Landau (1973). A philosophy of self-correction placed flexibility at the core of organisational reliability. Two main loops of interaction were identified, and they both supported the social construction of organisational reliability. The first involved foremen and their staff. Their cooperation facilitated most adaptations of regulations and procedures, and was thus a key to building organisational reliability. At first glance, it may appear that the role North Anna's journeymen and foremen played in the organisation closely resembles that of Bugey's foremen and journeymen (i.e. being in charge of field adaptations). However, the group of people executing procedures at Bugey succeeded only in tacitly compensating for organisational and procedural flaws, whereas at North Anna adjustment processes were explicitly part of the job of executing procedures. The second loop involved upper management and the group of people who executed procedures. The Security and Nuclear Safety and Operations Committee, a distinctive feature of North Anna, offered a continuous opportunity for exchange and provided for rapid feedback and dialogue between upper management and lower ranks.

The conjunction of these two loops enabled North Anna to centralise and decentralise at the same time. Final decisions were taken by upper management, but a clear delegation of power was left to the foremen and their crews when field issues were concerned. Organisational reliability was built upon this capacity to ensure such bottom-up and top-down movement.

However, some shortcomings were also visible, as exemplified by the vicious circle that was building between maintenance foremen and the supervisors of operations shifts. The atmosphere of competition between supervisors, the amount of responsibility falling to the foremen, and the possibility of individual sanction provoked strategies that are potentially harmful to organisational reliability.

2.6.4 Organisational reliability through compliance and exit at Nogent

At Nogent the social system was in transition, a fact that complicated assessment of the plant's organisational reliability. Nevertheless, it appeared that the plant was approaching a new equilibrium, one where organisational reliability was supported by contractors. Indeed, careful reporting by the contractors at this plant, probably a good feature of organisations seeking reliability, was a by-product of their self-protection strategy, which they were likely to continue as long as it remained in their own interest. For this reason I believe that assessing organisational reliability cannot be separated from the assessment of power relations and groups interests.

Contractors played a key role in enhancing organisational reliability at Nogent, but they were confronted with EDF agents suffering greatly from a loss of responsibility. At the organisational level, the result was a loss of coherence, or as Weick (1993, 1995) could say a 'loss of sense'. Two worlds were working side by side, largely ignoring each other: EDF's world was highly demotivating, favouring a strategy of 'exit' and, more rarely, 'voice' (through the union) (Hirschman, 1970), whereas the contractors' world was one in which staff were highly motivated because their contracts and jobs were at stake, and because they had found in contracts a way to protect themselves and develop their power by complying with established regulations.

2.7 SUMMARY

I have presented four portraits of organisational reliability. The intention has not been to rank the organisations studied, but to provide the means to evaluate the basis on which organisational reliability is built. Since organisational reliability is socially constructed, I maintain that an understanding of it cannot be separated from a minute study of social interactions and the uncovering of employees' strategies. As we know now, some of these strategies can be highly profitable, whereas others can be counterproductive and sometimes harmful. These two features are, nonetheless, often so closely linked to each other that they seem to be two sides of the same coin.

Notes

1   High-reliability organisations are organisations that can sustain a high level of performance and an impressive level of reliability (which conjointly permits and facilitates their
social acceptance).
2   Violation is a deviation from an existing rule. An improvisation is really what is also
called the 'expected task' (Terssac, 1992, p. 94), a series of gestures that do not exist on
paper but that are necessary to carry on the task (see Bourrier, 1996b).

References

ACKERMANN, W. and BASTARD, B. (1991) *Culture de sûreté et modalités d'organisation du travail* [Safety culture and forms of work organisation]. Unpublished manuscript. Paris: Centre de Sociologie des Organisations, Centre National de la Recherche Scientifique (CNRS).

AMALBERTI, R. (1988, September) *Savoir-faire de l'opérateur: aspects théoriques et pratiques en ergonomie* [The savoir-faire of operators: Theoretical and practical aspects of ergonomics]. Paper presented at the 23rd Congress of the Société d'Ergonomie de Langue Française (SELF), Models and practices of the analysis of work, Paris.

AMALBERTI, R. (1996) *La conduite de systèmes à risques* (Managing high-risk systems]. Collection *Le Travail Humain*. Paris: Presses Universitaires de France.

BOURRIER, M. (1994) *Compliance as a strategy* (Working paper No. 94-10). Berkeley, CA. Institute of Governmental Studies, University of California at Berkeley.

BOURRIER, M. (1996a) Organising maintenance work at two American power plants. *Journal of Contingencies and Crisis Management*, 4, pp. 104–12.

BOURRIER, M. (1996b) *Une analyse stratégique de la fiabilité organisationnelle: organisa-tion des activités de maintenance dans quatre centrales nucléaires en France et aux Etats-Unis* [A strategic analysis of organisational reliability: Organisation of activities and maintenance at four nuclear centres in France and the United States]. Unpublished doctoral dissertation, Institut d'Etudes Politiques de Paris.

CHABAUD, C. and TERSSAC, G. DE (1987) Du marbre à l'écran: rigidité des prescriptions et régulations de l'allure de travail [From galleys to screen: The rigidity of monitoring and regulating the work pace]. *Sociologie du travail*, 3, pp. 305–22.

CHISHOLM, D. (1989) *Coordination without hierarchy: Informal structures in multi-organisational systems*. Berkeley: University of California Press.

CLARKE, L. (1989) *Acceptable risk? Making decisions in a toxic environment*. Berkeley: University of California Press.

CONINCK, F. DE (1995) *Travail intégré, société éclatée* [Integrating work, splitting society]. Paris: Presses Universitaires de France.

CROZIER, M. (1963) *Le phénomène bureaucratique* [The bureaucratic phenomenon]. Paris: Seuil.

CROZIER, M. and FRIEDBERG, E. (1977) *L'acteur et le système* [Actor and system]. Paris: Seuil.

CROZIER, M. and THOENIG, J.-C. (1975) La régulation des systèmes organisés complexes, le cas du système de décision politico-administratif local en France [The functioning of complex organisations, the case of local politico-bureaucratic decision-making system in France]. *Revue Française de Sociologie*, **16** (1), pp. 3–32.

DE KEYSER, B. and NYSSEN, A. S. (1993) Les erreurs humaines en Anesthésie [Human errors in anaesthesia]. *Le Travail humain*, **56** (2–3), pp. 243–66.

DOWNS, A. (1967) *Inside bureaucracy*. Boston: Little, Brown.

FAVERGE, J. M. (1970) L'homme, agent de fiabilité et d'infiabilité [The human being, agent of reliability and unreliability]. *Ergonomics*, **13**, pp. 301–27.

FAVERGE, J. M. (1980) Le travail en tant qu'activité de récupération [Work as a recuperative activity]. *Bulletin de Psychologie*, **33**, pp. 203–06.

GIRIN, J. and GROSJEAN, M. (eds) (1996) *La transgression des règles au travail* [Rule violation at work]. Paris: L'harmattan.

GOULDNER, A. W. (1954a) *Patterns of industrial bureaucracy*. Glencoe, Ill: Free Press.

GOULDNER, A. W. (1954b) *Wildcat strike*. Yellow Springs, OH: Antioch Press.

GRANOVETTER, M. (1985) Economic action and social structure: The problem of embeddedness. *American Journal of Sociology*, **91**, pp. 481–510.

GRAS, A., MORICOT, C., POIROT-DELPECH, S. and SCARDIGLI, V. (1990) *Le pilote, le contrôleur et l'automate* [Pilots, controllers, and robots]. Paris: Editions de l'IRIS.

GUILLERMAIN, H. and MAZET, C. (1993) *Tolérance aux erreurs, sur-fiabilité humaine et sûreté de fonctionnement des systèmes socio-techniques* [Error tolerance: On human reliability and the functional safety of sociotechnological systems] (CNRS report No. LAAS 93418). Paris: Centre National de la Recherche Scientifique (CNRS).

HABER, S. B., BARRIERE, M. T. and ROBERTS, K. H. (1992, June) *Outage management: A case study*. Paper presented at the Institute of Electrical and Electronic Engineers (IEEE), Monterey, CA.

HABER, S. B., SHURBERG, D. A., BARRIERE, M. T. and HALL, R. E. (1992) *The nuclear organisation and management analysis concept methodology: Four years later*. Unpublished manuscript. Upton, Long Island, NY: Brookhaven National Laboratory, Associated Universities.

HALE, A. and BARAM, M. (in press) *Safety management and the challenge of organisational change*. Amsterdam: Elsevier.

HIRSCHHORN, L. (1993) Hierarchy versus Bureaucracy: The case of a nuclear reactor. In ROBERTS, K. (ed.), *New Challenges to understanding organisations* (pp. 137–51). New York: Macmillan.

HIRSCHMAN, A. (1970) *Exit, voice, and loyalty. Response to decline in firms, organisations and states*. Cambridge, MA: Harvard University Press.

LANDAU, M. (1973) On the concept of a self-correcting organisation. *Public Administration Review*, **33**, pp. 533–9.

LANDAU, M. (1991) On multiorganisational systems in public administration. *Journal of Public Administration Research and Theory*, **1**, pp. 5–18.

LA PORTE, T. R. and CONSOLINI, P. (1991) Working in practice but not in theory: Theoretical challenges of 'high-reliability organisations'. *Journal of Public Administration Research and Theory*, **1**, pp. 19–47.

LA PORTE, T. R., ROBERTS, K. and ROCHLIN, G. (1987, December) *High reliability organisations: The research challenge* [Revision of LA PORTE, T. R., High reliability organisations: The problem and its research dimensions] (Working Paper). Berkeley, CA. Institute of Governmental Studies, University of California at Berkeley.

LA PORTE, T. R. and THOMAS, C. (1995) Regulatory compliance and the ethos of quality enhancement: Surprises in nuclear power plant operations. *Journal of Public Administration Research and Theory*, **5**, pp. 109–37.

LEPLAT, J. and TERSSAC, G. DE (eds) (1990) *Les facteurs humains de la fiabilité* [The human factors of reliability]. Marseille: Octares.

LINHART, R. (1978) *La division du travail* [The division of labour]. Paris: Galilée.

OSTY, F. and UHALDE, M. (1993) *La régulation sociale en centrale nucléaire, les conditions sociales de l'extension des centrales deux tranches: le cas de Flamanville* [Social regulation at a nuclear centre and the social conditions and extension of the two main units: The case of Flamanville] (Sociological Working Paper, No. 26 of the Laboratoire des Sociétés et du Changement institutionnel). Paris: Institut de Recherches et d'Etudes sur les Sociétés Contemporaines.

PERROW, C. (1984) *Normal accidents: Living with high-risk technologies.* New York: Basic Books.

POYET, C. (1990) L'homme agent de fiabilité dans les systèmes informatisés [The human being as an agent of reliability in the information system]. In TERSSAC, G. DE and LEPLAT, J. (eds), *Les facteurs humains de la fiabilité* (pp. 223–41). Marseille: Octares.

REASON, J. (1987) The Chernobyl errors. *Bulletin of the British Psychological Society*, **40**, pp. 201–6.

REASON, J. (1990) *Human error, causes and consequences.* Cambridge: Cambridge University Press.

ROBERTS, K. H. (1993) *New challenges to understanding organisations.* New York: Macmillan.

ROCHLIN, G. I. (1988) *Technology and adaptive hierarchy: Formal and informal organisation for flight operations in the US Navy* (Working Paper No. 88-18). Berkeley, CA. Institute of Governmental Studies, University of California at Berkeley.

ROCHLIN, G. I. (1993) Defining 'high reliability' organisations in practice: A taxonomic prologue. In ROBERTS, K. H. (ed.), *New challenges to understanding organisations* (pp. 11–32). New York: Macmillan.

ROCHLIN, G. I. and VON MEIER, A. (1994) Nuclear power operations: A cross-cultural perspective. *Annual Review of Energy and the Environment*, **19**, pp. 153–87.

SAGAN, S. D. (1993) *The limits of safety: Organisations, accidents, and nuclear weapons.* Princeton, NJ: Princeton University Press.

SCHULMAN, P. (1993) Negotiated order of organisational reliability. *Administration and Society*, **25**, pp. 353–73.

SELZNICK, P. (1949) *T.V.A. and the grass roots.* Berkeley, CA: University of California Press.

SETBON, M. (1993) *Pouvoirs contre Sida* [Politics against AIDS]. Paris: Seuil.

SHRIVASTAVA, P. (1987) *Bhopal: Anatomy of a crisis.* Cambridge, MA: Ballinger.

TERSSAC, G. DE (1992) *Autonomie dans le travail* [Autonomy at work]. Paris: Presses Universitaires de France.

VAUGHAN, D. (1996) *The* Challenger *Launch Decision.* Chicago: University of Chicago Press.

WEICK, K. (1987) Organisational culture as a source of high reliability. *California Management Review*, **29**, pp. 112–27.

WEICK, K. (1993) The collapse of sensemaking in organisations: The Mann Gulch disaster. *Administrative Science Quarterly*, **38**, pp. 628–52.

WEICK, K. (1995) *Sensemaking in organisations.* Thousand Oaks, CA: Sage.

WHYTE, W. F. (1948) *Human relations in the restaurant industry.* New York: McGraw Hill.

WILDAVSKY, A. (1989) *Searching for safety.* New Brunswick, NJ: Transaction Publishers.

CHAPTER THREE

Finnish and Swedish practices in nuclear safety

BJÖRN WAHLSTRÖM

Technical Research Centre of Finland
VTT Automation

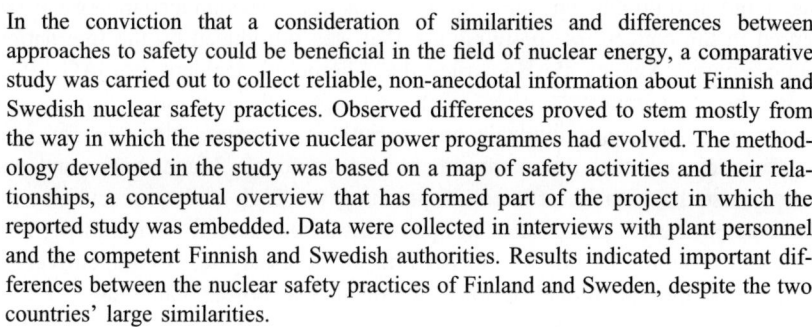

In the conviction that a consideration of similarities and differences between approaches to safety could be beneficial in the field of nuclear energy, a comparative study was carried out to collect reliable, non-anecdotal information about Finnish and Swedish nuclear safety practices. Observed differences proved to stem mostly from the way in which the respective nuclear power programmes had evolved. The methodology developed in the study was based on a map of safety activities and their relationships, a conceptual overview that has formed part of the project in which the reported study was embedded. Data were collected in interviews with plant personnel and the competent Finnish and Swedish authorities. Results indicated important differences between the nuclear safety practices of Finland and Sweden, despite the two countries' large similarities.

Nuclear power accounts for an important share of the electricity generated in many countries. Nonetheless, there is also opposition to nuclear power in the same countries, a major concern being that nuclear power plants cannot be operated safely over extended periods of time. However, experience from plants that have performed well provides evidence that safe operation is possible. Finding out which characteristics make nuclear power plants operate safely at acceptable levels of output has thus become a key issue in the debate about this form of energy production.

Because operational experience has shown organisation and management to be central to safety and acceptable performance, some research has focused on identifying organisational factors that can affect safety and operation. It has, however, been difficult to make the findings specific enough for application to the daily routines at nuclear power plants. It has also been difficult to assess the extent to which findings are exchangeable between different cultural environments.

The study reported in this chapter was carried out as a part of a project conducted within the Nordic research cooperation on nuclear safety (NKS). The aim of the project, RAK 1: 'A survey and an evaluation of safety practices', is to identify possible deficiencies and evaluate the effectiveness of safety practices, especially for activities with a large impact on safety. One part of the project was concerned with mapping a conceptual model of nuclear safety. The map illustrates the path between the requirements for various safety activities and the solutions to the problems they pose. The second part of the project, based on interviews, was intended to validate the map of nuclear safety practices and to provide non-anecdotal information about differences in safety practices. Data for that study were collected in 62 interviews in Finland and Sweden with the personnel at nuclear power plants, the relevant authorities, and one reactor vendor. The total effort spent in the project from 1995 to 1997 has been approximately half a person year per year, not counting the effort of the project reference group and the persons interviewed.

3.1 BENEFITS OF A COMPARISON

The safety of nuclear power depends not only on technically successful plant designs, but also on management and work practices. This understanding has been captured in the concept of *safety culture* that was introduced by the International Atomic Energy Agency (IAEA) after the Chernobyl accident (INSAG, 1991). Perhaps the most important aspect of the concept is the realisation that the commitment to safety by individuals and society is a major precursor for safety and efficiency, but that this commitment is not enough. Management and work practices also have to be efficient in pursuing the high-quality execution of all tasks. Nuclear power plants have to operate in a commercial environment, which means that it should be possible to perform all necessary tasks competitively.

Studies on the operational performance of nuclear power plants have indicated wide differences (see Beckjord *et al.*, 1987, IAEA, 1989, Marcus *et al.*, 1990). Some plants have been able to achieve high energy availability year after year, whereas other plants have trouble reaching the relatively modest average level of the nuclear power plants in the world. The difference in income from sold electricity gives an immediate impression of the problems poorly performing plants have to struggle with. Typically, they have a large backlog of modifications and improvements to be carried out. Sometimes urgent investments may add up to a major share of the annual turnover. By contrast, well-performing plants typically have their operation in order and can invest in fine-tuning the plant for optimal performance.

Findings from several operational safety review teams (OSART) missions of the IAEA highlight practices of well- and not-so-well performing plants. 'At each plant visited there are some practices or ideas which could be adopted by others ...

and some areas where improvements could be made have been identified even at the best plants' (IAEA, 1988, p. 4). This observation indicates that an exchange of sound practices between plants could benefit the whole industry. But is an exchange of safety practices possible? A blind exchange may not be feasible, for such practices are anchored in a cultural environment. Suitable interpretation and translation of them, however, should make it possible to transfer the principles rather than exact scripts. There is also an incentive to transfer sound operational practices: the industry as a whole is judged by its worst performers (Rees, 1994). Only openness and self-regulation will be able to neutralise the political hostility toward the nuclear power industry in many countries and possibly even build support for nuclear power.

Experience from the business world indicates that comparisons and benchmarking are efficient ways for a plant to improve its performance. An in-depth comparison of practices, performance indicators, and company cultures permits an assessment of a plant's strengths and weaknesses. A feasible way to achieve a transfer of sound operating practices is thus to use bench marks for two or more actors in the field. The benefit of the comparison then becomes the awareness of other practices and the necessity of assessing their potential benefits. Awareness of other practices makes it easier to see the advantages and disadvantages of a plant's own practices and can help fine-tune them without major changes.

3.2 THE NUCLEAR POWER PROGRAMMES IN FINLAND AND SWEDEN

The nuclear era in Sweden began in 1956 when the company AB Atomenergi filed an application to build and operate a materials research reactor in Studsvik (a small village near Nyköping, about 50 miles southeast of Stockholm). AB Atomenergi was also involved in the construction of the first nuclear power reactor in Sweden, the Ågesta, which delivered district heat to a suburb of Stockholm from 1964 to 1974. An upgraded version of this plant, the Marviken reactor, was abandoned when an analysis revealed problems in maintaining core stability.

The first electricity-producing reactor in Sweden, Oskarshamn 1, was ordered in 1965 by Oskarshamnsverkets kraftgrupp AB (now OKG) from Asea Atom (now ABB Atom). That facility was followed by four more Swedish reactor orders to ABB Atom, each one with improved constructions to be built in Ringhals, Barsebäck, and Oskarshamn. The first reactors ordered by the Swedish State Power Board (Vattenfall) were a boiling water reactor (BWR) from ABB Atom and a pressurised water reactor (PWR) from Westinghouse in Ringhals. The Ringhals PWR reactor was followed by three more PWRs. At the fourth nuclear site in Sweden, Forsmark, two BWRs, Forsmark 1 and 2, were built by ABB Atom. These reactors were all improvements on the earlier generations, and another round of improvement was made by ABB Atom in its Oskarshamn 3 and Forsmark 3 reactors.

In Finland serious consideration of nuclear power for generating electricity was initiated in 1963, when the council of IVO (the largest Finnish power utility) authorised its board to ask for tenders for a nuclear power plant. The tenders were opened in 1965, but no order was placed at that time. In 1968 the Finnish government decided to postpone the whole project for building nuclear power plants. In 1969 discussions were opened anew, and one PWR plant was ordered from the Soviet Union, together with an option on a second installation. Soon after, private industry in Finland formed a new power company, TVO, which ordered one BWR plant from ABB Atom with an option on an additional one. Orders for the two options were placed soon afterwards.

The situation of nuclear power in Finland and Sweden today is different from that of the early 1970s. In keeping with the decisions after the referendum on nuclear power in 1980, Sweden has decided to phase out nuclear power by the year 2010. In Finland a consortium between IVO and TVO applied for a decision in principle to build a fifth nuclear power unit, but the application was narrowly defeated in parliament in 1993. The discussion in Finland on further reactors has stalled, but at regular intervals voices are heard in favour of reopening the question. The discussion in Sweden has been more concentrated on when, how, and which units are to be shut down ahead of schedule. Both industrial management and labour unions are lobbying for reconsideration of the decision to phase out nuclear power, but so far without result.

3.3   REGULATORY ENVIRONMENTS IN FINLAND AND SWEDEN

Nuclear power in Finland and Sweden is regulated by national laws and regulations. The first Atomic Energy Act in Sweden was passed in 1956 and in Finland 1958. Both Acts stipulated that the operation of nuclear facilities requires an operating license awarded by the authorities. The Acts have been revised in both countries several times. A major amendment of the Nuclear Energy Act in Finland was passed in 1987 after a long discussion dating as far back as 1978. The Act is rather detailed and contains technical safety requirements. A decree in 1988 and a decision by the Finnish government in 1991, laid down further details on how to ensure the safety of nuclear power plants.

Laws and regulations in Finland and Sweden empower a national authority to act as a regulator. In Sweden this authority is split between two bodies, the Swedish Nuclear Power Inspectorate (SKI) and the Swedish Radiation Protection Institute (SSI). In Finland the Finnish Center for Radiation and Nuclear Safety (STUK) regulates both reactor safety and radiation protection. Both in Finland and Sweden the responsibility for safety is placed squarely on the operator of the nuclear facilities. The authorities have the power to require the facilities to shut down if there is any doubt about their safety. Operating licenses in Finland have always been issued for a limited period, but in Sweden such restriction is only the exception.

Finland and Sweden differ greatly when it comes to the details of the requirements for operating a nuclear power plant. In Finland STUK has written a comprehensive set of safety manuals, the YVL guides, which take a stand on various issues. Salminen (1997) gives a recent presentation on the use of these guides. Their creation goes back to the Loviisa NPP project, in which the agreement between the utility and the vendor called for the licensing of nuclear power plants to be subject to Finnish requirements. The YVL set presently consists of about 70 guides arranged into eight technical areas. They are updated regularly and go through a detailed reviewing process before they are adopted. The YVL guides are not mandatory but do represent a strong recommendation. They are relatively general, giving a frame in which the designers of nuclear power plants are granted relative freedom. The YVL guides have a legal status lower than the decisions of STUK. That is, a new guide will not automatically be applied to an old plant, but new guides are known to have been made mandatory after a certain transfer period.

The Swedish licensing and inspection systems are quite different from the Finnish. The system in Finland is to a large extent governed by the YVL guides, whereas this kind of a system does not exist in Sweden, where American regulations are used instead and reference is made to the US Nuclear Regulatory Commission requirements. In Sweden a license to operate a nuclear power plant is awarded on the basis of an application, to which a final safety analysis report (FSAR) is attached. For the most part, the substance of an FSAR is based on a common understanding between SKI and the applicant. When the application is accepted, the FSAR is seen as an integral part of the license and, in a way, as an agreement between the authority and the utility on the construction and operation of the plant. Inspections are performed to ensure that actual practices are in compliance with the requirements of the FSARs. When new requirements are adopted, they are communicated to the licensees in regulatory letters. The situation with regard to more general requirements is going to change in Sweden. In the most recent update of the nuclear energy act in Sweden SKI was given the authority to issue its own regulations, and this process has begun.

3.4 SIMILARITIES AND DIFFERENCES

In this comparison of similarities and differences, it is fair to say that the former outnumber the latter. The differences are easy to trace back to historical variations in the way in which nuclear power was introduced in Finland and Sweden. The largest difference is perhaps the fact that Sweden has its own national vendor for reactors. The idea of developing internal Swedish requirements was considered inappropriate, because Sweden hoped to receive export orders for her reactors. The decision to use American licensing guidelines is thus understandable. The rapid development of the BWR concept of ABB Atom has reflected great flexibility in approaches to new solutions, and a willingness to solve problems as they emerge. In retrospect, a problem with this model has been that the documentation of the designs has sometimes lagged behind and has not always been explicit enough.

The other major difference between Finland and Sweden is the present political climate surrounding nuclear power. In Finland the option to build additional nuclear power plants is still open, but Sweden is adhering to the date of 2010 for the phase-out of nuclear power. In Sweden this threat for the whole industry has been part of day-to-day life for more than ten years. Discussions with representatives of the nuclear industry reveal frustration with the situation and a fear that it may become difficult to attract young people to the industry. A similar concern for the age structure of the people employed by the industry is shared in Finland, and the construction of a new plant is considered important for the long-term development of competence in this field.

A major similarity between the plants in Finland and Sweden is their commercial success. All plants have been able to achieve high power-availability figures for many years, and they can be judged as very efficient with respect to all performance indicators. There are several reasons for this success. From the outset, the nuclear industry attracted the best and brightest young engineers, who are the ones most likely to have the innovative outlook upon which a nuclear power plant's design depends. The first important advances in the nuclear industry had already been taken in the major countries of the world, a fact that made it possible to achieve good results with reasonable effort. Both Finland and Sweden have an industrial infrastructure that is able to support large, high-tech projects. The practices associated with fossil fuel power plants had not become deeply entrenched either in Finland or Sweden, so it was possible to consider the needs of nuclear power with an unbiased mind.

Despite the fact that the nuclear power programme in Sweden is three times larger then that in Finland, SKI has fewer resources than the corresponding parts of STUK. This difference in resources is also reflected in working practices in that SKI channels its inspection efforts more to the safety processes of the utility companies than to technical details. This focus is reflected in SKI's continued pursuit of plans for risk-based regulation designed to help direct its attention to the most important activities.

There is efficient communication, both formal and informal, between the authorities and the utilities both in Finland and Sweden. Regular meetings are held at each hierarchical level, and difficult issues are confronted openly in discussions. Scientific societies, committees, steering groups for research projects and similar kinds of contact fora provide the opportunity for representatives of the organisations involved to meet and exchange views on current problems.

Swedish and Finnish procedures for considering human factors and organisational issues differ. In Sweden all nuclear power plants and SKI have their own 'man-technology organisation' or MTO groups. These groups have been given the responsibility of supplying their respective organisations with expertise in the behavioural sciences. The members of an MTO group analyse events and incidents for human errors and organisational deficiencies, participate in audits and reviews, and advise on major modifications in the plants. These groups also have a large influence and are valued by technically oriented people. In Finland general opinion has been that such specialised groups are not needed, and that the issues they address can be dealt with as a part of technical considerations.

The authorities and the utility companies in both Finland and Sweden have close contact with international bodies. Working groups of the IAEA and OECD's Nuclear Energy Agency (NEA) have proved to be important fora for communication with international experts. The utility companies have engaged in user-group co-operation. The contact network of the World Association of Nuclear Operators (WANO) provides a significant channel for both international experience and exchange programmes between nuclear power plants.

The training of control room operators in Sweden is centralised in Studsvik, where KSU, a company owned by the Swedish utilities, operates simulators for all the Swedish plants. In Finland simulator training has been decentralised, with simulators being operated by the utility companies themselves at their plant sites.

Education and research are central to maintaining long-term safety in nuclear power plants. In recent years the number of university graduates with a specialisation in nuclear engineering has been decreasing both in Finland and Sweden, but so far it has been possible to meet the nuclear power industry's most urgent needs. ABB Atom and the power companies in Sweden have initiated programmes for young university graduates to be trained in various skills considered important by the industry. In Finland no such programmes are operated, but a pool of skills is maintained by the Technical Research Centre of Finland (VTT). In Finland public research on nuclear power is funded by the Ministry of Trade and Industry, but in Sweden SKI administrates the research directly. Much of the research on nuclear power in Finland is conducted by the VTT, in cooperation with STUK and the utility companies. In Sweden no similar research organisation exists, and the projects are scattered between consultant companies and university groups.

3.5 ONGOING ACTIVITIES

In Finland and Sweden major restructuring of the electricity supply has been initiated. The development follows the path already taken by the United Kingdom and Norway, where an electricity market has been established. This restructuring implies that competition is being brought into electricity production. Electricity utilities are being broken up into production companies and companies responsible for the operation of the electricity grid. Electricity transmission and distribution is considered an infrastructure where a monopoly is acceptable. Upon payment of a connection fee, a producer is allowed to use this infrastructure to deliver power to customers. The restructuring process has influenced the ownership structure of the whole industry, in which many mergers and new contacts have taken place since 1995. Among the nuclear utilities there have been fears that increased competition will make it more difficult to exchange information important to safety.

The nuclear power plants both in Finland and Sweden are going through major modernisations. The projects involved stem partly from a concern to ensure safe and uninterrupted operation for additional decades, and partly by a need to replace solutions that for various reasons are difficult to sustain. In Finland an additional motive for the modernisation projects is to increase the plants' electric power output. In Sweden there have also been political motives, reflecting a willingness to invest in continued operation of the plants.

In Sweden large projects have been initiated to reconstitute the design base of the country's nuclear power plants. This work was prompted partly by the incident in 1992 at Barsebäck where strainers were clogged far more rapidly than expected. This incident pointed to deficiencies in the design of the five older ABB Atom constructions. Two large projects, one for the Ringhals plant and the other for the Barsebäck and Oskarshamn plants, have been contracted to ABB Atom. A similar but smaller project has been launched by Forsmark. A deliberate decision has been to involve young people as much as possible, in order to facilitate the generational change of the nuclear industry in Sweden.

3.6 A MAP OF SAFETY PRACTICES

Information for comparing Finnish and Swedish activities in nuclear safety was collected on those activities which were regarded as the most important. However, the original idea of constructing a map of safety practices showing a path between requirements and solutions eventually proved somewhat difficult. A comprehensive picture of the safety activities involved can be achieved only through a combination of many different views.

Perhaps the most important components of all activities are the concepts of goals, planning, and feedback (see Figure 3.1). A goal is defined for all activities, a planning process is used to search for ways to achieve defined goals, and the feedback of actual outcomes provides inputs for improvements in the next round. Safety activities should be described and operational. The description of safety

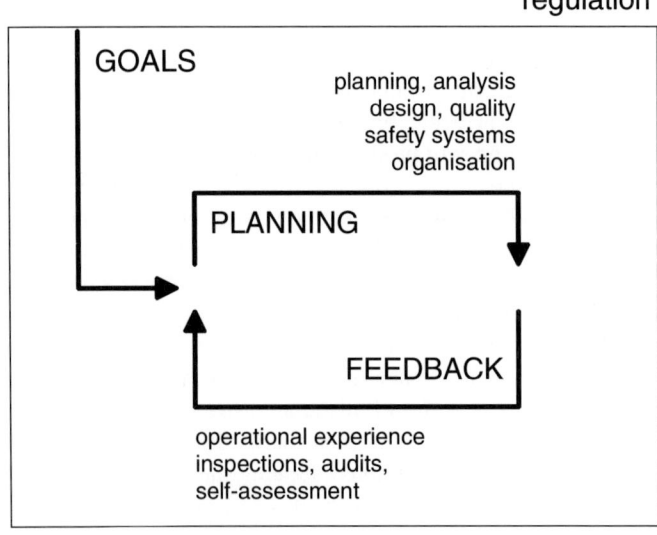

Figure 3.1 The generic components of safety practices.

activities should be accurate enough to serve as a kind of procedure for how to do the work. Described safety activities are operational if actual practices match the description. Inspection can therefore be seen as a twofold process: described arrangements of the safety activities are compared with an ideal model, and it is then verified that activities are performed as described.

It is proposed that the model of safety activities should include various angles of view. Systematic planning is an important part of all safety activities. Safety control can be seen as three interacting control systems: technical, administrative, and societal (see Figure 3.2). The administrative and societal controls involved are allocated to certain organisations, which are given the responsibility for their efficiency. The technical and administrative system of a plant can be described as in Figure 3.3, along the two axes of abstraction (e.g. goals, functions, and design) and aggregation (ranging from the system as a whole to its constituent parts), with the design of the plant typically advancing in the direction of the arrows. Three major processes interact in building a plant and its operational practices: design and construction, verification and validation, and licensing. Systematic *quality assurance* is a special aspect to be integrated in other activities, and it involves a description of work practices and regular audits. Safety precautions are structured around *threats and barriers*, an approach that also lays the foundations for the defence-in-depth principle (INSAG, 1997). Lastly, *quantification* of the efficiency of various safety precautions makes it possible to set priorities for possible improvements.

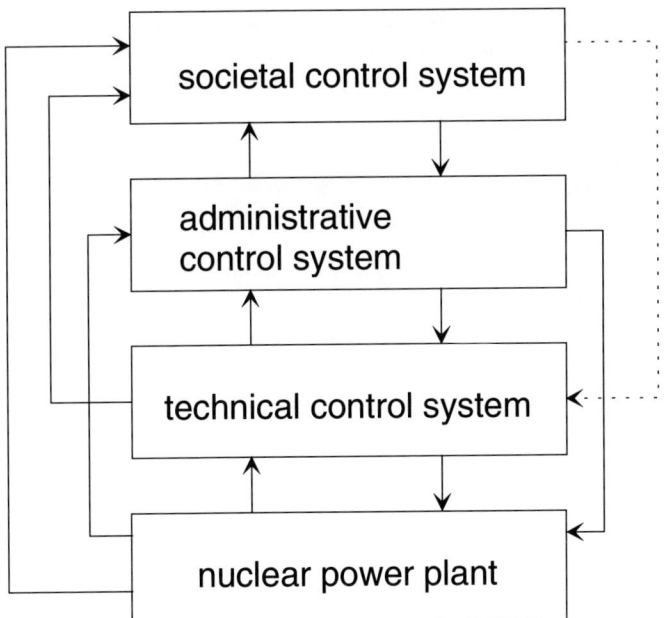

Figure 3.2 Three interacting control systems ensure the safety of a nuclear power plant.

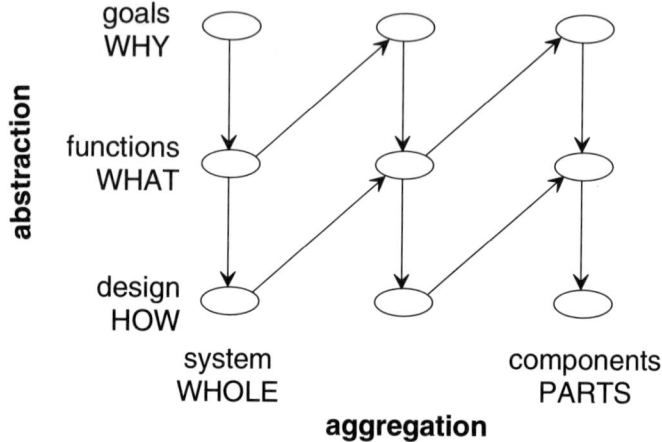

Figure 3.3   Relationships between abstraction and aggregation.

An early version of these views on safety activities was used to structure inter-views with various experts within the industry and with the safety authorities. The information collected in those discussions became the basis for refining the model of safety activities. The next step in developing the ideal model will be to formalise the concepts to make them more consistent and more accurately described. The model has been used for an assessment of the modification activities at nuclear power plants in Finland and Sweden. It can also be used to structure inspections and self-assessments at nuclear power plants, and can provide a scheme for a more systematic comparison of different approaches to nuclear safety.

3.7 CONCLUSION

This review of Finnish and Swedish practices pertaining to nuclear safety has made several aspects of the topic clear. First, nuclear power has a political dimension that makes it unlike other industries. This fact has to be recognised when admin-istrative and societal control systems are set up. Secondly, it is evident in retrospect that the effort needed to operate and maintain nuclear power plants has been underestimated in that industry. Nonetheless, nuclear power is still economically competitive compared with other methods of generating electric power. Thirdly, two explanations for the success of Finnish and Swedish nuclear power plants were given repeatedly in the interviews with safety authorities and experts. One was that engineers, authorities, and personnel in the Finnish and Swedish nuclear power industry have demonstrated a degree of farsightedness and a proactive approach to dealing with hazards. The other reason is that they have been prepared to learn from experience, to think for themselves, and to remain open-minded. These explanations stress the importance of systematic planning and the involvement of all personnel in the effort to optimise the use of skills and resources. They are also well in line with the safety concepts of feed-forward and feedback control suggested in Rasmussen (1987) (see also Wilpert in this volume). Fourthly, large similarities between Finland and Sweden must not be permitted to obscure the existence of important differences, which have been observed between Finnish and Swedish safety practices. Those differences are mostly of historical origin: that is, they have arisen from the ways in which the respective nuclear power programme developed. Fifthly, the results of the project seem to indicate that the concept of safety culture is immediately understood by people in the nuclear community. To operationalise the concept, however, it may be necessary to include additional definitions and methods that will enable authorities and experts in the nuclear power industry to use it in their audits and self-assessments. If this concept is embedded in the map of safety practices and combined with appropriate organisational and cultural theory, it can provide insights into this process of redefining safety culture.

A well-designed and efficient plant, a strong economy and well-trained personnel certainly strengthen the likelihood of high performance. Neglecting these factors risks a downward spiral in operating conditions, a vortex of deterioration from which it is difficult to escape. Even for a plant that is performing well it is crucial not to stretch these resources too thinly, since the resulting lack of flexibility may endanger the working spirit of the plant personnel if something unforeseen happens. These reserve resources can in the meantime be put to practical use supporting international activities, such as creating new standards or providing help to countries that are struggling to enhance the performance of their nuclear power plants.

References

BECKJORD, E., GOLAY, M., GYTOPOULOS, E., HANSEN, K., LESTER, R. and WINJE, D. (1987) *International comparison of LWR performance* (MIT Energy Laboratory Report No. MIT-EL 87–004). Cambridge, MA: Massachusetts Institute of Technology.

IAEA (International Atomic Energy Agency) (1988) *OSART results: A summary of the results of operational safety review team missions during the period August 1983 to May 1987* (IAEA-TECDOC-458). Vienna: IAEA.

IAEA (International Atomic Energy Agency) (1989) *Good practices for improved nuclear power plant performance* (IAEA-TECDOC-498). Vienna: IAEA.

INSAG (International Nuclear Safety Advisory Group) (1991) *Safety culture* (Safety Series No. 75 – INSAG-4). Vienna: IAEA.

INSAG (International Nuclear Safety Advisory Group) (1997) *Defence in depth in nuclear safety* (INSAG-10). (Manuscript in preparation).

MARCUS, A. A., NICHOLS, M. I., BROMILEY, P., OLSON, J., OSBORN, R. N., SCOTT, W., PELTO, P. and THURBER, J. (1990) *Organisation and safety in nuclear power plants* (NUREG/CR-5437). Washington, DC: US Nuclear Regulatory Commission.

RASMUSSEN, J. (1987, May) *Safety control and risk management*. Paper presented at the IAEA/NPPCI Specialists Meeting on The human factor information feedback in nuclear power: Implications of operating experience on system analysis, design and operation, Roskilde, Denmark.

REES, J. V. (1994) *Hostages of each other: The transformation of nuclear safety since Three-Mile Island*. Chicago: University of Chicago Press.

SALMINEN, P. (1997) Renewal of operating licences of Finnish nuclear power plants. In International Atomic Energy Agency (IAEA) (ed.), *Reviewing the safety of existing nuclear power plants* (pp. 139–50). Vienna: IAEA.

CHAPTER FOUR

The cultural context of nuclear safety culture: a conceptual model and field study

NAJMEDIN MESHKATI

University of Southern California, Los Angeles

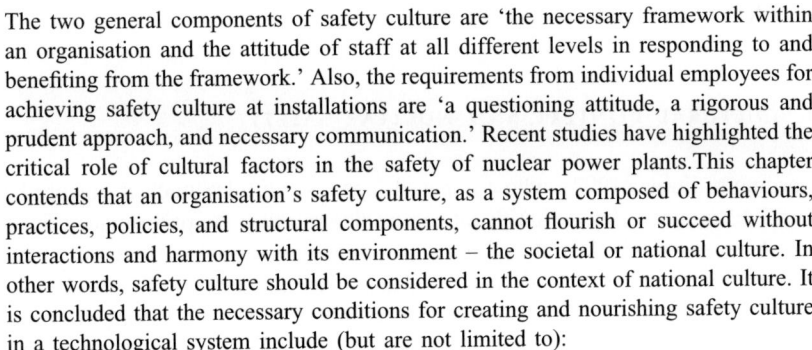

The two general components of safety culture are 'the necessary framework within an organisation and the attitude of staff at all different levels in responding to and benefiting from the framework.' Also, the requirements from individual employees for achieving safety culture at installations are 'a questioning attitude, a rigorous and prudent approach, and necessary communication.' Recent studies have highlighted the critical role of cultural factors in the safety of nuclear power plants. This chapter contends that an organisation's safety culture, as a system composed of behaviours, practices, policies, and structural components, cannot flourish or succeed without interactions and harmony with its environment – the societal or national culture. In other words, safety culture should be considered in the context of national culture. It is concluded that the necessary conditions for creating and nourishing safety culture in a technological system include (but are not limited to):

- an understanding of systems-related factors affecting human performance;
- determination of the extent to which systems-related factors interact with factors of organisational culture and the national culture;
- promotion of a questioning attitude and openness in the organisation;
- development of conducive regulations and a supportive regulatory environment.

Human and organisational factors play a vital role in the safety of large-scale technological systems (Meshkati, 1988; 1989a, b, c; 1991a, b). Fortunately, this fact has been almost universally recognised by the nuclear industry around the world. Nevertheless, it was the accident at the Chernobyl nuclear power plant that formally introduced the concept of *safety culture* to the vocabulary of nuclear safety. According to the International Atomic Energy Agency (IAEA), 'the term "safety culture" came into use after the Chernobyl accident' (*Nuclear Safety Review 1992*, p. D24). The expression is used nowadays quite extensively in the safety-related context, particularly in the field of nuclear energy. For instance, the former chairman of the US NRC, Ivan Selin, has referred to safety culture as one of four major sources of concern about the safety of the former Soviet-designed reactors (*Nuclear Engineering International*, July 1994). The concept of safety culture has also been embraced by other industries (Geller, 1994), such as chemical processing (Miner, 1991) and commercial aviation (Rodney, 1991; Lauber, 1993).

This chapter contends that an organisation's safety culture, as a system composed of behaviours, practices, policies, and structural components, cannot flourish or succeed without interaction and harmony with its environment – the societal or national culture. In other words, safety culture should be considered in the context of national culture.

The objectives of this chapter are threefold:

- to identify the element of a safety culture;

- to delineate a conceptual model of the interactions of dimensions of national culture with those of the safety culture, and to report on an experimental study on how the safety and productivity of US-owned manufacturing plants in other countries are affected by both the national culture of the host country and the organisational culture of the subsidiary plant;

- to identify important, culturally based parameters affecting nuclear safety.

4.1 CULTURAL CONTEXT AND NUCLEAR SAFETY

According to Fujita (1992), who compared Japanese with US nuclear power plants, cultural differences play a significant role in determining the performance of operators. A recent seminal study of nuclear power operations in the United States, Germany, France, Switzerland, and Sweden by Rochlin and von Meier (1994) has highlighted the *critical* role of cultural factors in the safety of nuclear power plants. Their findings were that 'cultural differences were central and functional to operational safety and reliability' (p. 160) and that 'regulators almost always insist that their actions are meant to encourage or nourish the safety culture, but [that] many of the operators we interviewed pointed out that awkward or clumsy regulation can also interfere with the safety culture' (p. 181). Rochlin and von Meier stated that 'care must be given to the group dynamics when designing computer-based control rooms; and because the dynamics of the control room differ from country to country (and possibly from plant to plant), the design must take cultural differences into account' (p. 175). The authors also noted that regulations should be sensitive to cultural norms and they cautioned the blind 'transfer of US or other regulatory approaches developed in a radically different technical and cultural context' (p. 181). In conclusion, they raised an important warning flag for the nuclear industry around the world:

> To tinker with staff and departmental organisation, assignments of responsibility, delegation of authority, operator discretion, or even control-room design and integration without understanding the role of culture is to perform a real-time experiment with possibly irreversible consequences. (p. 182)

4.2 SAFETY CULTURE AND ORGANISATIONAL CULTURE

Since the Chernobyl accident, several studies of safety culture have emerged from the nuclear industry around the world (e.g. General Accounting Office [GAO], 1990; Saji, 1991; Ostrom, Wilhelmsen, and Kaplan, 1993; Joksimovich, 1992; Dien, Llory, and Montmayeul, 1992). According to the International Atomic Energy Agency (IAEA),

> the [Chernobyl] accident can be said to have flowed from deficient safety culture, not only at the Chernobyl plant, but throughout the Soviet design, operating, and regulatory organisations for nuclear power that existed at the time. Safety culture requires total dedication, which at nuclear power plants is primarily generated by the attitudes of managers of organisations involved in their development and operation. (INSAG, 1992, p. 24)

In a report by the International Nuclear Safety Advisory Group (INSAG) of the IAEA, safety culture is defined as 'that assembly of characteristics and attitudes in organisations and individuals which establishes that, as an overriding priority, nuclear plant safety issues receive the attention warranted by their significance' (INSAG, 1991, p. 4).

According to the US NRC, nuclear safety culture is a prevailing condition in which each employee is always focused on improving safety, is aware of what can go wrong, feels personally accountable for safe operation, and takes pride and 'ownership' in the plant. Safety culture is a disciplined, crisp approach to operations by highly trained staff, who are confident but not complacent, follow sound procedures, and practice effective teamwork and effective communication. Safety culture is an insistence on a sound technical basis for actions and a rigorous self-assessment of problems (GAO, 1990).

According to Pidgeon (1991), safety culture can be characterised as the set of beliefs, norms, attitudes, roles, and social and technical practices that are concerned with minimising the exposure of employees, managers, customers, and members of the public to conditions considered dangerous or injurious. Also, the principal cultural unit within which a safety culture is assumed to be located is that of the organisation (Pidgeon and O'Leary, 1994). In other terms, safety culture is a product of a larger concept of organisational culture (Joksimovich, 1992), an assertion verified by a study supported by the NRC (Orvis, Moieni and Joksimovich, 1993).

Organisational culture through its operative norms, significantly affects system safety and reliability (Weick, 1987). Furthermore, many factors, such as the 'shared experience' of the organisational members, are formed and nurtured by the organisational culture. According to Rochlin, La Porte, and Roberts (1987), this factor contributes to organisational adaptability, 'self-designing', and self-adjusting and eventually results in higher organisational reliability. Therefore, the organisational culture is a key determinant in the formation of the safety culture of a technological system.

According to Schein (1985), organisational culture is 'a pattern of basic assumptions – invented, discovered, or developed by a given group as it learns to cope with its problems of external adaptation and internal integration – that has worked well enough to be considered valid and, therefore, to be taught to new members as the correct way to perceive, think, and feel in relation to those problems' (p. 9). Kotter and Heskett (1992) contend that organisational culture has two levels that differ in terms of their visibility and their resistance to change.

> At the deeper and less visible level, culture refers to values that are shared by the people in a group and that tend to persist over time even when group membership changes. . . . At the more visible level, culture represents the behaviour patterns or style of an organisation that new employees are automatically encouraged to follow by their fellow employees. Each level of culture has a natural tendency to influence the other.
>
> (p. 4)

Operationally, organisational culture is defined as a set of shared philosophies, ideologies, values, beliefs, expectations, attitudes, assumptions, and norms (Mitroff and Kilmann, 1984). Cultural norms refer to the set of unwritten rules that guide behaviour (Jackson, 1960). Use of this concept allows the capturing of those dimensions of organisational life that may not be visible in the more rational and mechanical aspects of the organisation.

4.3 DEFINITION AND DIMENSIONS OF NATIONAL CULTURES

Culture, according to anthropologists, is the way of life of a people – the sum of their learned behaviour patterns, attitudes, customs, and material goods. According to Azimi (1991), the culture of a society consists of a set of ideas and beliefs. These ideas and beliefs should have two principal characteristics or conditions. First, they should be accepted and admitted by the majority of the population. Secondly, the acceptance of these beliefs and ideas should not necessarily depend upon a scientific analysis, discussion, or convincing argument. In the context of technology transfer and utilisation, culture could also be defined operationally as the 'collective mental programming of peoples' minds' (Hofstede, 1980).

This view might be anathema to many scholars of 'hard' sciences and other engineering-dominated fields, but according to Stephen Jay Gould, the renowned Harvard University professor of geology, biology, and the history of science, even *scientific* theories are strongly culturally based (Gould, 1981, p. 22):

Facts are not pure and unsullied bits of information; culture also influences what we see and how we see it. Theories, moreover, are not inexorable inductions from facts. The most creative theories are often imaginative visions imposed upon facts; the source of imagination is also strongly cultural.

4.3.1 Dimensions of national culture

Culture, defined earlier as the collective mental programming of the people who work in technological systems, affects not only the safety but also the success and survival of any technology. National cultures differ in at least four basic dimensions: power distance; uncertainty avoidance; individualism–collectivism; and masculinity–femininity (Hofstede, 1980). *Power distance* refers to the extent to which a society accepts the fact that power in institutions and organisations is distributed unequally. It is an indication of the interpersonal power or influence between two entities, as perceived by the least powerful of the two (Boeing Commercial Aircraft Group [BCAG], 1993). *Uncertainty avoidance* refers to the extent to which a society feels threatened by uncertain and ambiguous situations. It also refers to attempts to avoid these situations by providing greater career stability, establishing more formal rules, not tolerating deviant ideas and behaviours, and believing in absolute truths and the attainment of expertise. *Individualism* refers to a 'loosely knit' social framework in which people are supposed to take care of themselves and their immediate families only, whereas *collectivism* is characterised by a tight social framework in which people distinguish between in-group and out-group. They expect their in-group members (e.g. relatives, clan, organisation) to look after them and, in exchange, they owe absolute loyalty to the group. The *masculinity* dimension expresses the extent to which the dominant values in a society are 'masculine,' as evidenced by decisiveness, interpersonal directness, and machismo (Johnston, 1993). Other characteristics of masculine cultures include assertiveness, the acquisition of money and material goods, a relative lack of empathy, and a lower perception of the importance of quality-of-life issues. This dimension can also be described as a measure of the need for ostentatious manliness in the society (BCAG, 1993). *Femininity*, the opposite pole on this continuum, represents lower assertiveness and greater empathy and concern for issues regarding the quality of life.

The four cultural dimensions discussed above also have significant implications for nuclear safety and control-room operations. For instance, according to Helmreich (1994a) and Helmreich and Sherman (1994), there is evidence that operators with high power distance and high uncertainty avoidance prefer and place a 'very high importance' on automation. Furthermore, it is known that the primary purpose of regulations is to standardise, systematise, and impersonalise operations. This is done, to a large extent, by ensuring adherence to (standard and emergency) operating procedures. On many occasions it requires replacing operators' habits with more desirable ones that are prescribed in procedures or enforced by regulations. However, according to several studies, an operator's culturally driven habit is a more potent predictor of behaviour than his or her intentions, and there could be occasions on which intentions cease to have an effect on operators' behaviour (Landis, Triandis, and Adamopoulos, 1978). This places in doubt the effectiveness of those regulations and procedures that are incompatible with operators' culturally driven habits.

4.4 FIELD STUDY AND ITS IMPLICATIONS FOR THE SAFETY CULTURE IN NUCLEAR INSTALLATIONS

A major, though subtle, factor affecting the safety and performance of a technological system is the degree of compatibility between its organisational culture and the national culture of the host country. It is an inevitable reality that groups and organisations within a society also develop cultures that significantly affect how the members think and perform (Schein, 1985). Demel and Meshkati (1989) and Demel (1991) conducted an extensive field study to explore how the performance of US-owned manufacturing plants in other countries is affected both by the national culture of the host country and the organisational culture of the subsidiary plant. A manufacturing plant division of a large American multinational corporation was examined in three countries: Puerto Rico, the United States, and Mexico. Hofstede's (1980) *Values Survey Module* for national culture and Reynolds' (1986) *Survey of Organisational Culture* were administered. Performance measures (i.e. production, safety, and quality) were collected through the use of secondary research.

The purpose of this investigation was threefold:

- to determine if there were any differences between the national cultures of Puerto Rico, the United States, and Mexico;
- to find out if there were any differences between the organisational cultures of the three manufacturing plants;
- to establish whether there was any compatibility between the organisational culture of the plants and the national culture of the three countries, and to examine if the compatibility or incompatibility affected their performance in terms of production yields, quality, safety, and cycle time.

This study's aim was to examine the relationship, if any, between the compatibility of national and organisational cultures on the one hand, and performance on the other. The results of this study indicated that there were differences between the national culture dimensions of Puerto Rico, the United States, and Mexico. However, no significant differences were found between the organisational cultures of the three plants, perhaps because of selection criteria, according to which candidates may have been carefully screened for behavioural styles, beliefs, and values that 'fit in' to the existing organisational culture. Additionally, socialisation may have been another factor. For example, the company may have had in-house programmes and intense interaction during training, which could have created a shared experience, an informal network, and a company language. These training events often include songs, picnics, and sporting events that provoke feelings of togetherness. Also, the company may have had artifacts (the first level of organisational culture) such as posters, cards, and pens that remind the employees of the visions, values, corporate goals of the organisation, and that help to promote its culture.

Therefore, it seems that a 'total transfer' has been realised by this multinational corporation. These manufacturing plants produce similar products, so they must achieve uniform quality in their production centres. To do so, this company has transferred both its technical installations and machines as well as its organisation. Moreover, to fulfill its purpose, the company chooses its employees according to highly selective criteria.

These results notwithstanding, Hofstede's (1980) research demonstrated that even within a large multinational corporation famous for its strong culture and socialisation efforts, national culture continued to play a major role in differentiating work values. Differences were found between the national cultures of Puerto Rico, the United States, and Mexico.

There are concepts in the dimensions of organisational culture that may correspond to the same concepts of the dimensions of national culture. The 'power distance' dimension of national culture addresses the same issues as the perceived oligarchy dimension of organisational culture, in that they both refer to the nature of decision-making. In countries where power distance is large, only a few individuals from the top make the decisions; uncertainty avoidance and perceived change address the concepts of stability, change, and risk-taking. One extreme is the tendency to be cautious and conservative (e.g. to avoid risk and change when possible) in adopting different programmes or procedures. The other extreme is the predisposition to change products or procedures, especially when confronted with new challenges and opportunities – in other words, to take risks and make decisions. Uncertainty avoidance may be related to perceived tradition in the sense that, if the employees have a clear perception of 'how things are to be done' in the organisation, their fear of uncertainties and ambiguities will be reduced. Agreement on a perceived tradition in the organisation goes well with a country with high uncertainty avoidance. Individualism–collectivism and perceived cooperation address the concepts of cooperation between employees and of trusting and assisting colleagues at work. In a collectivist country cooperation and trust among employees is keenly pursued; in an individualist country, less so.

One could say that perceived tradition (of the organisational culture) may also be related to individualism–collectivism in the sense that, if members of an organisation have shared values, know what their company stands for and what standards they are to uphold, they are more likely to feel as if they are an important part of the organisation. They are motivated, because life in the organisation has meaning for them. Ceremonies (of the organisational culture) and rewards to honour top performance are very important to employees in any organisation. However, the types of ceremony or reward that will motivate employees may vary across cultures. In other words, rewards and ceremonies should vary depending on whether the country has a masculine orientation (where money and promotion are important), or a feminine orientation (where relationships and working conditions are important). If given properly, they may keep the values, beliefs, and goals uppermost in the employees' minds and hearts.

Because cultural differences may play significant roles in achieving the success of corporate performance, the findings of this study may have important managerial implications. First, an organisational culture that fits one society might not be readily transferable to other societies. In other words, the organisational culture of the company should be compatible with the culture of the society to which the company is transferring. There needs to be a comfortable match between the internal variety of the organisation and the external variety (coming from the host country). When the cultural differences are understood, the law of requisite can then be applied as a concept for systematic investigation of the influence that culture has on the performance of the multinational corporation's manufacturing plants. This law may be useful when examining environmental variety in the new cultural settings. Secondly, the findings of the present study have confirmed that cultural compatibility between organisational cultures of the multinational corporations and the cultures of the countries they are operating in play a significant role in the performance of their manufacturing plants.

It can be suggested, therefore, that the decision about which management system to promote should be based on specific human, cultural, social, and deeply rooted local behaviour patterns. For the success of their operations it is critical for multinational corporations operating in cultures different from their own to ensure and enhance cultural compatibility. As a consequence, it can be recommended that no organisational culture should be transferred without prior analysis and recommendations for adjustment and adaptation to the foreign country's cultures and conditions. This research has given a clear view of the current potential for supervising and evaluating cultural and behavioural aspects of organisations as affected by their external environment and their relation to the performance of the organisations. Culture, both national and organisational, will become an increasingly important concept for technology transfer.

Results of this study showed that, whereas there were differences between the national cultures of the three countries, there were no significant differences between the organisational cultures of the three manufacturing plants. It is noteworthy that the rank order of the performance indicators for these plants was in exact concordance with the rank order of the compatibility between the organisational culture and the national culture of the host country: Mexico had the highest overall cultural compatibility and the highest performance, Puerto Rico had high overall compatibility and the next highest overall performance, and the United States had the lowest cultural compatibility and the lowest overall performance.

4.5 NATIONAL CULTURE: ATTENUATOR OR AMPLIFIER OF THE SAFETY CULTURE?

The following issues and examples are an attempt to demonstrate some important culturally based attitudes affecting organisational functioning, technology utilisation, and, in particular, safety culture in nuclear power plants (discussed in detail in Meshkati, 1994):

- Risk perception.
- Attitude toward work.
- Work group dynamics.
- Attitude toward technology.
- Attitude toward organisation, hierarchy, procedure, and working habits.
- Attitude toward time and time of the day.
- Religious duties and their effects on work.
- Achievement motivation and orientation.
- Population stereotype (e.g. colour association).
- The 'If-it-ain't-broke, don't-fix-it' attitude.

For instance, according to Otway and von Winterfeldt (1982), people's perceptions of technologies and of risk depend upon the information to which they have been exposed, the information they have chosen to believe, the values they hold (including religious and ideological beliefs), the social experiences to which they have had access, the dynamics of stakeholder groups, the vagaries of their political process, and the historical context within which all of the aforementioned have taken place. The perception of the risk posed by a given technology is also influenced positively by, among other things, whether the technology increases the standard of living, creates new jobs, enhances national prestige, and/or creates greater independence from foreign suppliers. According to Douglas and Wildavsky (1982), risk perception is influenced or moderated in society by a 'cultural bias' that 'elevates some risks to a high peak and depresses others below sight' (p. 9).

The response of nuclear reactor operators to nuclear power plant disturbances is shown in Figure 4.1. The operators are constantly receiving data from the displays in the control room and looking for change or deviation from standards or routines in the plant. It is contended that their responses during transition from the rule-based to the knowledge-based level of cognitive control, especially at the knowledge-based level, are affected by the safety culture of the plant, and are also moderated or influenced by their cultural background. Their responses could start a vicious circle which, in turn, could lead to inaction, which wastes valuable time and control room resources. Breaking this vicious circle requires boldness to make or take over decisions, so that the search for possible answers to the unfamiliar situation does not continue unnecessarily and indefinitely. It is contended that the boldness is culturally driven to a high degree, and that it is a function of the plant's organisational culture, reward system, and regulatory environment. Boldness, of course, is also influenced by operators' personality traits, aptitude for risk-taking, and perception (as mentioned before), which are also strongly cultural. Other important aspects of the national culture include *hierarchical power distance* and *rule orientation* (Lammers and Hickson, 1979), which govern acceptable behaviour and could determine the upper boundary of operators' boldness.

According to INSAG (1991), 'two general components of the safety culture are the necessary framework within an organisation [whose development and

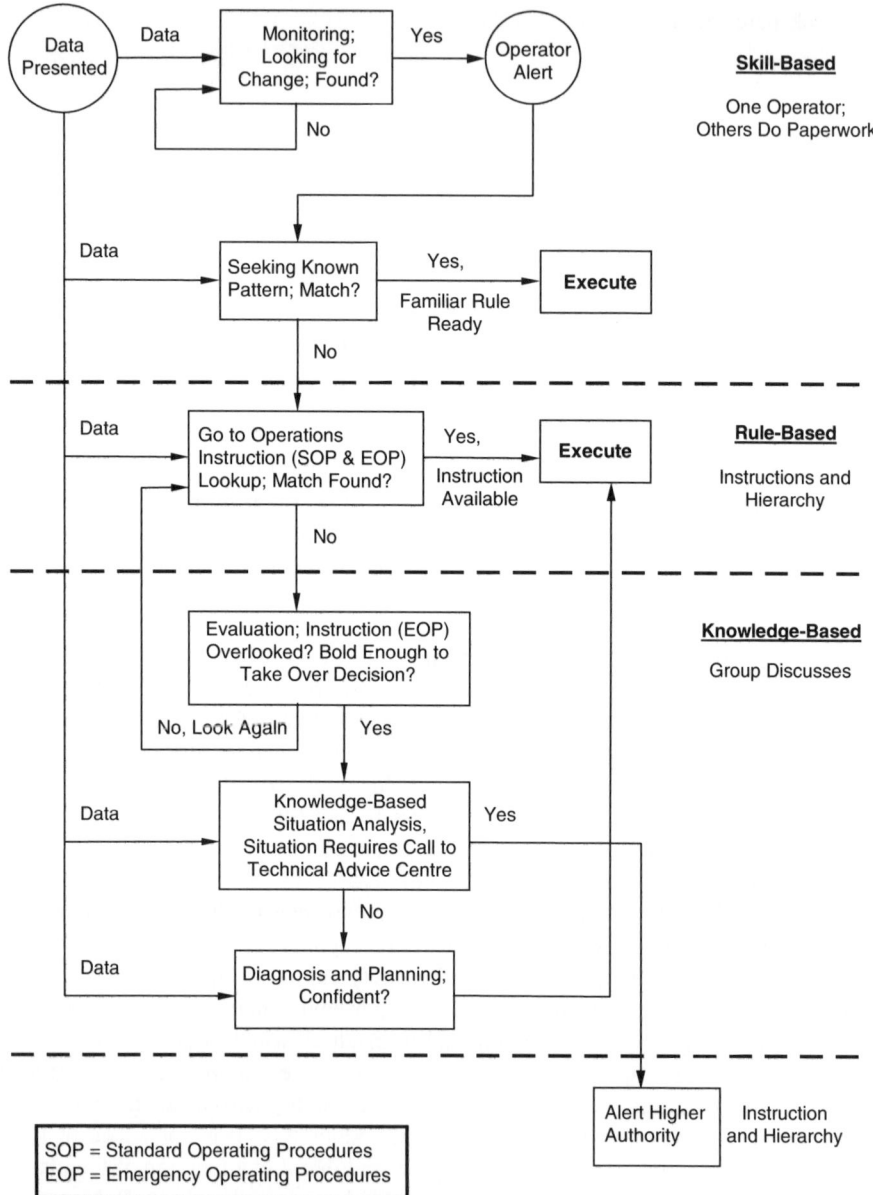

Figure 4.1 Model for nuclear power plant operators' responses to disturbances.

From: *Integration of workstation, job, and team structure design in the control rooms of nuclear power plants: Experimental and simulation studies of operators' decision styles and crew composition while using ecological and traditional user interfaces* (p. 42), N. Meshkati, B. J. Buller, and M. A. Azadeh, August 1994, Grant Report, prepared for the US Nuclear Regulatory Commission (Grant No. NRC-04-91-102), Los Angeles: University of Southern California. Reprinted with permission. Adopted from J. Rasmussen, March 1992.

maintenance is the responsibility of management hierarchy] and the attitude of staff at all different levels in responding to and benefiting from the framework.' Also, the requirements of individual employees for achieving safety culture at the installation are 'a questioning attitude, a rigorous and prudent approach, and necessary communication' (pp. 13–14). However, other dimensions of national culture – uncertainty avoidance, individualism-collectivism, and masculinity–femininity – could either resonate with and strengthen or attenuate safety culture, while interacting with these general components and requirements. For instance, a questioning attitude (on the part of operators) is greatly influenced by the power distance, rule orientation, and uncertainty avoidance of the societal environment and the openness in the organisational culture of the plant. A rigorous and prudent approach, which involves understanding the work procedures, complying with procedure, being alert for the unexpected, and so forth, is moderated by power distance and uncertainty avoidance in the culture. It is also moderated by the sacredness of procedures, the critical nature of step-by-step compliance, and a definite organisational system at the plant. Communication, which involves obtaining information from others, transmitting information to others, and so on, is a function of all the dimensions of national culture as well as the steepness and rigidity of the hierarchical organisation of the plant.

The nuclear industry shares many safety-related issues and concerns with the aviation industry, and there is a continuous transfer of information between them (e.g. Electronic Power Research Institute [EPRI], 1984). Cultural and other human factors are considerations affecting the performance of a cockpit crew and are, to a large extent, similar to those affecting nuclear plant control-room operators. It is worth, therefore, recalling the crash of a passenger aircraft, an accident to which cultural factors within the cockpit and between it and the air-traffic control tower were found to have contributed significantly (National Transportation Safety Board [NTSB], 1991). Flight 052 (AV052), a Boeing 707 of the Airline of Columbia, crashed in Cove Neck, New York, on 25 January 1990, fatally injuring 73 of the 158 persons aboard. According to Helmreich (1994b):

> In a culture where group harmony is valued above individual needs, there was probably a tendency to remain silent while hoping that the captain would 'save the day.' There have been reported instances in other collectivist, high power distance cultures where *crews have chosen to die in a crash rather than disrupt group harmony and authority* and bring accompanying shame upon their family and in-group.
> (emphasis added, p. 17)

> High Uncertainty Avoidance may have played a role (in this accident) by locking the crew into a course of action and preventing discussion of alternatives and review of the implications of the current course of action. High Uncertainty Avoidance is associated with a tendency to be inflexible once a decision has been made as a means of avoiding the discomfort associated with uncertainty. (p. 17)

Moreover, the importance of the cultural factors *vis-à-vis* automation in the aviation industry has been further highlighted by two recently published studies. Sherman and Helmreich (in press), in their study of national culture and flightdeck automation, surveyed 5,705 pilots across 11 nations and have reported that 'the lack of consensus in automation attitudes, both within and between nations, is disturbing'. They have concluded that there is a need for clear explication of the philosophy governing the design of automation. Most recently, the US Federal Aviation Administration Human Factors Study Team (FFA, 1996) issued a report entitled *The Interfaces Between Flightcrews and Modern Flight Deck Systems*. This team identified several 'vulnerabilities' in flightcrew management of automation and situation awareness that are caused by a number of interrelated deficiencies in the current aviation system, such as 'insufficient understanding and consideration of cultural differences in design, training, operations, and evaluation' (p. 4). They have recommended a host of further studies under the heading of Cultural and Language Differences. They included pilots' understanding of automation capabilities and limitations, differences in pilot decision regarding when and whether to use different automation capabilities, the effects of training, and the influence of organisational and national cultural background on decisions to use automation.

4.6 CONCLUSION

Safety culture in a nuclear installation should be considered in the context of its local national culture. Necessary conditions for creating and nourishing safety culture in a nuclear installation include (but are not limited to):

1 a thorough understanding of the dimensions of local national culture;

2 a determination of the extent of their match with those of the organisational culture of the installation;

3 their compatibility with the prescribed requirements for safety culture;

4 a further understanding of the effects of cultural variables on the interactions between human operators and automation in control rooms.

References

AZIMI, H. (1991) *Madarhaaye Toseah-Nayafteghi in Eqtesad-e-Iran* [Circles of underdevelopment in the Iranian economy]. Teheran, Iran: Naey.

BCAG (Boeing Commercial Aircraft Group) (1993) Crew factor accidents: regional perspective. In *Proceedings of the 22nd technical conference of the international air transport association (IATA) on Human Factors in Aviation* (pp. 45–61). Montreal: IATA.

DEMEL, G. (1991) Influences of culture on the performance of manufacturing plants of American multinational corporations in other countries: A macroergonomics analysis. Unpublished master's thesis. Los Angeles, CA: Institute of Safety and Systems Management, University of Southern California.

DEMEL, G. and MESHKATI, N. (1989) Requisite variety: A concept to analyse the effects of cultural context for technology transfer. In *Proceedings of the 33rd annual meeting of the human factors society* (pp. 765–9). Santa Monica, CA: Human Factors Society.

DIEN, Y., LLORY, M. and MONTMAYEUL, R. (1992) Operators' knowledge, skill, and know-how during the use of emergency procedures: Design, training, and cultural aspects. In *Proceedings of the fifth conference on human factors and power plants* (pp. 178–81). New York: Institute of Electrical and Electronics Engineers (IEEE).

DOUGLAS, M. and WILDAVSKY, A. (1982) *Risk and culture: An essay on the selection of technical and environmental dangers.* Berkeley, CA: University of California Press.

EPRI (Electric Power Research Institute) (1984, January) *Commercial aviation experience of value to the nuclear industry (EPRI NP-3364).* Prepared by Los Alamos Technical Associates. Palo Alto, CA: EPRI.

FAA (Federal Aviation Administration) (1996, June 18) *Federal aviation administration human factors team report on the interfaces between flightcrews and modern flight deck systems.* Washington, DC: FAA.

FUJITA, Y. (1992) Ebunka: Do cultural differences matter? In *Proceedings of the 1992 fifth conference on human factors and power plants* (pp. 188–94). New York: Institute of Electrical and Electronics Engineers (IEEE).

GELLER, E. S. (1994, September) Ten principles for achieving a total safety culture. *Professional Safety*, pp. 18–24.

GAO (General Accounting Office) (1990, April) *Nuclear safety: Concerns about reactor restart and implications for DOE's safety culture* (GAO-RCED-90-104). Washington, DC: GAO.

GOULD, S. J. (1981) *The Mismeasure of Man.* New York: Norton.

HELMREICH, R. L. (1994a, May 18) *Operational Personnel Perception on Automation.* Presentation at the International Civil Aviation Organisation (ICAO) Flight Safety and Human Factors Seminar, Amsterdam.

HELMREICH, R. L. (1994b) Anatomy of a system accident: Avianca Flight 052. *International Journal of Aviation Psychology*, **4**, pp. 265–84.

HELMREICH, R. L. and SHERMAN, P. (1994) Flightcrew perspective on automation: A cross-cultural perspective. *Report of the Seventh ICAO Flight Safety and Human Factors Regional Seminar* (pp. 442–53). Montreal: International Civil Aviation Organisation (ICAO).

HOFSTEDE, G. (1980) *Culture's Consequences.* Beverly Hills, CA: Sage.

INSAG (International Nuclear Safety Advisory Group) (1991) *Safety Culture* (Safety Series No. 75-INSAG-4). Vienna: International Atomic Energy Agency (IAEA).

INSAG (International Nuclear Safety Advisory Group) (1992) *The Chernobyl Accident: Updating of INSAG-1* (INSAG-7). Vienna: International Atomic Energy Agency (IAEA).

JACKSON, J. M. (1960) Structural characteristics of norms. In HENRY, M. B. (ed.), *The dynamics of instructional groups: Socio-psychological aspects of teaching and learning.* Chicago: University of Chicago Press.

JOHNSTON, A. N. (1993) CRM: Cross-cultural perspectives. In WIENER, E. L., KANKI, B. G. and HELMREICH R. L. (eds), *Cockpit Resource Management* (pp. 367–97). San Diego: Academic Press.

JOKSIMOVICH, V. (1992) Safety culture in nuclear utility operations. In *Proceedings of the 1992 fifth conference on human factors and power plants* (pp. 182–7). New York: Institute of Electrical and Electronics Engineers (IEEE).

KOTTER, J. P. and HESKETT, J. L. (1992) *Corporate Culture and Performance.* New York: Free Press.

LAMMERS, C. J. and HICKSON, D. J. (1979) A cross-national and cross-institutional typology of organisations. In LAMMERS, C. J. and HICKSON, D. J. (eds), *Organisations alike and unlike: International and interinstitutional studies in the sociology of organisations* (pp. 420–34). London: Routledge & Kegan Paul.

LANDIS, D., TRIANDIS, H. C. and ADAMOPOULOS, J. (1978) Habit and behavioural intensions as predictors of social behaviour. *The Journal of Social Psychology*, **106**, pp. 227–37.

LAUBER, J. K. (1993) A safety culture perspective. *Growing a Safety Culture*. In *Proceedings of the 38th Annual Corporate Aviation Safety Seminar* (pp. 11–17). Arlington, VA: Flight Safety Foundation.

MESHKATI, N. (1988, October 18–20) An integrative model for designing reliable technological organisations: The role of cultural variables. Paper prepared for the World Bank Workshop on Safety Control and Risk Management in Large-Scale Technological Operations, Washington, DC.

MESHKATI, N. (1989a) An etiological investigation of micro- and macroergonomic factors in the Bhopal disaster: Lessons for industries of both industrialised and developing countries. *International Journal of Industrial Ergonomics*, **4**, pp. 161–75.

MESHKATI, N. (1989b, November 6–11) Self-organisation, requisite variety, and cultural environment: Three links of a safety chain to harness complex technological systems. Paper for the World Bank Workshop in Risk Management [in Large-Scale Technological Operations], Karlstad, Sweden.

MESHKATI, N. (1989c) Technology transfer to developing countries: A tripartite micro- and macroergonomic analysis of human-organisation-technology interfaces. *International Journal of Industrial Ergonomics*, **4**, pp. 101–15.

MESHKATI, N. (1991a) Human factors in large-scale technological systems' accidents: Three Mile Island, Bhopal, Chernobyl. *Industrial Crisis Quarterly*, **5**, pp. 133–54.

MESHKATI, N. (1991b) Integration of workstation, job, and team structure design in complex human-machine systems: A framework. *International Journal of Industrial Ergonomics*, **7**, pp. 111–22.

MESHKATI, N. (1994) Cross-cultural issues in the transfer of technology: Implications for aviation safety. *Report of the Flight Safety and Human Factors Regional Seminar and Workshop* (pp. 116–37). Montreal: International Civil Aviation Organisation (ICAO).

MESHKATI, N., BULLER, B. J. and AZADEH, M. A. (1994, August) *Integration of workstation, job, and team structure design in the control rooms of nuclear power plants: Experimental and simulation studies of operators' decision styles and crew composition while using ecological and traditional user interfaces*, (Vol. 1). Grant Report prepared for the US Nuclear Regulatory Commission (Grant No. NRC-04-91-102). Los Angeles: University of Southern California.

MINER, S. G. (1991, August) Creating the safety culture. *Occupational Hazards*, pp. 17–21.

MITROFF, I. I. and KILMANN, R. H. (1984) *Corporate tragedies: Product tampering, sabotage, and other catastrophes*. New York: Praeger.

NTSB (National Transportation Safety Board) (1991) *Aircraft accident report: Avianca, the Airline of Columbia, Boeing 707–321B, HK 2016 fuel exhaustion Cove Neck, New York, January 25, 1990* (Report No. NTSB-AAR-91-04). Washington, DC: NTSB.

Nuclear Engineering International Promoting a change in safety culture: Selin's view. (1994, July) pp. 16–17.

Nuclear Safety Review (1992). Vienna: International Atomic Energy Agency.

ORVIS, D. D., MOIENI, P. and JOKSIMOVICH, V. (1993, April) *Organisational and management influences on safety of nuclear power plants: Use of PRA techniques in quantitative and qualitative assessments* (NUREG/CR-5752). Prepared for the US Nuclear Regulatory Commission. San Diego, CA: Accident Prevention Group.

OSTROM, L., WILHELMSEN, C. and KAPLAN, B. (1993, April–June) Assessing safety culture. *Nuclear Safety*, **34** (2), pp. 163–72.

OTWAY, H. J. and VON WINTERFELDT, D. (1982) Beyond acceptable risk: On the social acceptability of technologies. *Policy Sciences*, **14**, pp. 247–56.

PIDGEON, N. F. (1991) Safety culture and risk management in organisations. *Journal of Cross-Cultural Psychology*, **22**, pp. 129–40.

PIDGEON, N. F. and O'LEARY, M. (1994). Organisational safety culture: Implications for aviation practice. In JOHNSTON, N., MCDONALD, N. and FULLER, R. (eds), *Aviation psychology in practice* (pp. 21–43). Hants: Avebury Technical.

REYNOLDS, P. D. (1986) Organisational culture as related to industry, position, and performance. *Journal of Management Studies*, **23**, pp. 333–45.

ROCHLIN, G. I., LA PORTE, T. R. and ROBERTS, K. H. (1987, Autumn) The self-designing high-reliability organisation: Aircraft carrier flight operations at sea. *Naval War College Review*, pp. 76–90.

ROCHLIN, G. I. and VON MEIER, A. (1994) Nuclear power operations: A cross-cultural perspective. *Annual Review of Energy and Environment*, **19**, pp. 153–87.

RODNEY, G. A. (1991) Rebuilding a safety culture. In *Proceedings of the 44th International Air Safety Seminar* (pp. 38–43). Arlington, VA: Flight Safety Foundation.

SAJI, G. (1991, July–September) Total safety: A new safety culture to integrate nuclear safety and operational safety. *Nuclear Safety*, **32**, pp. 416–23.

SCHEIN, E. H. (1985) *Organisational culture and leadership*. San Francisco: Jossey-Bass.

SHERMAN, P. J. and HELMREICH, R. L. (in press) National culture and flightdeck automation: Results of a multi-nation survey. *The International Journal of Aviation Psychology*.

WEICK, K. E. (1987, Winter) Organisational culture as a source of high reliability. *California Management Review*, **29**, pp. 112–27.

CHAPTER FIVE

Implicit social norms in reactor control rooms

MARIN IGNATOV

Bulgarian Academy of Sciences, Sofia

All control and steering activities in a reactor control room are highly structured. The explicit work-task structuring is based on omnipresent procedure manuals and other regulations. There is much unsystematic evidence, however, that the explicit rules are not always followed even in such extremely regulated environments. The author examines the interrelationship between explicit and implicit behavioural norms. The iterative research process that is recommended includes collecting and analysing data, planning of procedures, and facilitating the confidential exchange of information. Results show support for the initial hypothesis that implicit norms are predictors of safe performance at the individual level. The violation of implicit norms is punished in a very specific way – through social ridicule in the sub-unit and rejection. In contrast, the observance of implicit norms leads to acknowledgment, popularity, and a gain in status. Implicit norms evolve through a repeated reinforcement of behaviour in specific situations. Hence, implicit norms are not easy to recall without also recalling the specific situation. They seem to be cognitively represented only in connection with information about situations that have been experienced.

Improving system safety in high-hazard, low-risk systems such as nuclear power plants is not only a problem of technology. To an increasing extent it encompasses psycho-organisational inquiries into the social behaviour and social interaction of the operational personnel. Taking into account this trend, the International Nuclear Safety Advisory Group (INSAG, 1986, 1991) introduced the concept of 'safety culture.' It was understood as an assembly of characteristics and attitudes in organisations and individuals, which establishes that, as an overriding priority, nuclear plant safety issues receive the attention warranted by their significance.

Wilpert (1991) has argued, however, that the initial INSAG understanding of safety culture restricts the impact area to predominantly cognitive matters. He stresses the need for investigations of the relationships between values, beliefs, and attitudes on the one side, and patterns of behaviour on the other. In a parallel development the Study Group on Human Factors of the Advisory Committee on the Safety of Nuclear Installations (ACSNI, 1993) has come to similar conclusions, and has defined the safety culture of an organisation as the product of individual and group values, attitudes, perceptions, competencies, and patterns of behaviour that determine the commitment to, and the style and proficiency of, an organisation's health and safety management.

From this perspective the steady growth of interest in cultural and cross-cultural analyses appears understandable. A review study by Rochlin and von Meier (1994) indicated a broad band of acceptable strategies in implementing elements of the concept of safety culture. Similar conclusions are drawn in this volume's chapters by Rochlin, Reason, Meshkati, Bourrier, and Semmer. One promising approach to answering the challenge of understanding safety behaviour is to develop research tools and methods for a deeper psychological comprehension of the organisation of activities in the reactor control room, and to combine this effort with the cross-cultural understanding of norms and beliefs. This chapter describes some of the results of a qualitative pilot study that is part of a bigger research programme investigating the regulatory importance of 'seen-but-unnoticed', implicit norms and rules of behaviour in the nuclear reactor control room.

The explicit normative bases of all activities in the control room are the written procedures and rules. All control and steering activities in a reactor control room are highly structured. The explicit work-task structuring is based on the omnipresent procedure manuals and other regulations. Practical observations show, however, that explicit rules are not always followed. More or less formalised implicit standards or specific schemes of conduct play an important role in the overall regulation of operators' activities. The actual behaviour of the operators derives from the explicit structuring, but is modified through psychological processes of 'redefinition.' There is much empirical, though unsystematic, evidence for the existence of specific or individual preference rules, which derive from implicit principles, such as group or social mores or even ethnic folkways. Such rules define the 'proper' conduct of operators in reactor control rooms.

A main research problem is that both explicit and implicit preference rules give rise to patterns of behaviour that in some cases might be contradictory in nature. Sometimes, following implicit norms might lead to a better or more efficient control style than would strict adherence to written procedures, but it might also diminish efficiency and even lead to incidents. The still unanswered research question is how such implicit norms based on real features of life and work, such as social responsibility, self-image, and social tact, 'redefine' written procedures and other explicit safety rules.

The subjective process of redefining work tasks has been analysed by Hackman (1970). His research concentrated on regulated patterns of action, which are redefined in order to suit individual value systems and to correspond to the level of trust the operator has in his or her partners in the working environment. Hackman stated that the redefinition process depends on three psychological regulatory systems:

- adequate or inadequate perception of the task and the circumstances;
- the individual's readiness for specific action;
- the individual's experiences with the same or similar tasks.

One may assume that implicit social norms, rules, and beliefs can influence all three regulatory systems. They modify the perception of a given task, influence the individual readiness for action, and shape the individual's experiences to bring them into accord with group or social expectations.

The purpose of this pilot study was to extend prediction of safe performance to the individual and sub-unit level of analysis. Implicit norms, rules, and 'hidden' beliefs were checked as predictors of safe performance on a preliminary, qualitative basis. The main hypothesis was that safe performance is regulated and can therefore be predicted by implicit safety norms.

The pilot study was aimed also at examining the necessary organisational prerequisites for the subsequent larger investigation of implicit norms. For this purpose a cooperation agreement with the management of an eastern European nuclear power plant was signed.

5.1   METHOD

During the pilot study, a review of safety documents and procedures was carried out. Psychological and managerial knowledge was transmitted in several meetings, discussions, and seminars in order to increase the acceptance of human factors research methods and approaches among operators and managers of the power plant. Specific team-building and group-discussion techniques were systematically applied. Consequently, closer and more fruitful relationships between the researchers and the power plant's personnel were established.

In accordance with the pilot phase of the project, a qualitative and descriptive approach was chosen. Information came from 12 open-ended interviews with control room operators. The interviews were carried out in the last quarter of 1995. The interviewed operators were not considered high or low incident-prone by the management or by the psychological counsellors of the nuclear power plant. My interest centred on the interviewees' versions of reality in the control room situation. Questions of self-image, preferred behaviour, lifestyle, and risk-taking within and outside the working environment were discussed. The aim of this initial effort was to produce descriptions of basic implicit norms, so that the results can be used in the design of the subsequent larger investigation.

5.2 FINDINGS

Several implicit norms were discerned in the qualitative data and are given below as examples. These norms have not yet been cross-validated, but serve solely as a basis for further hypothesis-building.

5.2.1   Efficiency norms

For almost all interviewees good operator performance is synonymous with efficiency. As one interviewee put it: 'Good operators know how to maximise output and minimise risks'. The ultimate proof of such efficiency is a personal history of incident-free work in the reactor control room.

5.2.2   Norms about sharing of operational responsibility

Some operators stressed that they were good in the job, not because they never violated a written rule, but because they could distinguish between those situations in which responsibility should be taken individually and those in which responsibility should be shared. One knows only afterwards, of course, whether the situation was serious or unimportant, so implicit norms about responsibility-sharing evolve slowly and in a very contradictory way. Interestingly, there were reports about cases where operators would not wake up the sub-unit or plant manager at night because they were afraid that the problem might turn out to be harmless.

The mechanisms of implicit norms about sharing responsibility can be very subtle. The suggestion to call the sub-unit manager in cases that contradict the implicit norm might not be rejected in a harsh way. Just a single word or a gesture might be enough to give the initiator the impression of being incompetent for wishing to call the sub-unit manager because of a 'trifle.'

5.2.3   Norms concerning the correctness of electronically displayed information

Some experienced operators could be convinced that a given operational situation might be safe even if the relevant display is out of order. Someone who suggests further examination runs the risk of ridicule. It does not appear to be so important whether this fear of rejection is justified or not. There are situations in which such interpersonal fears of ridicule might keep operators from reacting, even though the majority in the reactor control room consider action meaningful. In one case an operator reported that at the beginning of his career he used to ask repeatedly about the correctness of electronically displayed information, even though for the other operators everything seemed to be under control. The same person reported that for a period of time he was considered anxious and overcautious by the other operators.

5.2.4 Norms about different ranges of tolerance for explicit rules

Two younger operators spoke repeatedly of older operators showing a disregard for some of the written rules. As a social value or behaviour pattern, this outspokenness was not considered inappropriate as long as the following implicit rule applied to the more experienced operator: 'He has never had an incident, so he knows what he is doing'. This rule did not apply to younger operators, however.

Another finding was that, when older operators taught the young 'trainees' how to work in the control room, they also passed on implicit social norms. The newcomers were instructed on how to become confident operators, so that they could take the occasional risk while maintaining control. For the older operators the real expertise was not found exclusively in operating manuals. It was rather a process of the continuous enlargement and enrichment of personal experience.

5.2.5 Norms about passing on information to superiors

There was also interview evidence for the existence of implicit norms about whether superiors were to be informed about problems. It might be considered uncooperative to report innocuous minor violations of rules.

5.2.6 Norms of self-presentation

Operators work in the reactor control room in the presence of others. They manage two processes simultaneously: the technological process in the reactor or the turbine, and the socio-psychological process of passing on information about themselves, their feelings, their character, and social status. The latter process conveys particular impressions that operators wish to make. It appeared to be a very strong implicit norm for operators to give the outward impressions of competence and self-confidence regardless of the doubt they might feel inwardly. Most of the operators considered themselves to be patient, alert, cautious, and technically able.

5.2.7 Gender-specific ('macho' norms) about sharing personal emotional experience

Some operators felt that they would encounter contempt and antipathy or become objects of ridicule or scorn if their fear, anxiety, or doubt were to be discovered by their co-workers in the reactor control room. There appeared to be no mechanism for sharing fears in the specifically 'macho' environment of the control room. The researcher assessed the presence of women operators in control rooms as positive in this context. As representatives of the opposite sex, the women were not expected to comply fully with this implicit norm, so they were in a much better position to make feelings of fear and anxiety a 'legitimate' matter, thus modifying the whole structure of implicit 'macho' norms for the male operators as well.

5.3 CONCLUSIONS

Some evidence of implicit norms as predictors of safe performance at the individual and the sub-unit level was found, though it was not cross-validated. It was also concluded that implicit safety norms might be more pervasive in this nuclear power plant, because of the special social situation in the country of its location. It also became clear that the implicit norms of the sub-units are not distinct enough to yield significant differences at the sub-unit level of analysis.

The results of the interviews supported the initial hypothesis that implicit norms determine what is regarded as allowed/not allowed, and important/not important. The unwritten rules might differ from the formal, official ones, or even contradict them, but they would not necessarily hamper the efficiency of the cooperative effort in reactor control rooms.

Another aspect of implicit norms became evident when discussing models of penalisation of unwanted behaviour. The violation of implicit norms might also be punished, but in a very different way – through social ridicule in the sub-unit and also rejection. In contrast, the observance of implicit norms leads to acknowledgment, popularity, and a gain in status.

Conclusive evidence of the strength of norm-enforcing mechanisms were the concrete examples of responses to a violation of explicit or implicit rules (see also Reason, in this volume), not the official proclamation of the safety culture. 'Negative' implicit norms evolve with great subtlety, 'step by step'. In an extremely structured and formalised organisation such as a nuclear power plant, great importance should be attributed to the behaviour of models, especially that of sub-unit leaders. For example, a superior who talks about safety as the highest priority but who ignores clear violations of safety rules and even complains about existing rules as a 'stressful burden', will undermine explicit norms, in spite of repeating them incessantly.

Because of different motivational and cognitive prerequisites, the evolved implicit norms differ both in their content and regulatory strength from one individual, group, organisational sub-unit, and organisation to the next. To a certain extent these differences are natural characteristics of the human interaction. But in many situations relevant to safety, they may interfere with communication and decision-making.

Implicit norms evolve through repeated reinforcement of behaviour in specific situations, making implicit norms difficult to recall without recalling the specific situation. The norms seem to be cognitively represented only in connection with information about situations that have been experienced. Subsequent generalisation and verbalisation of implicit safety norms appear to be very difficult. Further research is needed in this direction.

Implicit norms in the reactor control room are more widely accepted and practiced than others: some are relevant for every kind of social interaction, whereas others are confined to rather specific areas. The differentiated knowledge pertaining to specific implicit norms would probably allow the researcher to predict behaviour in situations of rule conflicts. Often, implicit norms are operating in difficult and ambiguous situations and, hence, can be a factor in diminishing or fostering the efficiency of actions taken by operators.

The overall conclusion from the pilot study is that the specifics of safety practices as well as the implementation process for new or 'imported' safety measures can be understood only if the social researcher is in a position to uncover and observe the corresponding implicit social norms. This finding stresses the need for a deeper understanding of the cultural and psychological basis of safety behaviour, and is consistent with findings of recent cross-cultural studies on safety practices in nuclear power plants (Rochlin, von Meier, 1994; Bourrier, in this volume).

A typically inappropriate and inefficient strategy for coping with implicit norms would be to try to suppress them by developing more and more procedures and rules. Often, this strategy coincides with a tendency to deny the very existence of implicit norms. Such defensive and naive positions neither foster nor deepen the individual organisation's understanding of safety-related activities in the nuclear control room.

In conclusion, the results allowed an overall positive feasibility evaluation of the larger investigation that is planned. The survey showed that it was possible to infer implicit norms from empirical data through an iterative research process of collecting and analysing data, planning procedures, and facilitating the confidential exchange of information in an interview situation.

References

ACSNI (Advisory Committee on the Safety of Nuclear Installations, Study Group on Human Factors) (1993) *Third report: Organising for safety*. London: HMSO.

HACKMAN, J. R. (1970) Tasks and task performance in research on stress. In MCGRATH, J. E. (ed.), *Social and Psychological Factors in Stress* (pp. 202–37). New York: Holt, Rinehart & Winston.

INSAG (International Nuclear Safety Advisory Group) (1986) *Safety Culture*, Safety Series, Report 1 (INSAG-1). Vienna: International Atomic Energy Agency (IAEA).

INSAG (International Nuclear Safety Advisory Group) (1991) *Safety Culture*, Safety Series, Report 4 (INSAG-4). Vienna: Atomic Energy Agency.

ROCHLIN, G. I. and VON MEIER, A. (1994) Nuclear power operations: A Cross-Cultural Perspective. *Annual Review of Energy Environment*, **19**, pp. 153–87.

WILPERT, B. (March 1991) System Safety and Safety Culture. Paper presented at IAEA & IIASA Meeting on the influence of organisation and management on the safety of Nuclear Power Plants and other industrial systems, Vienna.

CHAPTER SIX

Situational assessment of safety culture

NORBERT SEMMER AND ALEX REGENASS

University of Bern, Department of Psychology

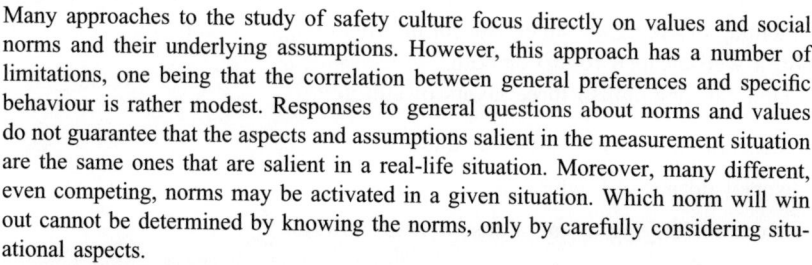

Many approaches to the study of safety culture focus directly on values and social norms and their underlying assumptions. However, this approach has a number of limitations, one being that the correlation between general preferences and specific behaviour is rather modest. Responses to general questions about norms and values do not guarantee that the aspects and assumptions salient in the measurement situation are the same ones that are salient in a real-life situation. Moreover, many different, even competing, norms may be activated in a given situation. Which norm will win out cannot be determined by knowing the norms, only by carefully considering situational aspects.

These observations suggest the necessity for an approach that starts out from situations rather than norms. Situational approaches are used successfully in human resource management. Various experienced people are interviewed about incidents in which something went wrong or almost went wrong. They are asked about the situation and about the reactions, feelings, and thoughts that they and others had. The answers elicited in such interviews make it possible to construe 'typical situations'. People can then be asked how they would respond in a given typical situation, how others would respond, and so forth. It is crucial that the situations and a sample covering all important aspects are typical of real-life cases. If these situations are carefully developed, the situational approach can be a significant complement to other methods of assessing safety culture.

6.1 CULTURE AND SITUATIONS

6.1.1 Elements of culture

Safety culture is seen as a salient factor in shaping effective safety regimes in nuclear power plants. The interest in such a holistic concept as culture is inspired by the insight that not everything can be governed by formal rules and regulations (INSAG, 1991). At the same time, the concept is an elusive one, and trying to assess it proves quite difficult (Hofstede, 1991; Rochlin and Meier, 1994).

To narrow down the scope of culture as a concept, one can define organisational culture as 'a pattern of basic assumptions . . . [of the] correct way to perceive, think, and feel' (Schein, 1990, p. 111; see also Meshkati, in this volume). The function of culture is to coordinate the actions of its members by aligning or coordinating their perceptions, thoughts, feelings, and, above all, actions. A stable culture can be comforting to the individual member. It reduces uncertainty by giving interpretation and meaning to events and actions. This culture is a product of a long learning process, during which a group or an organisation develops accepted ways of dealing with its problems (Schein, 1990).

Culture helps the individual appraise situations and events. This appraisal is twofold, consisting of an interpretation of reality (how things are) and a comparison of this interpretation with a social norm (how things should be). Directly or indirectly, this appraisal leads to observable behavioural results (e.g. going and checking something or asking the supervisor for information), physical products (e.g. a finished tool, a well-cleaned room) or both.

One can distinguish three layers of culture:

- observable artifacts and behaviours;
- social norms and values;
- basic underlying assumptions (Schein, 1985).

Artifacts and behaviour are the first things an outsider can see, feel, and hear upon entering an organisation. This category includes everything visible (dress codes and the tidiness of the workplace) as well as the availability and appearance of manuals, company records, products, and annual reports. The main problem with artifacts is, as Schein (1985) has warned, that they are hard to decipher accurately. If a forklift operator drives too fast and is not stopped by his supervisor, the lapse may be a significant indicator that safe behaviour is not being ensured, but it also could be an isolated event.

Values and social norms are the main target of many approaches to the study of safety culture. Values can be understood as preferred states, and social norms 'make explicit the forms of behaviors appropriate for members of the [organization]' (Katz & Kahn, 1978, p. 385). The distinction between values and norms is somewhat blurred, however. Norms stress the concrete formulation of a required behaviour, whereas values tend to give a rationale for requiring a particular behaviour in the first place. Many existing survey instruments that are designed directly

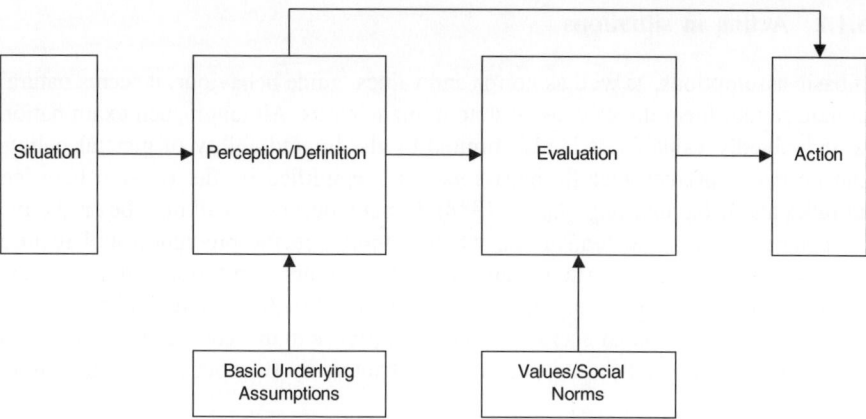

Figure 6.1 Perception, evaluation and selection of an action in a given situation influenced by basic underlying assumptions, values and social norms.

to assess safety focus on such norms and values. One example is the safety survey developed at the Institute of Work Psychology at the Swiss Federal Institute of Technology (Grote and Künzler, 1993, p. 61). It contains items such as 'In cases of conflict, safety has priority over production' (item O2) and 'For each step in the working process it is clear who is responsible' (item O6). The respondents are asked, how 'true' these statements are. Similarly, responses about personal norms of safety are elicited in the Idaho Safety Norm Survey (Ostrom, Wilhelmsen, and Kaplan, 1993) by items like 'In our company, the employees are aware of their part in safety'.

Basic underlying assumptions are defined by (Schein, 1990, p. 112) as 'taken-for-granted, underlying, usually unconscious assumptions that determine perceptions, thought processes, feelings, and behavior'. They often are so deeply rooted that they are rarely spoken about in the organisation, and often people are even not aware that they are operating (this is what Schein means by the term *unconscious*). 'Hard to access' may be a better term. In other words, basic underlying assumptions are deeply rooted beliefs about how things are. Their influence on the perception and definition of a situation is especially relevant. These beliefs act as rules that tell a person what to attend to and what to ignore. These selection rules operate very early during the intake and processing of the information present in a given situation, and they operate to a great extent automatically (see also Ignatov, in this volume).

Basic underlying assumptions often stem from personal experience, but they are also passed on through tradition as part of socialisation. A statement of such an assumption could be: 'The boss gets irritated if we call him in the middle of the night and things later turn out to be of minor importance, even if he tells us the opposite'. Or: 'Of these two instruments, the first is more reliable; in case of doubt, the second one must not be trusted'. A basic underlying assumption on a more personal level could be: 'I am the only one in our team who does not know exactly how to do this task, and no one must ever find that out!' These assumptions sometimes seem so obvious and self-evident that they are never tested, or they are ego-threatening and are therefore not talked about (see Figure 6.1).

6.1.2 Acting in situations

If basic assumptions, as well as norms and values, guide behaviour, it seems natural to inquire into them directly, as is done in many cases. Although such examination is undoubtedly valuable, it is also limited by the limited ability of general beliefs and norms to predict specific behaviour, as exemplified by the relation between attitudes and behaviour (e.g. Ajzen, 1988). General questions will only be predictive for general behavioural tendencies, whereas more specific prediction will require consideration of more specific features of the behaviours in question and the context in which they take place (the principle of compatibility) (Ajzen and Fishbein, 1980).

Six aspects in particular seem of special importance in this context. (For examples supporting some of our points, see also the chapters by Bourrier and by Ignatov in this volume):

1. Not all assumptions and social norms are official

Groups have informal norms and values, which may well be in conflict with official ones. For example, the existence of production norms in teams is well-documented. Production norms regulate the amount of work that is considered legitimate (e.g. a fair day's work). If someone produces significantly more (a 'rate buster') or less (a 'chiseller'), he or she is sanctioned by the group. Similarly, a norm of group solidarity may forbid 'squealing' which in the present case would imply informing management of possible safety problems caused by other members of the group (or even its supervisor).

Such group norms modify, and may even undermine, the way in which organisational norms are implemented. Such local social norms will not be detected easily in surveys, and special care has to be taken to uncover them. With regard to safety, unofficial social norms could be 'Don't be a coward,' or 'Don't make a fool of yourself'. (In the Swiss construction industry there is a term, *Sicherheitsstündeler*, that may best be translated as *safety fanatic*. Needless to say the term is derogatory.) Similarly, groups develop shared beliefs about reality that define, for instance, how dangerous certain actions are considered to be, whether it is safe to violate certain regulations (see the private rules mentioned in Bourrier, this volume), or whether certain information can be trusted.

2. Organisations may communicate conflicting norms

Official policy papers and speeches may state that 'security has an overriding priority' (INSAG, 1991, p. 1), but the everyday behaviour of managers may signal the opposite. Typically, behaviour will have more impact than the officially espoused values. A plant manager's evident signs of irritation in response to a delay in reactor start-up may shift the priorities of the team from safety to speed, even if that manager repeatedly emphasises the absolute priority of safety concerns. These shifts can happen very subtly, as when a manager praises his team for meeting a deadline that could only be met by taking short cuts and bending rules. It is these 'theories in use' (Argyris and Schön, 1974) that are most important, not proclamations.

3. Social norms are general prescriptions, how to do things 'right.' Specific situations, however, typically involve several norms that have to be balanced

For instance, a norm that errors must be discussed openly may coexist with a solidarity norm that forbids one to report errors of a colleague to supervisors when there is a risk that the colleague may be punished. In reality, the question may not be whether to conform to a norm, but rather how to balance different and potentially conflicting norms.

4. Behaviour is governed by many calculations about the desirability and probability of various outcomes (Ajzen, 1988; Vroom, 1964)

These considerations involve social norms (which may, as has been pointed out, conflict) and calculations of personal gains, losses, and costs. Many people, for example, have and endorse a strong social norm of helping when asked. Nevertheless, it is an open question whether that norm would overrule other considerations when, for example, one is in a hurry and comes across a shabbily dressed man lying in the street and reeking of alcohol. With regard to safety someone who would prefer a more cautious strategy in a given situation may calculate the probability of being seen as a coward (especially if the risky strategy proves successful), and may therefore refrain from voicing his or her concerns.

Norms, therefore, are one thing, but applying them in specific situations where there may be many additional considerations might raise quite a different conflict between norms and behaviour. This can go so far that people hold beliefs that they themselves realise to be irrational, but that nevertheless are influential in determining their behaviour. Research on the operation of the experiential and the rational system of information-processing has shown many examples of such irrationality in probability judgment (Epstein, 1994).

5. Many social norms are difficult to describe verbally

Just as many people can detect a grammatical error, yet may be unable to describe the underlying rule, many norms cannot easily be described and can only be activated in specific situations, usually when violations of these norms are detected. This difficulty in describing social norms is particularly marked with the many exceptions that usually arise when appropriate but conflicting norms come into play. The problem is probably best exemplified by noting how hard it is to teach norms – and their limitations – to children. Children tend to take norms literally, and when they are scolded for saying quite loudly 'Doesn't that man look funny?' they are apt to reply 'But you told me I should always tell the truth!' When defining the norm, one usually does not think of more than a few exceptions, and without situational cues one is often unable to think of additional ones, much less define them in a generic way.

6. Norms and beliefs are not always active

In many cases it may not be evident to an individual or a group that a given norm is relevant in a given situation, or that a certain issue is at stake. For example, it is well known that the risks associated with certain actions tend to be underestimated, if those actions have been carried out successfully many times (e.g. Reason, 1990). This underestimation may lead to the assumption that danger does not really exist or, if it does, that it exists only for beginners. Similarly, if members of a crew feel certain about a diagnosis (e.g. that they can afford to ignore a display's high temperature warning because the gauge has shown inaccurately high values on several occasions), they will perceive things to be normal. In these cases, the situation may not be defined as one in which safety is an issue. Safety norms might therefore not become salient in the minds of the people involved. Only hindsight, after it has become clear that safety was a problem, will suggest that it would have been natural to activate safety norms in the given situation. But at the moment when the relevant decisions are made, other issues are salient, such as efficiency self-definitions, or identities (e.g. as an experienced, non-anxious operator). The fact that different situations activate different goals, norms, and identities has been seen as leading to situational ethics – that is, circumstances under which norms are violated because they do not appear relevant to the situation at hand (Cropanzano, James, and Citera, 1993). Similar points have been made by authors concerned with attitudes, for example, Fazio (1986), who distinguishes attitudes in terms of their 'accessibility', stating that 'attitude activation may occur as a result of some situational cue that defines attitudes as relevant to the immediate situation' (p. 213). Wilson and Hodges (1992) have suggested 'that people often have a large, conflicting "data base" relevant to their attitudes on any given topic, and the attitude they have at any given time depends on the subset of beliefs to which they attend' (p. 38). Clearly, it is necessary to consider the situation and the process of its definition, and all the complex implications for the activation of quite different and potentially conflicting social norms, and for the costs and benefits involved in different behaviours.

It remains important, of course, to inquire directly about organisational and social norms. But the assessment of safety culture must go a step further. It must address the issue of how actual situations activate norms.

6.2 THE SITUATIONAL APPROACH

Many social norms, especially their basic underlying assumptions, are not able to be uncovered through direct questioning. Acknowledging that values and assumptions are expressed in situations, we propose a situational approach, in which subjects are not directly questioned about values or norms, but are confronted with a dilemma that stems from conflicting social norms and various costs and benefits associated with different types of behaviour. The subjects are asked what they would do in such a situation, what they think others would do, what reactions they would expect their behaviour to elicit from others, and so forth. The value of such an approach is well-established in the area of personnel selection, where it serves as the foundation for such interview techniques as the Situational Interview (Latham, 1989), the Patterned Behaviour Description Interview (Janz, 1989), or the Behavioural Event Interview (Spencer and Spencer, 1993), all of which have proven superior to classical forms of interviewing (see Eder and Ferris, 1989). This finding is consistent with personality research showing that people's behaviour tends to be quite stable and predictable in relation to specific situations (Shoda, Michel, and Wright, 1994).

The following examples are intended as illustrations of our method. (Note that the situation we use has not been found through interviews, but invented.)

Table 6.1(a)   The situational approach

Situation 1
Imagine you have to restart the reactor after a SCRAM. You have only 20 minutes left to do this. After that period you will have to wait for two days because of xenon build-up. A number of security checks are advisable. It is unclear whether they can be done in such a short time.

- What do you do?
- How do you proceed?
- What would the others think of your reaction?

Approach
The response could be analysed with respect to safety procedures, knowledge of regulations, assessment of the riskiness of the situation, or conflicting norms or values. A situation of this kind could then be further developed in order to assess more specific questions, including how difficult decisions of this type are taken in the team, whether everybody's opinion is asked for, whether dissenting voices are likely, and how they are dealt with.

Table 6.1(b)

Situation 2
Imagine you have to restart the reactor after a SCRAM. You have only 20 minutes left to do this. After that period you will have to wait for two days because of the xenon build-up. A number of security checks are advisable. It is unclear whether they can be done in such a short time.

The shift supervisor decides to give it a try, even though he has to waive a security check. Experience shows that this check is not really necessary, because the conditions it monitors have always proved to be acceptable.

- Do you agree with the supervisor's decision?
- Do you think the other members of the team would agree with the decision?
- How would you react if someone raised doubts about the decision?

A further variant may explicitly introduce the aspect of group pressure:

Table 6.1(c)

Situation 3

Imagine you have to restart the reactor after a SCRAM. You have only 20 minutes left to do this. After that period you will have to wait for two days because of the xenon build-up. A number of security checks are advisable. It is unclear whether they can be done in such a short time.

The shift supervisor decides to give it a try, even though he has to drop a security check. Everybody knows that this check is not really necessary because the conditions it monitors have always proved to be acceptable.

The whole team therefore supports the supervisor. But you have your doubts.

- Will you make your doubts known?
- What reaction do you expect from your supervisors and your colleagues?
- Would you insist if they ridicule you, doubt your competence, or the like?

Through these types of questions, social norms and basic underlying assumptions, which are often implicit, may be activated through the situational context. Special attention should be given to aspects that involve threats to one's self-esteem and positive identity, as in situations where someone might be afraid to look foolish, anxious, incompetent, stubborn, and the like.

Responses to the questions in these scenarios can have different formats. The responses may be left to the interviewee (free-response format). Such responses often provide valuable information about what the key aspects of the situation are for the individual, what associations come up, and so forth. The disadvantage is that answers to free-response format questions are difficult to compare across individuals, groups, or organisations. Comparison is easier in multiple-choice formats, as in the following example.

Table 6.1(d)

Possible Responses to Situation 2

1 I have my doubts, but it is *his* responsibility.
2 I have my doubts, but protests are useless.
3 I think he is wrong and I will tell him so.

Of course, one could combine the two formats by starting with free-response questioning and then eliciting possibilities that have not been mentioned. These kinds of interviews can be conducted with individuals as well as with groups. The form does not necessarily have to involve interviews; after some experience with this type of assessment one could perhaps develop situational questionnaires as well.

6.2.1 Developing scenarios: the critical incident interview

The quality of situational interviews depends to a large extent on the quality of the situations on which they are based. If the situations can be regarded as prototypical of those encountered in everyday life, and if they entail dilemmas that people accept as realistic, then situational interviews yield informative responses. Identifying such situations is, therefore, crucial. In our field of study, situational questions are typically based on the critical-incident method (Flanagan, 1954). Various people with experience in the relevant field are interviewed about incidents that were critical in the sense that something did go wrong or almost went wrong. (This definition does not necessarily pertain to dramatic failures such as accidents; smaller events that may be precursors are also of interest!) These people are then asked about the characteristics of this situation, how it developed, how other people reacted to it, what actions made the situation go wrong or get worse or, if the situation was mastered, exactly what actions were responsible for this mastery. Actions not taken, such as the failure of a group's members to communicate the collective doubts they had had about the situation, are included as well, along with the thoughts and feelings involved.

Extensive interviewing, usually combined with observation, is necessary to collect a number of situations. The goal is not principally to collect situations of major importance, such as those with consequences that made it obligatory to report them to the authorities, but to include the little dilemmas and the coping behaviours which are typical of a given organisation (or group) in everyday operations. This goal requires attention to situations of a much smaller scale. The description of these situations is then discussed with experts (such as operators, supervisors, and security personnel). The result is a set of situations that seem prototypical and realistic both in substance and wording. Similarly, possible reactions of various people or groups can be discussed with these experts, in order to generate a set of responses that may be used in multiple-choice alternatives or as probes in interviews.

Another possible source for critical situations is incident reports. Incident reports do often have limitations, one being that often only major incidents are reported. But, as mentioned above, the approach advocated in this chapter also focuses attention on situations of a much smaller scale. The second disadvantage is that incident reports often do not contain the information needed to draw inferences about the psychological processes involved (see Wilpert *et al.*, 1994).

6.2.2 Other variants of the situational approach

The type of scenario just described follows the logic of the situational interview (Latham, 1989) in that it presents situations that have been developed by means of the critical-incident interview technique. Respondents are asked to imagine one of those situations; they may or may not have experienced one like it.

Another variant is an interview in which respondents are requested to describe behaviour in situations that they have actually experienced (Janz, 1989; Spencer and Spencer, 1993). The cues for these situations are specific behaviours or personal reactions. For example:

- Describe a situation in which you have detected and corrected a mechanic malfunction.

- Describe a situation in which you have openly discussed an error which you committed.

- Describe a situation in which you have insisted on being on the conservative side.

- Describe a situation in which you have felt uneasy about how to proceed.

- Describe a situation in which you had doubts about the decision proposed by the shift supervisor.

- Describe a situation in which you felt that regulations unnecessarily hindered effective action.

When answering follow-up questions, respondents must give specific details of what happened, must state verbatim what was said rather than report in general terms ('I made my doubts known'), and, if necessary, must enact an *ad hoc* role-play or some other activity (see Spencer and Spencer, 1993, for a more detailed description of such an approach).

6.2.3   Prerequisites and limitations of the situational approach

Using the situational approach to develop instruments requires careful planning and considerable skill. Expertise in interviewing and behavioural observation is therefore indispensable, as is experience in the field under study. Collaboration between social scientists and technical experts and between people within an organisation and experts outside it seems advisable. As with any type of questionnaire or interview method, this type of assessment is subject to distortion by responses that are more socially desirable than accurate. Be that as it may, validity is expected to be higher than with more general interviews or questionnaires. First, the beliefs and norms that are actually relevant are more likely to be activated by situational cues than by general questions, which may activate general cognitions that might not be relevant in a given situation. Secondly, it is much easier to make the general assertion that safety is of utmost importance, than to describe specific behaviours in specific situations in a socially desirable way. Nevertheless, careful wording of questions and skilful application of interview procedures are crucial, as is a climate of trust between the interviewer and the interviewee.

Apart from these issues of validity, the situational approach may be a key tool for self-assessment. It provokes reflection and communication about safety considerations and the adequacy of underlying assumptions.

Situational assessment is not meant to replace other approaches. Interviews and questionnaires that focus directly on norms, assumptions, and habits are by no means dispensible. Nor are analyses of such indicators as the number of accidents and plant availability. But given that norms and goals are activated by, and interpreted in the light of specific situations, a situational approach seems a significant complement to other methods of assessing safety culture.

References

AJZEN, I. (1988) *Attitudes, personality, and behaviour*. Milton Keynes: Open University Press.

AJZEN, I. and FISHBEIN, M. (1980) *Understanding attitudes and predicting behaviour*. Englewood Cliffs, NJ: Prentice-Hall.

ARGYRIS, C. and SCHÖN, D. A. (1974) *Theory in practice: Increasing professional effectiveness*. San Francisco: Jossey-Bass.

CROPANZANO, R., JAMES, K. and CITERA, M. (1993) A goal hierarchy model of personality, motivation, and leadership. *Research in Organisational Behaviour*, **15**, pp. 267–322.

EDER, R. W. and FERRIS, G. R. (eds) (1989) *The employment interview: theory, research, and practice*. Newbury Park, CA: Sage.

EPSTEIN, S. (1994) Integration of the cognitive and the psychodynamic unconscious. *American Psychologist*, **49**, pp. 709–24.

FAZIO, R. H. (1986) How do attitudes guide behaviour? In SORRENTINO, R. M. and HIGGINS, E. T. (eds), *Handbook of motivation and cognition* (pp. 204–43). Chichester, UK: Wiley.

FLANAGAN, J. C. (1954) The critical incident technique. *Psychological Bulletin*, **51**, pp. 327–58.

GROTE, G. and KÜNZLER, C. (1993) *Sicherheit in soziotechnischen Systemen. Zwischenbericht des Polyprojektes 'Risiko und Sicherheit technischer Systeme'* [Safety in sociotechnological systems: Interim report of the polyproject on Risk and safety of technological systems]. (Polyprojektbericht 05/93). Zürich: Eidgenössische Technische Hochschule, Institut für Arbeitspsychologie (IfAP).

HOFSTEDE, G. (1991) *Culture and organisations: Software of the mind*. New York: McGraw-Hill.

INSAG (International Nuclear Safety Advisory Group) (1991) *Safety culture* (Safety series 75-INSAG-4). Vienna: International Atomic Energy Agency (IAEA).

JANZ, T. (1989) The patterned behaviour description interview: The best prophet of the future is the past. In EDER, R. W. and FERRIS, G. R. (eds), *The employment interview: Theory, research, and practice* (pp. 158–68). Newbury Park, CA: Sage.

KATZ, D. and KAHN, R. L. (1978) The social psychology of organisations. (2nd edn). New York: Wiley.

LATHAM, G. P. (1989) The reliability, validity, and practicality of the situational interview. In EDER, R. W. and FERRIS, G. R. (eds), *The employment interview: Theory, research, and practice* (pp. 169–82). London: Sage.

OSTROM, L., WILHELMSEN, C. and KAPLAN, B. (1993) Assessing safety culture. *Nuclear safety*, **34**, 163–72.

REASON, J. T. (1990) *Human error*. Cambridge: Cambridge University Press.

ROCHLIN, G. I. and VON MEIER, A. (1994) Nuclear power plant operation: A cross-cultural perspective. *Annual review of Energy and Environment*, **19**, pp. 153–87.

SCHEIN, E. H. (1985) *Organisational culture and leadership*. San Francisco: Jossey-Bass.

SCHEIN, E. H. (1990) Organisational culture. *American Psychologist*, **45**, pp. 109–19.

SHODA, Y., MICHEL, W. and WRIGHT, J. C. (1994) Intra-individual stability in the organisation and patterning of behaviour: Incorporating psychological situations into the idiographic analysis of personality. *Journal of personality and social psychology*, **67**, pp. 674–87.

SPENCER, L. M. and SPENCER, S. M. (1993) *Competence at work*. New York: Wiley.

VROOM, V. H. (1964). Work and motivation. New York: Wiley.

WILPERT, B., FANK, M., FAHLBRUCH, B., FREITAG, M., GIESA, H. G., MILLER, R. and BECKER, G. (1994) *Weiterentwicklung der Erfassung und Auswertung von meldepflichtigen Vorkommnissen und sonstigen registrierten Ereignissen beim Betrieb von Kernkraftwerken hinsichtlich menschlichen Fehlverhaltens* [Improvement in reporting and evaluation of significant incidents and other registered events in nuclear power plants in terms of human errors]. (BMU-1996–457). Dossenheim: Merkel (ISSN 0724–3316).

WILSON, T. D. and HODGES, S. D. (1992) Attitudes as temporary constructions. In MARTEN, L. L. and TESSER, A. (eds), *The construction of social judgments* (pp. 37–65). Hillsdale, NJ: Lawrence Erlbaum.

CHAPTER SEVEN

Advanced displays, cultural stereotypes and organisational characteristics of a control room

NEVILLE MORAY

Department of Psychology
University of Surrey, Guildford, UK

Several initiatives have recently been taken to provide international cooperation in supplying human factors (ergonomics) resources to the nuclear industry worldwide. The aim of this chapter is to draw attention to the degree to which cultural, organisational, and even ergonomic differences may have to be overcome, if such transfer of knowledge and behavioural technology is to be successful. There is evidence of the remarkably wide effects of cultural interpretations of displays and controls even between what have been thought of as relatively homogeneous, high-technology countries. These differences may mean that even 'ecological' displays are not 'natural' and cannot be expected to be universally effective without extensive training. It is clear that with technology transfer to developing countries the effects may be much larger. In addition, it is important to consider failure modes at the organisational level in foreseeing the way in which international cooperation may fail to be effective. Of particular interest is the effect of choosing patterns of human–computer interaction in the design of control rooms.

Recently, several initiatives have occurred that have the aim of promoting international cooperation for the safe operation of the nuclear power industry. For example, International Ergonomics Association (IEA) has set up a group to offer human factors services to the power industry. Such initiatives are welcome and important. But when one considers how effective international cooperation can be, particularly during an accident or emergency, a number of questions need to be considered. In brief, because international cooperation involves communication between people of different languages, technological cultures, and expectations, to what extent are the unspoken assumptions of different groups likely to lead to misunderstanding rather than effective synergy? In this chapter I look at three sources of possible difficulties:

- the understanding of displays and controls in the context both of conventional and advanced displays;
- the great variety of control room organisation and hierarchy;
- the choice of patterns of human-computer interaction and organisation in advanced plants.

7.1 DISPLAYS, CONTROLS, AND CULTURAL STEREOTYPES

Recently, there has been an upsurge of interest in display design for complex human machine interaction. This change has been due largely to the development of cheap and extremely powerful computer graphics systems, large databases, and enhanced networking. Perhaps the most striking development has been the 'ecological interface' movement, which is based on the ideas of Gibson (1979) and supported strongly by Rasmussen and Vicente (1989), Vicente and Rasmussen (1992), and Rasmussen, Pejtersen, and Goodstein (1994).

The claim of the ecological interface movement is that one can provide operators with displays and controls that are particularly 'natural' and, hence, particularly effective for users. The displays and controls 'afford' opportunities for the correct behaviour. In a sense the aim is to provide an environment in which little or no thought is required, indeed one in which rules are not necessarily followed. Instead, interaction with systems approaches the level of a perceptual-motor skill. With well-practised skills people do not appear to need to think about the task. The physical configuration of the environment, which embodies cues and information, is processed automatically by the nervous system, so that correct, adaptive behaviour emerges. Often cited examples for humans are sporting activities; for other species predator–prey interaction. Such displays have been implemented experimentally (see, for example, Moray et al., 1993; Christofferson, Hunter, and Vicente, in press; Hansen, 1995). But few, if any, have been properly evaluated in real working environments, and it is not at all clear that the displays offer anything at all 'natural.' (It is also interesting that no work has been done on 'ecological' controls – everything has been on the display side.) In fact, most, if not all, of the ecological displays that have been implemented are completely opaque to the operator, unless the latter has had considerable training and possesses extensive domain knowledge. The ecology of ecological displays and advanced displays in general does not afford universal, cross-cultural, perceptual understanding of the information displayed, unlike the perceptual ecology of the natural world. Although such displays may be effective in the sense that a well-trained person can use them to deal with plant disturbances in a highly effective manner, it is not at all clear that someone viewing them for the first time will be able to understand

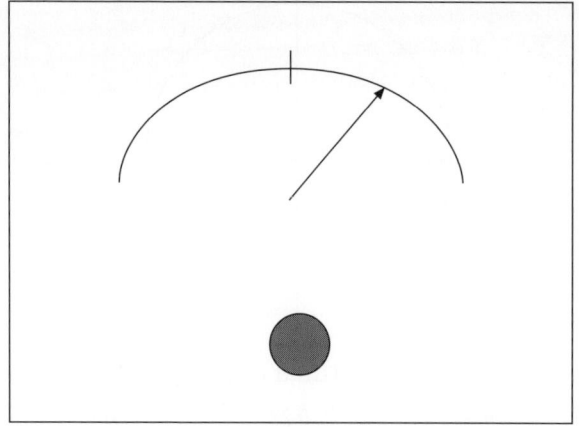

	USA	JAPAN	NETHERLANDS	GREECE	PORTUGAL
CLOCKWISE	0.00	0.09	0.00	0.08	0.00
ANTICLOCKWISE	1.00	0.91	1.00	0.92	1.00

Figure 7.1 The proportion of people from different cultures who answer 'clockwise' or 'anticlockwise' to the question, 'In which direction should the knob be turned in order to bring the pointer back to the midpoint of the scale?'

them – a prerequisite for cooperation by 'intervening' personnel who arrive during an emergency.

This problem raises a fundamental question: what is 'natural' and 'universal' in the way that people view displays and use controls? It is certainly true that people, and indeed animals of other species, can make use of information 'naturally' provided by the environment (although in many cases only after several months or years of maturation and learning, as with learning how to walk). Apparently, humans naturally perceive the 'meaning' of texture gradients in the environments, of distributions of light and shade, and perhaps even of invariants that signal the need to expend energy (Flach, Hancock, Caird, and Vicente, 1995), but there is little or nothing that is natural about the displays in any control room, whether they are the most advanced or the simplest. For example, it is obvious that a great deal of training is required to perform diagnosis using a Rankine cycle display (Moray *et al.*, 1993). But even much simpler displays and controls are not 'natural'.

Recently, there have been several papers investigating the extent to which there is agreement across cultures about stimulus-response cultural stereotypes for simple displays. Which way do people expect a pointer to move when a knob is turned or a lever displaced? In what order do people expect portions of a sequential display to be numbered? Examples of this work can be found in Courtney (1988), Sheng-Hsiung Hsu and Yu Peng (1993), and some of my own unpublished work. This research seems to indicate that there are very few cases where there is substantial agreement across cultures, but that there are many cases where there is either considerable disagreement or no strong expectation at all (see Figure 7.1). Clearly,

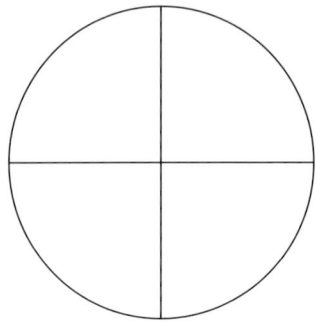

	USA	JAPAN	NETHERLANDS	GREECE	PORTUGAL
DA CB	0.25	**0.50**	0.28	0.17	**0.67**
AB CD	**0.45**	0.22	**0.44**	0.28	0.11
AB DC	0.10	0.17	0.28	**0.55**	0.11
Other	0.20	0.11	0.00	0.00	0.11

Figure 7.2 The proportion of people from different cultures who choose the given sequence for labelling sectors. The preferred ordering for each culture is shown in **bold**.

in this case there is a very strong cross-cultural agreement. On the other hand, in North America people expect an electric circuit to be 'on' or 'live' when the switch is 'up,' but in all of western Europe, most of the ex-British colonies, and much of Asia people expect a circuit to be live when the switch is pressed into the 'down' position. This difference in convention is obviously particularly dangerous in emergencies, where people are likely to intervene physically under time pressure when they cannot make themselves understood because they lack linguistic fluency.

Two further examples are of similar interest. Respondents were asked to use the letters A,B,C, and D to label the quadrants of Figure 7.2.

Clearly, there are very marked differences between observers in terms of the quadrant-letter correspondence they take for granted.

Finally, consider the differences in expectations for increasing flow by turning faucet handles (see Figure 7.3).

Thus, neither in perceptual expectancies nor in action expectancies is there much consistency across cultures, even when all the cultures are familiar with high technology systems. It seems to me highly improbable that much more complex displays, which are likely to be composed of units having properties that appear in these simple examples, will show any more compatibility and commonality across cultures.

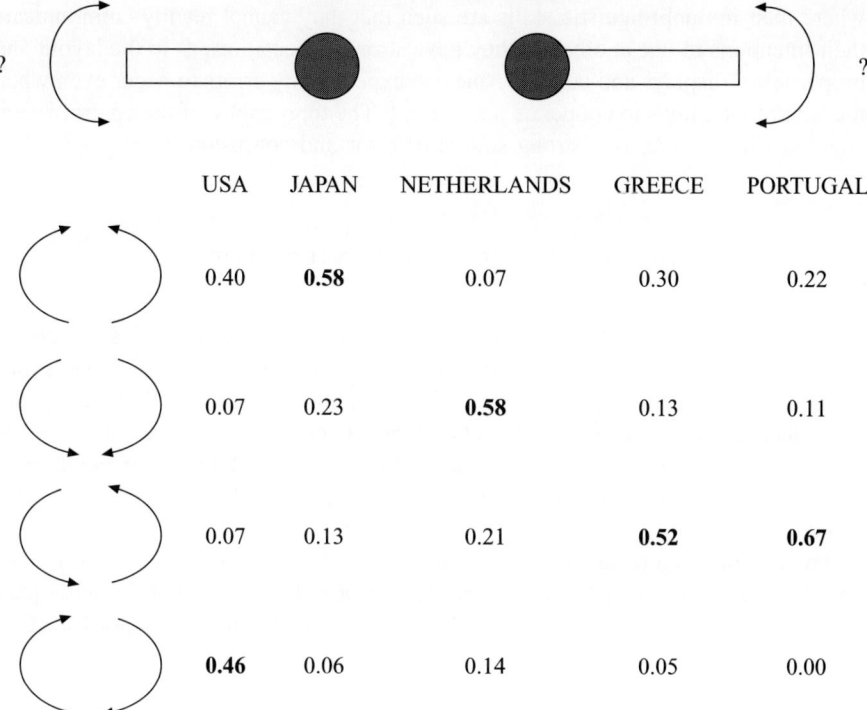

	USA	JAPAN	NETHERLANDS	GREECE	PORTUGAL
	0.40	**0.58**	0.07	0.30	0.22
	0.07	0.23	**0.58**	0.13	0.11
	0.07	0.13	0.21	**0.52**	**0.67**
	0.46	0.06	0.14	0.05	0.00

Figure 7.3 The proportion of people from different cultures who respond with the different patterns of expected movements to the question, 'In which direction would you turn the handles to increase the flow?'

By contrast, it is interesting to look at the example of an error caused by inadvertent 'trapping' of behaviour by a very strong stereotype. Moray (1993) reported a case where operators worked with a system involving two engines. The controls of the left engine were on the left of the control panel, those for the right engine on the right. There were three computer screens, and pages of information relevant to either engine could be brought up on any of the screens. Moray predicted a particular error, and that error in fact occurred on the first day of prototype-testing, fortunately in a simulator, not the real plant. The operator received a warning that the gearbox of the right engine was overheating. He called up the relevant information onto the left screen. Then, for nearly a minute, he tried to shut down the right engine using the left controls. The geometry of the display control relationships, with the left controls lying just below the image, were so strong that he was unable to resist using the left controls even though he himself had placed the right-hand information on the left screen. In the end he told the computer operator that the simulation must be faulty because he could not shut down the engine.

There is ample room for confusion when people from different cultures work together to solve a problem, particularly when they are under time pressure, and where their mutual linguistic skills are such that they cannot readily communicate their intentions to one another. If they have strong expectations as to the layout and properties of displays and controls, one can expect many errors to occur even when the best of intentions to cooperate are present. The topography of the control room, coupled with culture, is a strong source of error and confusion.

7.2   ORGANISATIONAL HIERARCHY IN CONTROL ROOMS

Just as there are culturally determined expectations about the physical layout of displays and the way in which controls will work, so are there cultural expectations about the hierarchical organisation of authority in the control room. Some years ago my colleagues and I analysed different patterns of control room organisation for the Atomic Energy Control Board of Canada. We were surprised by their variety and identified six main patterns in the ten countries we examined. The following figures show their variety.

There is no need to say which organisation comes from which country, for the point of this chapter is not to argue that one organisation is necessarily better than another. But I *can* say that the list is by no means exhaustive. It is not based on an examination of any eastern European installations, and it has only one Asian installation. The point which I wish to make is simply that a misunderstanding of the control room hierarchy can be expected to militate against efficient intervention by a visiting expert. Several control room organisational styles are shown in Figures 7.4 and 7.5. Consider for example the very fundamental difference between a control room of Type 1 and Type 6. Each has a specialist who is trained in nuclear engineering, a highly qualified person with theoretical knowledge above and beyond the engineering relevant to the particular plant. But, in Type 1 this specialist has no authority. He or she is on a sideline, available if required, but has no right to intervene. By contrast, in Type 6 the same person would have the absolute and final decision-making authority to declare an emergency, define strategy or tactics, and so forth. A foreign expert confronted by a person with such skills, and not appreciating the differences in hierarchical organisation, could waste valuable time in trying to intervene, if he or she misunderstood what one may call the political structure of authority in the control room. (See also the chapter by Bourrier in this volume.)

Related to this organisational issue is the question of the more subtle ways in which cultural stereotypes of behaviour may impinge on the working practices in the control room. In some cultures there is a strongly hierarchical attitude toward expressing opinions and questioning the actions of others. Indeed such behaviour may be inhibited even if there is no hierarchical structure, if the whole society puts a premium on certain patterns of social interaction. For example, in a different context I was puzzled by the lack of interaction and questions in scientific meetings in Japan, until I discovered that socially it is not normally acceptable to engage in

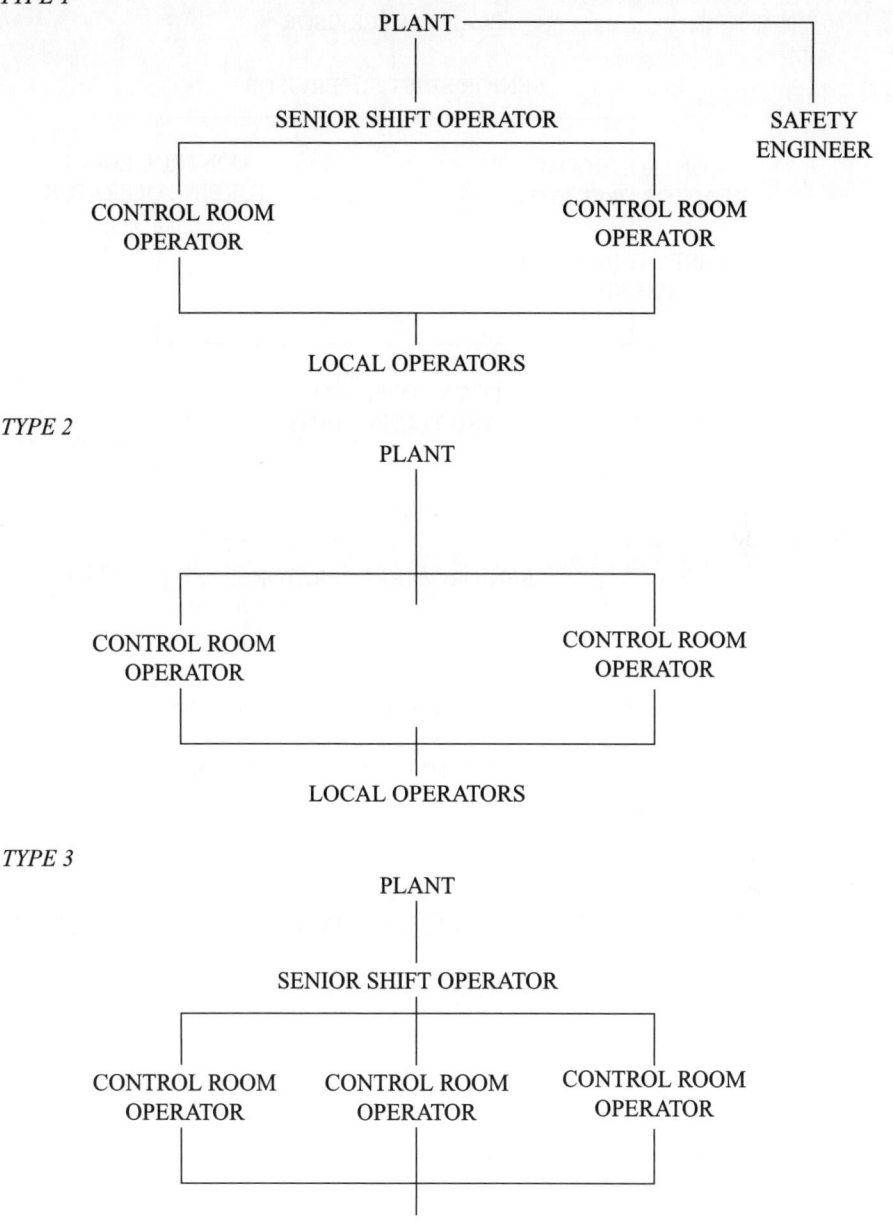

Figure 7.4 Examples of control room organisation (after Feher, Moray and Senders, 1988).

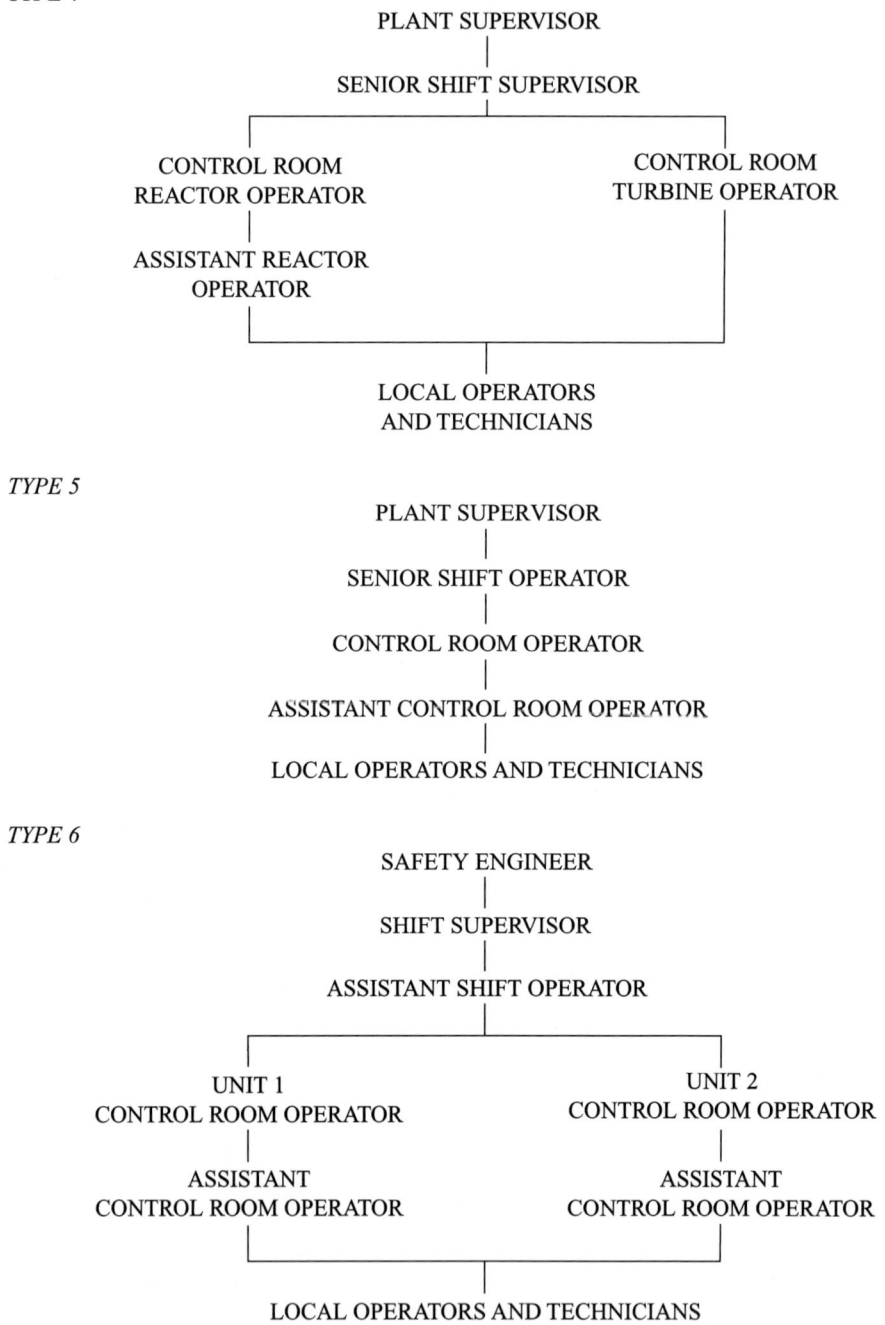

TYPE 4

PLANT SUPERVISOR

SENIOR SHIFT SUPERVISOR

CONTROL ROOM
REACTOR OPERATOR

CONTROL ROOM
TURBINE OPERATOR

ASSISTANT REACTOR
OPERATOR

LOCAL OPERATORS
AND TECHNICIANS

TYPE 5

PLANT SUPERVISOR

SENIOR SHIFT OPERATOR

CONTROL ROOM OPERATOR

ASSISTANT CONTROL ROOM OPERATOR

LOCAL OPERATORS AND TECHNICIANS

TYPE 6

SAFETY ENGINEER

SHIFT SUPERVISOR

ASSISTANT SHIFT OPERATOR

UNIT 1
CONTROL ROOM OPERATOR

UNIT 2
CONTROL ROOM OPERATOR

ASSISTANT
CONTROL ROOM OPERATOR

ASSISTANT
CONTROL ROOM OPERATOR

LOCAL OPERATORS AND TECHNICIANS

Figure 7.4 (cont'd)

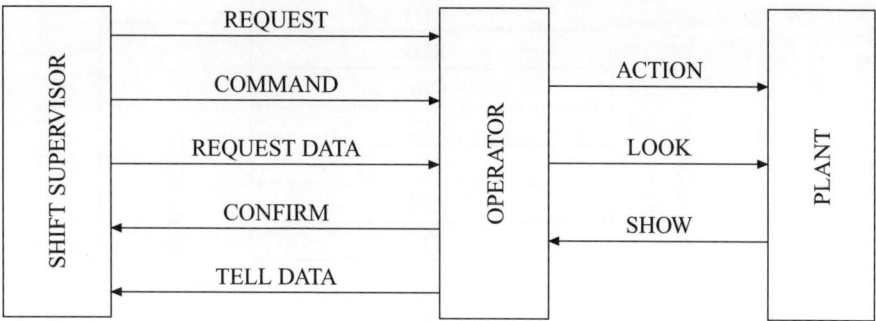

Figure 7.5 Manual control: hierarchical crew.

strong controversy in public. Again, the social style with which an expert expresses himself or herself, and indeed in some cases the gender of the expert, may have a very significant impact on the acceptability of their advice, or on the rapidity with which that advice will be considered. The interaction between such social and cultural stereotypes and the effectiveness of behaviour has great *potential* importance. It is simply not known to what extent it will have an impact in real emergencies, although social psychology and sociology, and the work in recent years on topics such as cockpit resource management suggests the effects will be far from negligible.

7.3 TASK ALLOCATION BETWEEN HUMANS, COMPUTERS, AND AUTOMATED SYSTEMS

Figures 7.5 to 7.12 show several possible configurations of control rooms, together with flows of information, communication, and control, as a function of the degree of automation, manual control, and human computer interaction. These figures can show the vast differences between the tasks, workload, and interactions required of operators under different control room designs. It is possible to develop a kind of sequential representation of the dynamics of communication and control from these graphs (although we will not do this in detail here).

For example, there is little need for interaction between the operator and the supervisor during normal operation (see Figure 7.7). The operator is coupled to the plant through a cycle in which the former looks at the plant (L), the plant shows (S) its state to the operator through the interface, and the operator takes appropriate action (A).

This arrangement continues until some abnormal situation arises. At that point the operator or the supervisor responds to alarms by retrieving and reading the emergency operating procedures (EOPs); issuing commands to the operator, who carries them out; asking the operator for confirmation that the commands have been executed; asking for data and other information while the operator, in addition

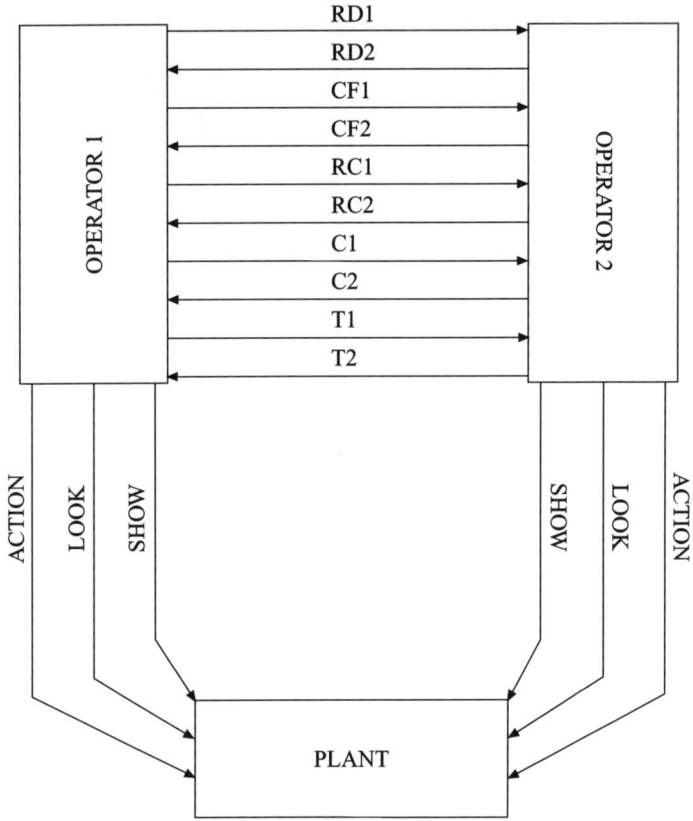

Figure 7.6 Manual control: heterarchical crew.
RD: Request data.   C: Command.
CF: Confirm.   T: Tell.
RC: Request confirmation.

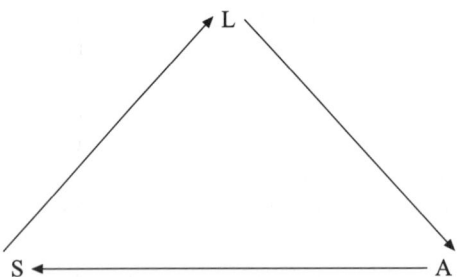

Figure 7.7 Cybernetics of normal situations in a nuclear plant.
L: The operator looks at the plant.
S: The plant shows its state to the operator.
A: The operator takes appropriate action.

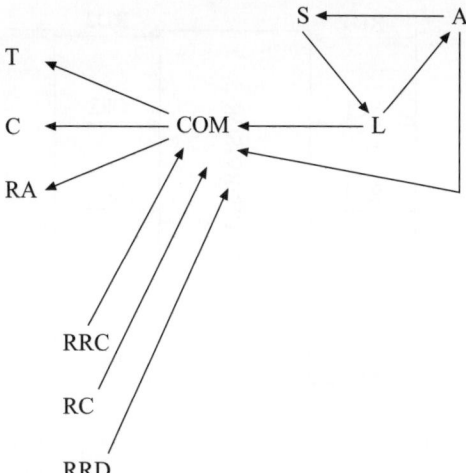

Figure 7.8 Cybernetics of abnormal event.
COM: Communication.
L: Operator looks.
S: Plant shows its state.
A: Operator asks for information from plant and relays request to crew.
T: Tells other members of crew.
C: Gives command.
RA: Requests action from others.
RRC: Reader of emergency operating procedures (EOPs) requests confirmation of plant
        state.
RC: Reader issues command based on EOPs.
RRD: Reader requests data.

to interacting with the plant, confirms that the supervisor's commands have been carried out, provides data to the supervisor, asks for advice, and so forth. The cybernetic graph now becomes much more complex, as shown in Figure 7.8.

The operator now has the additional burden of communication, while the supervisor has the problem of planning, integrating, requesting information, receiving confirmation, and so on. It is left as an exercise for the reader to develop similar flowcharts for the other systems.

What is the relevance of these configurations for safety? It is that most safety audits and Probabilistic Risk Analysis/Human Error Probability (PRA/HEP) estimates make certain implicit assumptions – for example, about the number and type of personnel present in the control room during an abnormal incident. It is clear that situations can arise that render any such estimates grossly inappropriate. Consider the case of Figure 7.9. If, for some reason, the shift supervisor were to be unavailable, all the work that that person normally performs – typically including finding and consulting the EOP manual – would have to be performed by one or other (or both) of the operators, and the workload would increase enormously. Similarly, if in Figure 7.10 there is a computer fault, the system would fail and would enter

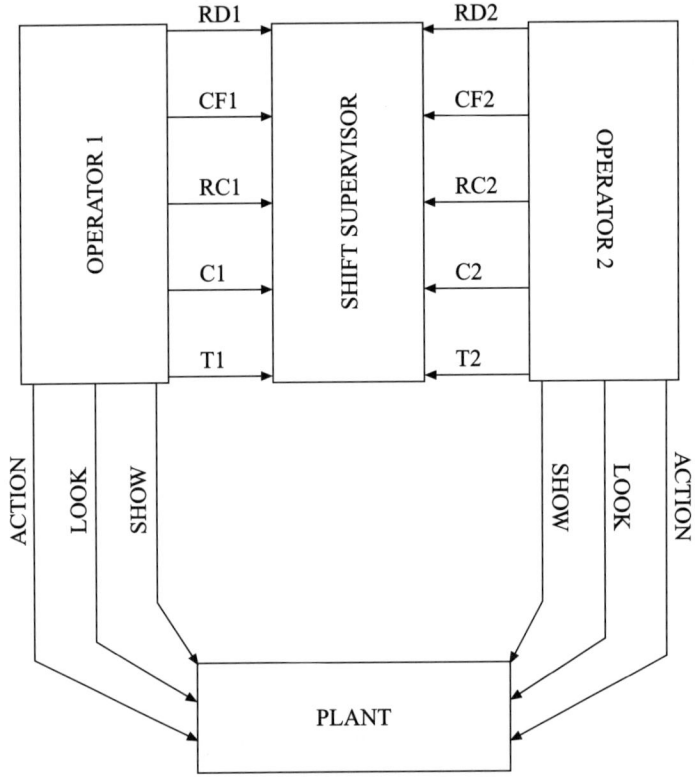

Figure 7.9 Manual Control: Heterarchical crew with supervisor.
RD: Request data.   C: Command.
CF: Confirm.   T: Tell.
RC: Request confirmation.

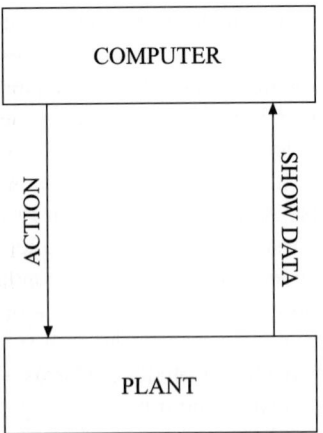

Figure 7.10 Full computer control.

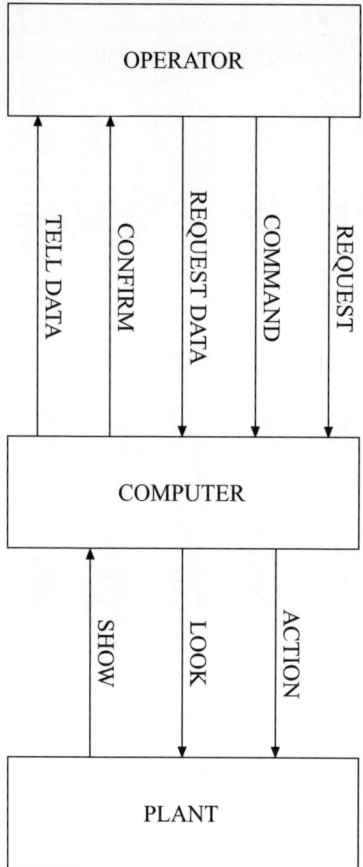

Figure 7.11 Computer control with human supervisory control.

a Figure 7.5 mode. Figure 7.11 shows a system in which the human operates in supervisory control mode. The process computer normally runs the plant, so that the human operator is no longer in direct contact with it, and sees the plant state only through the computer interface. Thus, the computer now filters the flow of information and control between the plant and the human. In Figure 7.12 the same system is shown during incident management. The human now has to communicate both with the shift supervisor who is in charge of emergency operations, and with the computer. Again, we see that should either the supervisor or the operator not be available, there will be a very heavy load on the other person, which may not be supported by an interface which makes strong assumptions about manning levels.

One is forced to the conclusion that, even when allocation of function and the introduction of automation are done well, one cannot rely on the results to lead to a stable system for which fixed safety parameters can be measured. It is not necessary to think in terms of dramatic events such as the death or incapacity

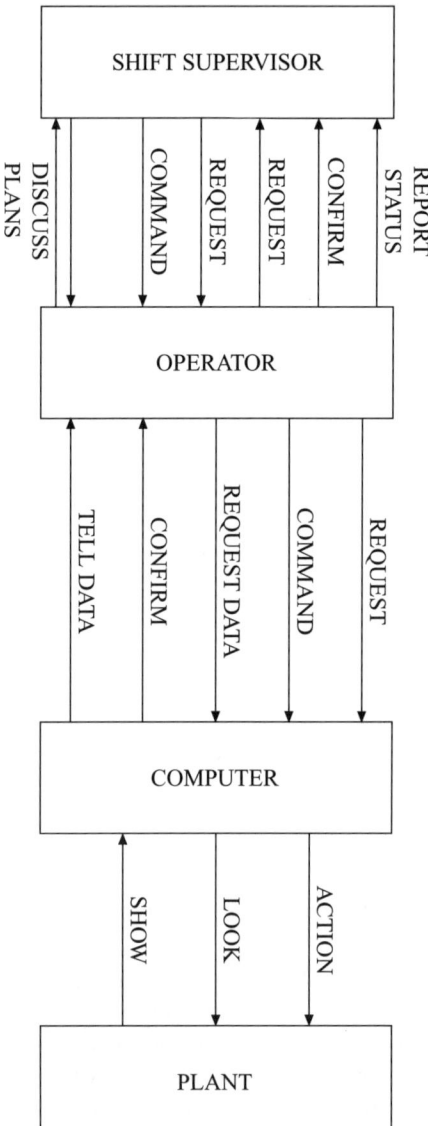

Figure 7.12 Computer control with human supervisory control and shift supervisor assistance.

of a crew member. Dysfunctional personal characteristics or social mismatches may well effectively change the cybernetic characteristics of a human machine system. Furthermore, the same point can be made about the question of how to optimise the intervention of an expert from outside the organisation that is experiencing the abnormal incident. Someone who is an expert at functioning in, say, a

Figure 7.9 mode may very well not function at all well in a Figure 7.11 mode, and vice versa.

7.3 SUMMARY

In recent years there has been much discussion about the use of expertise across cultural boundaries to increase the safety of hazardous high technology systems. Although it is hard to include HEP estimates into PRA, there has been even less consideration of the dynamics of intervention and the kind of difficulties that are to be expected when cross-cultural intervention occurs. As hazardous technologies become globalised, with the concomitant changes in cultural expectations, education, ergonomic stereotypes and social styles, only those who are sensitive to such matters are likely to be effective cross-cultural intervenors and to reduce the impact of error. While recent developments in interface design, such as the application of ecological psychology, may offer important improvements, it is not clear to what extent even they are free from the effects of learning and cultural stereotypes. Social dynamics (and political dynamics) are likely to be even more potent than ergonomics in many situations, although even the latter must not be neglected.

References

CHRISTOFFERSON, K., HUNTER, C. N. and VICENTE, K. J. (in press) A longitudinal study of the effects of ecological interface design on skill acquisition. *Human Factors*, **38**.

COURTNEY, A. J. (1988) Chinese response preferences for display-control relationships. *Human Factors*, **30**, pp. 367–72.

FEHER, M. P., MORAY, N. and SENDERS, J. W. (1987) *The development of criteria for Nuclear Power Plant Control Room Emergency Procedures*, Atomic Energy Control Board of Canada.

FLACH, J., HANCOCK, P., CAIRD, J. and VICENTE, K. (eds) (1995) *Global Perspectives on the Ecology of Human-Machine Systems*. Hillsdale, NJ: Erlbaum.

GIBSON, J. J. (1979) *The Ecological Approach to Visual Perception*. Boston: Houghton-Miflin.

HANSEN, J. P. (1995) Representation of system invariants by optical invariants in configural displays for process control. In FLACH, J., HANCOCK, P., CAIRD, J. and VICENTE, K. (eds), *Local Perspectives on the Ecology of Human Machine Systems*. Hillsdale, NJ: Erlbaum.

MORAY, N. (1993) Flexible interfaces can induce errors. In KRAGT, H. (ed.), *Enhancing Industrial Performance*. London: Taylor & Francis.

MORAY, N., JONES, B. G., RASMUSSEN, J., LEE, J. D., VICENTE, K. J., BROCK, R. and DJEMIL, T. (1993) *A performance indicator of the effectiveness of human-machine interfaces for nuclear power plants*. (NUREG/CR-5977.) Washington, DC: United States Nuclear Regulatory Commission.

RASMUSSEN, J. and VICENTE, K. J. (1989) Coping with human errors through system design: Implications for ecological interface design. *International Journal of Man-Machine Studies*, **31**, pp. 517–34.

RASMUSSEN, J., PEJTERSEN, A. M. and GOODSTEIN, L. (1994) *Cognitive systems Engineering.* New York: Wiley.

SHENG-HSIUNG HSU and YU PENG (1993) Control/Display relationship of the four-burner stove: A re-examination. *Human Factors,* **35**, pp. 745–9.

VICENTE, K. J. and RASMUSSEN, J. (1992) Ecological interface design: Theoretical foundations. *IEEE Transactions on Systems, Man and Cybernetics, SMC-22,* 589–606.

CHAPTER EIGHT

From theory to practice – on the difficulties of improving human-factors learning from events in an inhospitable environment

GERHARD BECKER

TÜV Rheinland e.V.

The goal of team effort by the Berlin University of Technology and TÜV Rheinland in Cologne in recent years has been to encourage organisational learning within the network of organisations involved in manufacturing, licensing, operating, supervising, and maintaining nuclear power plants. This collaboration has led to the proposal of a more systematic and comprehensive approach to event analysis as a way of improving the manner in which human contributions to events are dealt with in the existing system for reporting events in such installations.

Given the empirical findings and theoretical background of the concept used by the Technical University and TÜV Rheinland, the prospects for further improvements in safety were very promising. However, the initial steps to apply the concept and test the proposals in practice met resistance and brought many problems to light. The difficulties encountered, which prevented frank communication of organisational factors contributing to events, stem from the situation in those federal states of Germany where the political leaders want to close down nuclear power plants, and from issues of hierarchy and power within the utility organisation.

The author describes the problems that were discovered, investigates the reasons for them, and comments on the possibilities and strategies for overcoming the difficulties. A multilevel procedure was suggested for expanding what is learned from human contributions to events, near misses, and voluntarily reported weak points. The underlying strategies were aimed at respecting confidentiality and helping the parties involved to save face, an approach that enables the supervisory, managerial, and board levels of organisations to gain insights into their own involvement in bringing about events, a role that is often due to factors related to the safety culture of the organisation.

The relevance of human factors to the safe operation of complex technical facilities was recognised several years ago (Meister, 1962, 1971; Turner, 1978). Since then, the introduction of appropriate measures to defend against the threat revealed by these findings has taken a circuitous and sometimes devious route. Initial intervention strategies focused on first-end action at the human–machine interface. The resulting progress has been remarkable, but some of the measures had side-effects. For instance, automation has increased the complexity of the systems involved (Bainbridge, 1987). More and more people have come to realise that the complex interaction of a plant's organisation and the environment in which the plant is embedded may have a crucial influence on safety. In this context the importance of a sound safety culture has been emphasised in the last decade (INSAG, 1991).

In order that this more comprehensive understanding of human factors could be introduced into intervention strategies that are relevant to the daily practice of the nuclear industry in Germany, research projects (Wilpert *et al.*, 1994; Becker *et al.*, 1995) were conducted jointly by the safety research unit of the Technical University (TU) of Berlin and the Institute of Nuclear Technology and Radiation Protection (IKS) of TÜV Rheinland in Cologne. This chapter presents some of the key problems that we, the members of those research teams, encountered during the research project, as we tried to apply our scientifically convincing approach to a practical setting. The problems rooted in the political situation are illustrated, and the consequences of this situation demonstrated. The conclusion focuses on the undesired side-effects that the factors relating to the environment of an organisation may have on the organisation's behaviour. Proposals are drafted to overcome at least some of the problems.

8.1 EVENT-REPORTING AS AN ELEMENT OF ORGANISATIONAL LEARNING

Throughout the world, the approach taken to maintaining and enhancing the safety of nuclear power plants is the systematic collection and evaluation of operating experience. This type of feedback-control strategy is well accepted, and various event-reporting systems are established at both national and international levels. When combined with systematic procedures for analysing events, such an event-reporting system may be seen as an element of organisational learning (OL).

Because OL requires some kind of organised reporting system (Shrivastava, 1983), the German Event-Reporting System for Nuclear Installations was used as a starting-point in the research projects described in this chapter. That system, which is mandatory in Germany, has one feature in common with other licensee, event-reporting systems: the reporting and documentation of events is focused mainly on technical matters. Other features of the German Event-Reporting System for Nuclear Installations are as follows:

1   The reporting criteria are somewhat stricter than those used for the International Reporting Scale.

2   Evaluation at the federal level is undertaken by the Federal Agency for Radiation Protection (*Bundesamt für Strahlenschutz*, BfS) and an organisation acting as a consultant in reactor safety matters to federal authorities (*Gesellschaft für Anlagen- und Reaktorsicherheit*, GRS). Serious events of general concern result in critical event messages, which are circulated to all power plants, authorities, and TÜVs. The plants are obliged to check the relevance that the critical event messages may have to their own installation and to communicate their findings to the safety authorities.

3   Evaluation at the *Land* (state) level is performed by TÜVs on behalf of the *Land* authorities. The TÜVs also have to verify the nuclear power plants' responses to the critical event messages.

The information channels and the processes used are well-established. Within each utility the mandatory system is complemented by a reporting system that conveys information about safety and other relevant matters.

8.2   ENHANCING ORGANISATIONAL LEARNING – THE INCLUSION OF HUMAN FACTORS

The idea of improving traditional event-reporting and analysis in order to establish the framework for comprehensive OL requires a sound theoretical concept. To guide the development of our system, my colleagues and I have chosen the Sociotechnical System (STS) model originated by Emery and Trist (1960) and the model of the dynamics of accident causation presented by Reason (1990). The STS model has been proposed as an appropriate system for modelling the complex interactions of factors contributing to safety (and to other performance aspects of a system, such as product quality or availability).

This STS for the safety of nuclear power plants is intended to embrace all parties that contribute to the safety and reliability of these installations. Modifying an approach by Moray and Huey (1988), the STS may be modelled as a multilayered concept in which the relevant subsystems may be thought of as concentric circles. In the centre of the circles the *acting person* (the actor) as a member of a team interacts with the *technical system*. These units are circumscribed by the circle labelled the *organisation*. The organisation itself is embedded in the organisational *environment*, which also comprises such key institutions as the regulatory authority. As Wilpert *et al.* (1994) have shown, this approach to the learning system does not have the traditional bias of looking only at technical matters and the operator at the human-machine interface. Instead, it uses a broader perspective of human factors that avoids the danger of overlooking major contributing factors resulting from the outer STS layers.

The second model on which my colleagues and I have based the development of our approach is the model of the dynamics of accident causation as outlined by Reason (1990). For complex technical systems designed according to the defence-in-depth concept, it follows from this concept of accident evolution that several barriers must fail before an accident occurs. Single events, say, involving the operator in the nuclear power plant cannot lead to an incident with relevant consequences. Accordingly, analysts should search for several contributing factors. Furthermore, we follow Reason in his salient distinction between *active errors*, whose effects are felt almost immediately, and *latent errors*, whose adverse consequences may lie dormant within the system for a long time, becoming evident only when they combine with other factors (such as active errors) to breach the system's defences. An active error should therefore be seen as an opportunity to search for latent errors rather than blame the operator for being the cause of the incident.

Extending the event-reporting and analysis to include latent human factors to the extent proposed in this chapter means leaving the beaten track of analysis. The 'stop rule' that is usually applied when a human error by the operator has been noticed is outdated. Acknowledging that a failure by the active operator only invites the discovery of hitherto hidden and unsuspected conditions has far-reaching consequences. Because human failure is unavoidably connected to responsibility in modern society, this responsibility is shifted to other layers of the STS. If organisational and managerial factors in particular are identified as having contributed to events, responsibility is shifted to the managers themselves (Leplat, 1997).

8.3   THE ENVIRONMENT OF THE PLANT

The idea of having the possible contributions of all STS subsystems as a part of learning from events, near misses, and voluntarily reported weak points proved difficult to implement. The necessary intervention strategy must involve consideration of the complex interactions of the various participants in the German nuclear safety system. Achieving such scope is actually quite challenging, because no existing theoretical concepts and models apply to this inter-organisational field as detailed by Wilpert *et al.* in this volume.

Some of the traps in this inter-organisational field result from the peculiarities of the political situation in Germany. The legal system sets up a rather complex structure for the licensing of nuclear power plants and the supervision of their operation (see Wilpert *et al.*, in this volume). First, responsibility for enforcing the Federal Atomic Energy Act of 1985–97 falls on the states, whereas the Federal Ministry for the Environment, Nature Conservation and Nuclear Safety sees to it that the Act is enforced consistently. The main obstacles in the way of our aim of putting our methods into practice arise from the fact that some of the states have the political objective of closing down nuclear power plants. In these states, where the Green Party is a member of a coalition government, the authorities wish to abandon nuclear energy. Their attitude is expressed in their supervision of nuclear power plants, which is exemplified by:

1 Delays in granting permission to restart after outages.

2 The assignment of consultants willing and able to document flaws in plant-design objectively and thoroughly, thereby generating ever greater doubts and necessitating ever more regulations.

3 Demands for extra retrofitting (or backfitting).

4 Lengthy licensing procedures for such retrofitting.

The discretionary powers of *Land* authorities in Germany have been abused in many cases, as suggested by the fact that federal authorities have often issued directives overruling decisions of the states. The utilities, usually joint-stock companies, try to fight decisions that they consider arbitrary. One utility, for example, is conducting 35 lawsuits against the *Land* authorities, citing such grievances as delays in granting permission for start-ups and in licensing plant improvements.

These tensions between *Land* authorities and nuclear operators have serious consequences for the chances of our project helping to better OL. Since recognition of the importance of the factors enshrined in the concept of safety culture, the implications of that tension in Germany can be understood more clearly. It may be worthwhile, therefore, to reflect on the preceding observations in the light of the concept of safety culture and its implications.

According to the International Atomic Energy Agency (INSAG, 1991), 'Safety Culture is that assembly of characteristics and attitudes in organizations and individuals which establishes that, as an overriding priority, nuclear plant safety issues receive the attention warranted by their significance' (p. 1). The report emphasised that the behaviour of the workforce is determined by the requirements defined at the managerial level in the organisation. Consequently, the report contains requirements for management as well as for the organisational framework that should be established in order to achieve a real safety culture. It is pointed out that the requirements are valid not only for the utilities but for all organisations concerned with nuclear safety, including the regulatory authorities. Even though safety culture may not be a directly tangible thing, the report showed that it is possible to define various measurable indicators for assessing its effectiveness.

The concept of safety culture has meanwhile occupied the safety bodies of various countries (ACSNI, 1993; KSA, 1994) and has been the subject of a major conference organised jointly by the American Nuclear Society/Austrian Local Section (ANS/ALS), the International Atomic Energy Agency (IAEA), and the Nuclear Energy Agency (NEA) (ANS/ALS–IAEA–NEA, 1995). But the concept is still being analysed for all its implications, and its full meaning is still being interpreted. To some researchers it remains very vague (Wahlström, 1995), and others characterise safety culture as higher order organisational factors (Haber *et al.*, 1995). With the problem of improving OL in mind, I interpret safety culture as a further dimension of the STS for safety. This dimension can be illustrated if safety culture is

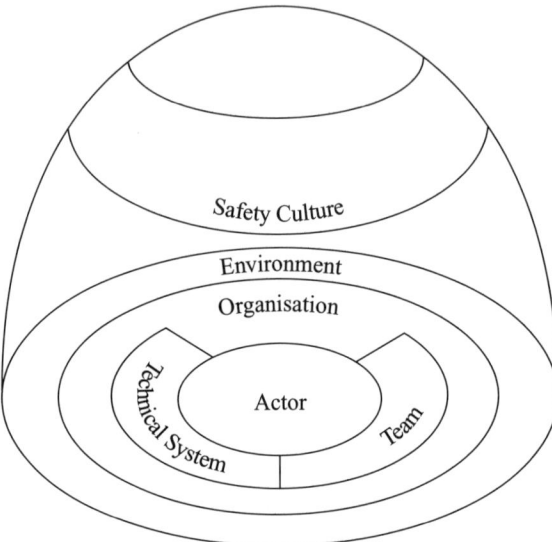

Figure 8.1 Safety Culture as the climate for the effective interaction of all parties involved.

assumed to be the climate that makes for effective cooperation between all parties involved (see Figure 8.1).

This 'dome' model is a metaphor for the idea that all participants of a system will be affected by any participant's behaviour that poisons the atmosphere of the organisation. This effect occurs even when all parties affirm that they have the same aims guiding their behaviour. The question to be answered is whether the desired safety culture can be achieved only if all parties not only have the same goals, but also agree on the means to achieve them. Because such consensus on the utilisation of nuclear power is unlikely in the near future, there is the issue of how to promote the intended learning process. Before these matters are considered here, however, the logic and motives guiding the behaviour of nuclear power plant managers in Germany should be examined more closely.

8.4   BEHAVIOUR OF NUCLEAR POWER PLANT OPERATORS

As I have already pointed out, the nuclear power plant employees whom my colleagues and I interviewed stiffly resisted our ideas of extended learning from events, when we tried to put them into practice. To these interlocutors, the behaviour of authorities in states intent on phasing out nuclear power plants was too unpredictable to risk adding organisational factors to event analysis. It is true that awareness had been growing that acknowledgment of human contributions to events should increase and that learning from events could be a way to ensure that it does. It was also true that the utilities had established a human-factors work group, whose members gave us the opportunity to present our findings and proposals and to exchange arguments. But we and the work group could not agree on the factors that should be incorporated into the system of learning from events. Instead, the utilities decided to implement their own approach to a more systematic consideration of human contributions to events. Based on concepts developed earlier by one of the utilities, it consisted only of factors relating to the human–machine interface and explicitly excluded all aspects resulting from management, supervision, and teamwork. This approach largely precluded the benefit that could be gained from deepening their understanding of the relevance that human-factors have for events. The utilities did extend the scope of their analysis of events, but new boundaries limiting the number of human factors to be considered were introduced.

8.5 UNDERSTANDING THE REASON FOR RESISTANCE

In order to understand the behaviour of the utilities, we must consider some of the implications resulting from our proposals. Our extended system for event analysis asked a great deal of the utility managers. It was the first time that they had been called upon to look at organisational shortcomings as an opportunity to learn from events. The scientific discussion about such questions may have a long tradition, but it has not caught the attention of managers in the German nuclear power industry. Even developments in the United States, like the Management Oversight Risk Tree (MORT; see Johnson, 1980), which emphasises management involvement in events, are not well known in Germany. Thus, the notion that organisational factors contribute to events is a revelation to all parties involved in Germany's nuclear safety system.

Authorities who are influenced by political pressures to phase out nuclear power plants, and who demand and closely check for absolutely strict compliance with regulations may interpret organisational weaknesses as management errors if they are documented in event reports. Such errors might foster doubts regarding the reliability of the plant management, which is one of the key criteria for granting or revoking the license of a nuclear power plant:

Section 7 – Licensing of installations
(2) A license may be granted only if
1. there are no known facts giving rise to doubts about the reliability of the applicant and of the persons responsible for the construction and management of the installation and the control of its operation and the latter have the requisite qualification for such construction, management, and operational control of the facility.

Given these legal conditions and the experience that some nuclear power plant operators have had with the authorities, one must accept that the fears they expressed were well-grounded. Some nuclear power plant operators are indeed in a difficult position, and no manager of a nuclear plant can be expected to risk the accusation of being unqualified. The resulting lack of consensus on the nuclear option has continued to hinder efforts to increase safety through the extension of OL. As shown elsewhere (Becker, 1997), there are also other obstacles to including organisational factors in the learning loop triggered by events. Such factors result from the hierarchical organisation, the distribution of power among the members of the hierarchy, and the management style established in the organisation.

The fear that federal and *Land* abuse of power may be suspected as an argument against disclosing internal deviations from the intended safety culture of a nuclear power plant. Such suspicions must be laid to rest through extended learning from events (and organisational factors). Given the adverse plant environment and the absence of a cooperative safety culture between authorities and operators, what are the options for achieving the expected results?

8.6   THE MULTILEVEL LEARNING APPROACH — ENHANCING HUMAN-FACTORS LEARNING

To overcome the problems outlined above, a multilevel learning approach has been proposed (Becker, in press). The idea behind it is to accept certain limits on the kind of information contained in the formalised event-reporting system, and to complement this system with a guided self-inquiry for use at the supervisory, managerial, and board levels in the organisation. Each of these levels in the organisation would analyse its own contribution to events in the same way, but the results of this level-by-level analysis would not be made known outside the organisation. The results of the analysis would be apparent only from the changes or improvements that are made from time to time in structures or management processes.

When the analysis of an event is to serve as the vehicle for discovering ways to improve at the supervisory and managerial levels, a bottom-up approach should be used. In other words, efforts to achieve better performance should begin with the report by the organisational unit that had been assigned to conduct event analysis, and should then be focused on the supervisory and managerial tasks that were found to be significant to the event under study. The staff member performing the event analysis is presumed to have been trained to work with the concept of human factors (thus making the person a human-factors evaluator), and that this person will use an aid designed for use in human-factors analysis. The scope of this aid is assumed to be limited to the purpose of analysis as proposed by the German utilities: the detection of measurable and documentable inadequacies points or flaws in such factors as display design or procedures.

To encourage the learning process at the supervisory and managerial levels in the organisation, the findings of the human-factors evaluator would then be used for self-inquiry at the next-highest management level. This approach is illustrated in Table 8.1, which presents some of the questions my colleagues and I proposed as an aid for probing a particular dimension of human factors.

It was not our intention to cast the analytical aid as a checklist, but to design a list of guide questions to be used during the problem-solving process of the event analysis. With these questions the human-factors evaluator must try to evaluate the factors influencing the event, preferably in a team session at which the operator involved in the event is also present.

Table 8.1 Excerpt from the aid for the human-factors specialist analysing human engineering

---

Displays

Is the display missing or defective?
Is the display design inadequate?
Is the arrangement or placement of the display unsuitable for the task?
Are labels inadequate?

Is there no second source of information for checking the correctness of the display or alarm?
Is the automatic action survey impossible or difficult?
Are there no available criteria or displays of the state of the component or system?
Is feedback of action results missing?

Is the display or alarm system unreliable (e.g. frequent false alarms)?
Is there alarm inflation or poor establishment of priorities?

Is the computer display design confusing (e.g. too many steps to obtain information/related information placed on different displays)?

Table 8.2 Excerpts from the analytical aid for the head of the production department

---

Human engineering: display of information or controls

Was a similar shortcoming observed earlier? If yes:
  Did the measures taken solve the problem?
  Is the process that led to the earlier decision still adequate in the light of the new experience?

Could the observed shortcoming be crucial in future events relevant for safety?

Does the shortcoming indicate a common mode problem?

The deficiencies detected formed the basis of a corresponding analytical aid for use by, say, the head of the production department. Keeping in mind the responsibilities defined in his job description, and knowing the possibilities and power he had to eliminate weak points, we were able to draft additional, appropriate, guide questions (see Table 8.2). Lists of corresponding guide questions can be drawn up for management tasks associated with vulnerabilities that the human-factors evaluator detects in procedures, training material, and maintenance instructions.

Another aspect that may be even more interesting to look into is the extent to which the proposed approach can also be applied to organisational factors that are more closely linked to the safety culture of the organisation than is the case with those factors addressed in Tables 8.1 and 8.2. *Communication* is a suitable example of the way in which an analytical guide for managers could cover precisely the

Table 8.3 Excerpts from the aid for the human-factors specialist analysing communication

Verbal communication	Written communication
Is communication non-existent or untimely? Is the transmission incomplete? Is the message misinterpreted? Was the interlocutor's understanding not checked? Were there difficulties with equipment? Was the environment noisy?	Is the information incomplete from shift to shift? Is the work instruction form incomplete, misleading, or not understood? Did communication fail between departments?

Table 8.4 Excerpts from the communication-related analytical aid for the head of the production department

Function	Verbal communication failure
Supervision	Does the supervisor ensure that: ■ standard terminology is used? ■ repeat back is used?
Equipment	Do problems with communication equipment occur frequently? Is it possible to take noise-abatement measures (e.g. insulated telephone booths) in important places in the plant?
Team	Do tensions frequently build up in the team? Could the information given to the human-factors specialist be incorrect because of group pressure or group conspiracy?
Organisation	Are there tensions between departments? ■ Is the workload distribution regarded as unfair? ■ Is the task distribution adequate? Are the safety officers – ■ easy to contact? ■ seemingly overburdened?
Training	Was misinterpretation of information influenced by: ■ lack of training? ■ deficiencies in the training programme?

factors that the human-factors evaluator may find difficult to deal with. Table 8.3 shows a list of typical guide questions with which the human-factors evaluator can discern communication-related factors that may have contributed to events.

The corresponding analytical aid for the manager of the production department might have the guide questions listed in Table 8.4. They are based on our research results and have been selected with an eye to the responsibilities that should be part of the job description for the head of the production department.

The sample questions in Table 8.4 show that the analysis performed on this management level is capable of identifying many indicators of the safety culture in the nuclear power plant. A similar list can be compiled for use by the head of the maintenance department. Of course, the analytical aid should be well geared to the task and competencies of the manager.

The approach illustrated so far covers only one or two levels of the hierarchy in the plant. But the idea of a multilayer learning system requires that one go further into the safety management system (SMS) and higher up in the hierarchy. Managers using questions such as those in Tables 8.2 and 8.4 are themselves dependent on the resources available for improvements, and on the decisions taken at a higher management level. Thus, when performing this self-assessment, the managers need to indicate the constraints under which they act. By doing so, they can give the top levels in the organisation an opportunity to question their own earlier decisions, and to judge the adequacy of the organisation's SMS and re-sources. A guide to facilitate the learning of this higher-level management can be drafted only after further work that takes into account the fact that the tasks of personnel in such positions have only rarely been documented. The implementation of a layered system for learning from events that encompasses all management levels will require a step-by-step approach based on continued research. However, the developments appearing to lead in the desired direction are reported by Hale, Bellamy, Guldenmund, Heming, and Kirwan (1997), who propose that the manage-ment delivery system for criteria and resources should be modelled.

8.7   CONCLUSION

The approach providing for multilevel learning from human contributions to events, near-misses, and voluntarily reported weak points has been devised in response to the problems that face the nuclear industry in Germany. The strained relations between the utilities and some regulatory authorities in that country make it nec-essary for the learning system to be adapted to the hierarchical structures of the typical nuclear power plant much more effectively than it has been.

In recommending an approach that calls for guided self-assessment of pertinent factors to the extent that the competence of the respective manager allows, I have assumed that those responsible for a nuclear power plant have internalised a questioning attitude as a prerequisite of a safety culture. Additional motivation for adopting the proposals put forward in this chapter may come from the desire to avoid events and unnecessary outages through an early response to voluntary reporting of weak points (no event – no unscheduled outage!). The assessment guides in this context need to lead to systematic analysis of, and providing information about, aspects and interrelations of human factors, because the managers are usu-ally not professionals in this field.

The proposals for guided event analysis that lends itself to self-assessment at the supervisory level are a prelude to a great step toward the systematic learning from events, near-misses, and voluntarily reported weak points. The strict confidentiality that this system provides for analyses performed at the supervisory and managerial levels of the organisation fosters improvement in learning, even in inhospitable organisational environments and thereby increases the likelihood that the assessment will afford insight into hitherto unrecognised factors of safety culture that contribute to events.

References

ACSNI (Advisory Committee on the Safety of Nuclear Installations) (1993) *Human factors study group third report: Organising for safety*. London: Health and Safety Executive.

ANS/ALS (American Nuclear Society/Austrian Local Section), IAEA (International Atomic Energy Agency), and Nuclear Energy Agency (1995) CARNINO, A. and WEIMANN, G. (eds), *Proceedings of the International topical meeting on safety culture in nuclear installations*. Vienna: IAEA.

BAINBRIDGE, L. (1987) Ironies of automation. In RASMUSSEN, J. and DUNCAN, K. (eds), *New technologies and human error* (pp. 271–83), Chichester: Wiley.

BECKER, G. (1997) Event analysis and regulation: Are we able to discover organisational factors? In HALE, A., FREITAG, M. and WILPERT, B. (eds), *After the event – from accident to organisational learning* (pp. 197–214). Oxford: Elsevier Science.

BECKER, G. (in press) Layer system for learning from human contributions to events: A first outline. In HALE, A. and BARAM, M. (eds), *Safety management and the challenge of organisational change*. Oxford: Elsevier.

BECKER, G., HOFFMANN, S., WILPERT, B., MILLER, R., FAHLBRUCH, B., FANK, M., FREITAG, M., GIESA, H. G. S. and SCHLIEFER, L. (1995) *Analyse der Ursachen von 'menschlichem Fehlverhalten' beim Betrieb von Kernkraftwerken* [Analysis of causes of 'human errors' in nuclear power plant operation] (BMU-1996-454). Dossenheim: Merkel. (ISSN 0724-3316)

EMERY, F. E. and TRIST, E. L. (1960) Socio-technical systems. *Management Sciences, Models and Techniques: Proceedings of the sixth international meeting of the Institute of Management Sciences*, Vol. 2 (pp. 83–97). New York: Pergamon Press.

HABER, S. B., SHURBERG, D. A., JACOBS, R. and HOFMANN, D. A. (1995) Safety culture management: The importance of organisational factors. In CARNINO, A. and WEIMANN, G. (eds), *Proceedings of the ANS/ALS-IAEA-NEA International Topical Meeting on Safety Culture in Nuclear Installations* (pp. 711–21). Vienna: IAEA.

HALE, A. R., BELLAMY, L. J., GULDENMUND, F., HEMING, B. H. J. and KIRWAN, B. (1997) Dynamic modelling of safety management. In SOARES, C. G. (ed.), *Advances in Safety and Reliability* (pp. 63–70). Oxford: Pergamon.

INSAG (International Nuclear Safety Advisory Group) (1991) *Safety culture* (IAEA Safety Series No. 75-INSAG-4). Vienna: International Atomic Energy Agency (IAEA).

JOHNSON, W. (1980) *MORT safety assurance systems: Occupational safety and health*. New York: Marcel Dekker.

KSA (Kommission für die Sicherheit von Kernanlagen) (1994, January, 20) *Sicherheitskultur im Kernkraftwerk* [Safety culture in nuclear power plants]. (Seminar Report No. KSA 7/64). Bern: KSA. [Swiss Confederate Nuclear Safety Commission]

LEPLAT, J. (1997) Event analysis and responsibility in complex systems. In HALE, A., FREITAG, M. and WILPERT, B. (eds), *After the event – From accident to organisational learning* (pp. 23–40). Oxford: Elsevier Science.

MEISTER, D. (1962) Individual and system error in complex systems. *Meeting of the American Psychological Association*, St. Louis.

MEISTER, D. (1971) *Human factors research and nuclear safety*. New York: Wiley.

MORAY, N. P. and HUEY, B. M. (1988) *Human factors research and nuclear safety*. Washington, DC: National Academy Press.

REASON, J. (1990) *Human error*. Cambridge: Cambridge University Press.

SHRIVASTAVA, P. (1983) A typology of organisational learning systems. *Journal of Management Studies*, **20** (1), pp. 7–28.

TURNER, B. A. (1978) *Man-made disasters*. London: Wykehan.

WAHLSTRÖM, B. (1995) Some cultural flavour of safety culture. In CARNINO, A. and WEIMANN, G. (eds), *Proceedings of the ANS/ALS-IAEA-NEA International topical meeting on safety culture in nuclear installations* (pp. 331–40). Vienna: IAEA.

WILPERT, B., FRANK, M., FAHLBRUCH, B., FREITAG, M., GIESA, H. G., MILLER, R. and BECKER, G. (1994) *Weiterentwicklung der Erfassung und Auswertung von meldepflichtigen Vorkommnissen und sonstigen registrierten Ereignissen bei Kernkraftwerken hinsichtlich menschlichen Fehlverhaltens* [Improvement of reporting and evaluation of significant incidents and other registered events in nuclear power operations in terms of human errors] (BMU-1996-457). Dossenheim: Merkel. (ISSN 0724-3316)

CHAPTER NINE

Inter-organisational development in the German nuclear safety system

B. WILPERT, B. FAHLBRUCH, R. MILLER, R. BAGGEN
AND A. GANS

Berlin University of Technology
Research Centre System Safety (FSS)

The authors describe an inter-organisational development project designed to institutionalise a programme called Safety through Organisational Learning in the German Nuclear Industry. Their analysis of the obstacles to this type of action research makes it clear that problems have to be tackled on the conceptual, methodological, and practical levels.

The notion of *system*, on which the research of the authors is based, embraces all levels that contribute to systems safety: the system's technology, the individual, the group, the organisation and its management, and the organisational environment. Conceptualising and modelling the dynamics of change in the whole system in this holistic fashion is a major challenge, particularly because theories of change in large-scale systems are generally wanting. The existing conceptual lacunae translate into methodological difficulties with initiating and sustaining systems change. What methods and approaches can successfully be applied at which pressure points of the system and at what time? The research presented in this chapter shows the manifold operational problems encountered in attempts to overcome structural and psychological barriers to change.

The authors describe their research strategy, which consists of a sequenced, concerted intervention programme focusing on German regulatory agencies; top management, employee representatives, and operational staff of German nuclear power plants; and private consulting agencies.

One of the ongoing projects of the Research Center for Systems Safety (FSS) at the Berlin University of Technology has been the task of introducing a methodical approach known as Safety through Organisational Learning (SOL) into the German nuclear safety system (GNSS) (Wilpert *et al.*, 1997). Because every actor in this system can contribute to its intended outcome – the safe production of nuclear energy in Germany – an organisational learning system has to encompass all of them: individual employees in nuclear installations, their organisation (nuclear installations and power plants), utilities, external consultants, research institutes, manufacturers, and regulatory authorities.

The project is based on the results and recommendations of previous work, in which researchers developed a comprehensive scheme for learning from experience by linking proper documentation of nuclear events (incidents, accidents, and near-incidents) to thorough analysis of the contribution made by all actors in the GNSS (Wilpert *et al.*, 1994). Consequently, the implementation of such proposals calls for an intervention strategy that includes all participants of that system. This chapter deals with the theoretical, methodological, and practical problems that an intervention project such as introducing SOL in the nuclear industry can encounter in a set of organisations interlinked by economic competition, conflictual goals, regulated relationships, and complex patterns of exchange.

9.1   THE GERMAN NUCLEAR SAFETY SYSTEM

9.1.1   Description

The first problem has been the fuzziness of what constitutes the GNSS. Traditionally, systems approaches include management of the focal organisation (e.g. a nuclear power plant) as relevant for the system safety. Factors outside the focal organisation are not, however, considered – a limited perspective that also characterises many approaches operating under the sociotechnical systems paradigm. But according to systems theory, *all* factors and components contributing to a system's output (e.g. product quality, reliability, and safety) must be considered part of the system. In other words, the GNSS as a total system also includes the pertinent elements of the focal organisation's organisational environment, such as regulatory bodies, consultant organisations, manufacturers, and the general public (see Figure 9.1). The major actors and their roles in the GNSS are described below.

The Federal Ministry for Environment, Nature Protection, and Reactor Safety

The German Atomic Safety Act of 1985–97 charges the Federal Ministry for Environment, Nature Protection, and Reactor Safety with overall responsibility for ensuring nuclear safety through national regulation and coordination. The ministry is assisted in this role by the Federal Agency of Radiation Protection, national scientific advisory committees (the Reactor Safety Commission [RSK] and the Radiation Protection Commission), and private research and consulting organisations

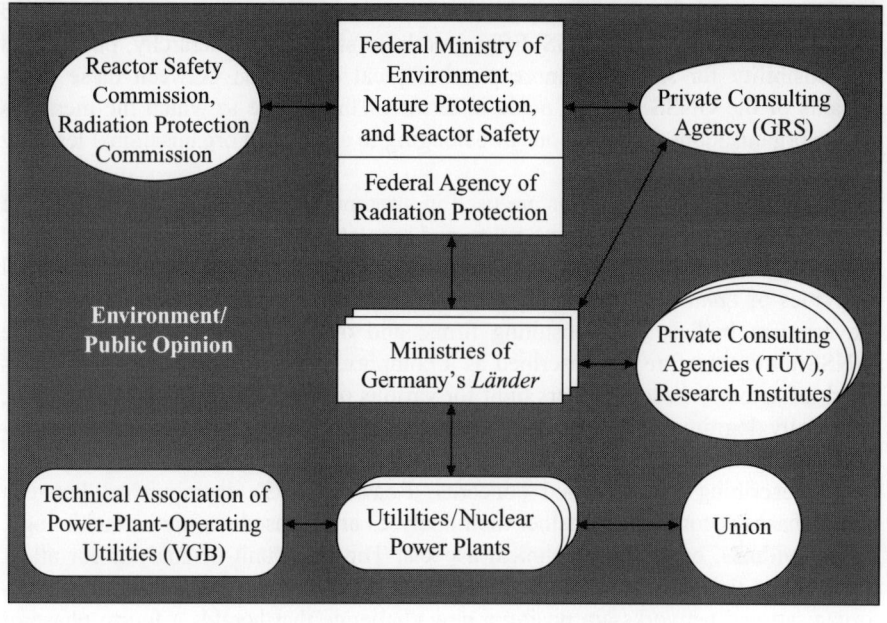

Figure 9.1   The relevant actors in the German nuclear safety system.

(e.g. *Gesellschaft für Anlagen- und Reaktorsicherheit* [GRS], an organisation acting as consultant in reactor safety matters to the Federal authorities).

Land ministries

Responsibilities for licensing, day-to-day inspection, and control of nuclear installations rest with the ministries of Germany's 16 *Länder* (federal states), which make frequent and extensive use of contracts with private consulting agencies (e.g. the *Technische Überwachungsvereine* [TÜV]), and ecological and metallurgical research institutes to help them in their decision-making.

Utilities

Organisations operating nuclear power installations are often linked to each other and to regional and local authorities through equity-sharing and mutual board representation. They have formed the Technical Association of Power-Plant-Operating Utilities (VGB), which provides joint services (e.g. operator training). In economic terms, they are competitors on the energy production market.

Stakeholder groups and the public at large

Here we have to consider articulate antinuclear movements, the media, and groups of private citizens that feel threatened by nuclear installations.

As suggested by the descriptions above, the individuals, organisations, and regulatory bodies of the GNSS differ greatly in size, goals, hierarchy, power, and responsibility for safety. A more psychological distinction between these participants of the GNSS may be made in terms of the degree to which the members of the organisations are conscious of belonging to a common organisational learning system.

Many different mechanisms are used in attempts to interlink and coordinate the actors in the GNSS: formal laws, rules, and regulations; court decisions; formal and informal information systems; informal decision-making processes; formal control activities of *Land*-level regulatory authorities; business contracts between utilities, government bodies, and consulting firms; and overlap in board personnel. The GNSS may, therefore, be described as a complex, often internally conflictual network of actors, whose members fight for various outcomes and different resources, especially legitimacy. These defining characteristics lead us to call it an *inter-organisational network*.

In describing safety-related concerns, Reason (1993) distinguished between three phases according to whether their analyses are focused mainly on technology, on individuals, or on sociotechnical aspects. The main unit of analysis for all of them is the individual organisation. However, goal-directed interventions in inter-organisational networks are posing a new challenge that heralds a fourth phase of safety topics: a safety-oriented inter-organisational network phase during which the need will be to address both intra-organisational and inter-organisational features in complex, partly conflictual settings.

9.1.2 Central conflict issues

In the heyday of nuclear power generation, all or most of the relevant actors agreed that nuclear energy production was desirable, a consensus fuelled to some extent by the first oil crisis. This consensus has broken down in recent years. The emergence of the Green Party as a potential coalition partner for Germany's Social Democratic Party (SPD) in the governments of some *Länder* has led to attempts to abandon the nuclear energy option in regional policy. In interpreting the shared national legal environment according to their political goals, these *Land* governments pursue administrative strategies intended to make nuclear energy production economically uninteresting for utilities.

This loss of consensus may be considered a crucial factor in the widespread distrust among members of the GNSS. The utility companies, for example, do not believe that the regulators' first priority is to increase safety. Indeed, they suspect that the ultimate goal of some regulators is to shut down nuclear power plants altogether. At the same time, the federal government is trying to keep the nuclear option open. The result has been administrative foot-dragging by some *Land* governments. Those *Länder* opposing the nuclear option delay licensing decisions, make increasingly frequent and tough demands on nuclear facilities, escalate regulatory controls, press court cases between federal and regional authorities and seek conflicts or legal battles pitting regional governments and local interest groups against utility companies. Scientific advisory groups and research institutes are drawn into the strife through attempts to enlist them for purposes of legitimising the positions of one adversary or the other. Lastly, the political scene in Germany is characterised by an articulate public minority opposed to nuclear energy production.

Although the overriding goal of all members of the GNSS must be to ensure safety, its members have additional, sometimes contradictory, objectives that may lead to conflicts. The regulatory bodies try to see to it that nuclear power plants obey regulations, but compliance may be costly for the utilities and may conflict with the efforts of nuclear power plants and utility companies to reduce their costs in order to maximise profit. In psychological terms, then, the activities of regulators often end up restricting the scope of action available to utilities. This restriction leads to an emotional and cognitive response that can be described as *reactance* (Brehm, 1966). In the present context reactance means that utilities try to secure or regain discretionary latitude that they perceive to be threatened by certain regulatory practices. The resulting distrust between regulators and utilities drives the spiral of resistance (e.g. information-hiding) and regulatory countermeasures ever higher. Despite the loss of consensus among the relevant actors in the GNSS, they are all intricately interlinked. There is a complex mix of hierarchy and symmetric, parity-based interaction, of partly formal and partly informal ties alongside associational and contractual relations, and of cooperative exchanges and conflictual encounters.

9.2 PROBLEMS OF INTERVENTION ON THE CONCEPTUAL LEVEL

Theoretical underpinnings for goal-directed interventions in large inter-organisational systems such as the GNSS are scant in the available literature. Problems arise because theoretical approaches originating in the field of economics focus mainly on material exchange, economic success, and voluntary membership. Sydow (1992) stressed the strategic nature of such systems by pointing out that they are established to maximise the participants' success in the face of external challenges from the market.

However, systems safety is only partly recognised as an external challenge, for the operators of utilities in the GNSS regard their work as inherently safe. Economic approaches to intervention are therefore not wholly relevant. Additional difficulties result from the fact that economic models deal with the performance of total systems in a descriptive, not explanatory, fashion. Consequently, advice for intervention remains very general or hypothetical, with no specific intervention strategies or methods being proposed.

In order to see what guidance various other current models can give for intervening in the GNSS, we now turn to selected theoretical approaches drawn mainly from political or organisational science.

9.2.1   Trans-organisational systems

Trans-organisational systems are composed of organisations with federative or coalitional structures joined by a common purpose (Aldrich and Whetten, 1981; Cummings, 1984; Warren, 1967). Cummings (1984) uses his Input–Process–Output Model to explain the performance of trans-organisational systems. Because such systems have a higher logical status than single organisations, all planned interventions to change trans-organisational systems require approaches commensurate with that higher status (Cummings, 1984). Note, however, that trans-organisational systems are conceived of as voluntary associations formed by different organisations for a common purpose.

Although there are different reasons for joining trans-organisational systems, one that seems relevant for the GNSS is *mandated motivation*, which is defined as the driving force for organisations to relate to each other as prescribed by rules or laws handed down by higher authorities. When, as with the GNSS, the level of interdependence is relatively low and the level of structuredness relatively high, the performance of the system is affected mainly by the level of effort that each organisation expends on tasks (in this context, meaning safety, competitive energy production, and the establishment of legitimacy). According to Cummings (1984), interventions should be aimed at structuring and organising the trans-organisational system. This recommendation is of little use to the GNSS, however, because the system is already highly structured, some parts of it, excessively so.

9.2.2   Organisation sets

Organisation sets consist of all organisations to which a focal organisation is directly related. The organisation set is a concept for analysis of the relation between an organisation and its relevant environment, which is composed mainly of other organisations. Therefore, the network of inter-organisational relations is characterised by a certain degree of social integration (Evan, 1965; Metcalfe, 1979). The focus of the concept is confined to the interaction between the focal organisation and the other relevant organisations of the set. It does not extend to the totality of interactions between all organisations in the set. The linkage between the organisations of the organisational set is established by the flow of information, products, and personnel.

The links between the members of the GNSS are no exception, at least as far as the flow of information is concerned. The precondition for an organisational set is the existence of an organisational network. Intervention should, therefore, attack the nodes of this network and increase the integration of expectations and requirements by means of cultural, normative, communicative, and functional measures. This proposal may indeed be considered highly relevant for interventions in the GNSS.

9.2.3 Inter-organisational relationships

Inter-organisational relationships are determined by two or more organisations that exchange resources and share collective and individual goals (Whetten, 1987). Inter-organisational relationships have interdependent processes resulting from the distribution of work and functions. Parties to the inter-organisational relationships take over certain roles and internalise certain performance expectations. To reach its goals, the social system builds and coordinates a structure and processes to organise the activities. Inter-organisational relationships have been described along dimensions of structure (formalisation, centralisation, and complexity) and processes (the intensity of the flow of resources and information) (Van de Ven, 1976).

The degree to which the concept of inter-organisational relationships is applicable to the GNSS is restricted because the participants of the GNSS share collective goals and exchange resources only to a limited extent (except perhaps the goal of safety). Interventions should help promote the coordination and compatibility of goals. This aim could be achieved through increased communication among the parties to inter-organisational relationships, a goal that must be considered relevant for the GNSS as well.

9.2.4 Organisational federations

Organisational federations consist of three or more organisations that pool their resources to achieve certain goals (D'Aunno and Zuckermann, 1987; Provan, 1983). Organisational federations are constituted with the intention of reducing the complexity of relations and activities within inter-organisational networks. The control over their activities is transferred to a management group.

In the GNSS the participating organisations pool their resources only partially (e.g. for personnel training), and they have no management control group in the strict sense. In the context of organisational federations, it is argued that an inter-organisational culture facilitates planned changes, so interventions should aim at establishing an inter-organisational culture. The argument seems to apply also to the GNSS, in which certain actors try to establish a safety culture embracing the GNSS as a whole.

9.2.5 Organisational communities

Organisational communities are conceived of as functionally integrated systems of interdependent populations of organisations that develop increasingly complex structures of coordination (Astley and Fombrun, 1987). This development is achieved through the construction of large organisations and an increasing density of inter-organisational relations. Organisational communities are described as substructures (activities for providing resources) and superstructures (social and political institutions guiding the process of providing resources). Economic competition and a short-age of resources are important preconditions for the emergence of organisational communities.

This aspect is different for the GNSS because it was predominantly constituted by law. Organisational communities develop increasingly large organisations and increasingly complex ways of coordinating between them, whereas the size of organisations in the GNSS is relatively stable. No valid recommendations for inter-ventions in large inter-organisational systems such as the GNSS are to be drawn from organisational communities.

9.2.6 Inter-organisational domains

Inter-organisational domains are functional systems and have a position between society as a whole and a single organisation (Trist, 1983). They consist of multiple different actors (i.e. individuals, groups, and organisations) that have a problem in common. In an inter-organisational domain, all activities and actors are interdepend-ent. A critical feature of inter-organisational domains is their voluntary nature, and the fact that their actors actively regulate the institutionalisation of the domain.

The GNSS differs from inter-organisational domains in that its actors are brought together and regulated by laws and rules. However, inter-organisational domains often have a reference organisation that has a leading or regulating function. In the GNSS the Federal Ministry for Environment, Nature Protection, and Reactor Safety serves as a kind of reference organisation with leading and regulating functions that are mandated by law. Interventions in inter-organisational domains, so the model suggests, should be aimed at establishing self-regulatory processes within those domains. As desirable as self-regulation may be, however, this goal may be diffi-cult to achieve in the GNSS because of trends toward overregulation. Overregulation reduces the very freedom of action that seems necessary for self-regulation.

9.2.7 Inter-organisational fields

Inter-organisational fields are composed of organisations from different populations, such as producer, consumer, and regulatory bodies (DiMaggio and Powell, 1983). Differences in the orientations of values held by constituents of inter-organisational fields often lead to conflicts. But formal and informal pressures exerted by powerful organisations (such as a strong regulator) or by expectations in society bring about what is called *coercive isomorphism* in inter-organisational fields (DiMaggio and Powell, 1983, p. 150). Coercive isomorphism makes organisations increasingly similar. Inter-organisational fields, according to this approach, lead to an adjustment of the participating organisations through processes that increase isomorphism (DiMaggio and Powell, 1983) and to a mutual adjustment of goals (Warren, 1967).

Although the loss of consensus in the GNSS does impede the adjustment of goals, the problem of incipient conflicts has an effect comparable to the coercive isomorphism described in the concept of inter-organisational fields. Furthermore, the theory of inter-organisational fields suggests that loosely coupled actors of such fields contribute to stability and flexibility of performance. The opposite is true for the GNSS, which is closely coupled and relatively inflexible. Thus, interventions in the GNSS, as in inter-organisational fields, must start from its structural diversity and must be aimed primarily at making GNSS actors aware that they are part of a common system.

Whatever differences exist between the characteristics of the GNSS and those of inter-organisational fields as conceptualised by DiMaggio and Powell (1983), the facts are that both systems consist of organisations embracing different populations and that conflicts within the GNSS, like those in inter-organisational fields, stem predominantly from value differences. Hence, the GNSS may be considered an inter-organisational field. The concept of field, as developed in field theory (Lewin, 1951) comprises the structures and dynamic forces that influence the behaviour of actors in the field.

The preceding review of various theoretical models of inter-organisational link-ages leads to the conclusion that none of the approaches referred to above is directly and fully pertinent to the GNSS, although they do overlap considerably. The proposed interventions aim either at improving coordination, communication, and integration or at structuring the system or establishing a common culture in the system. It is in such attempts at improvement and structuring that methodological problems arise.

9.3 METHODOLOGICAL PROBLEMS

As pointed out earlier, most of the reported approaches suggest some kind of intervention strategy. It is often recommended that coordination or communication should be improved or increased, but there are few ideas about how to translate these intentions into action. Many different questions emerge. For example, which methods should be applied to structure and organise a trans-organisational sys-tem or an inter-organisational field and make the participants aware that they are part of one system? How does one establish self-regulatory processes in inter-organisational domains and an inter-organisational culture in organisational federa-tions? What methods can promote the coordination of compatibility of goals by increasing communication in inter-organisational relationships or increase integration in organisation sets by introducing cultural, normative, communicative, and functional measures?

For the GNSS it was thus necessary to identify intervention approaches that can affect the entire system on all of its levels and subsystems. Furthermore, these interventions had to be able to guarantee desired change that is sustained and balanced. Faced with the dearth of concepts for changing systems in their entirety, we had to resort to traditional alternatives of organisational development (i.e. for single organisations) and had to try to adapt those concepts to the specific needs of the GNSS as an inter-organisational field.

Lewinian field theory (Lewin, 1951) suggests that an analysis of field forces should precede any concerted intervention designed to induce change. Such an analysis ought to identify change-resistant and change-inducing field forces as well as barriers and gatekeepers relevant to goal-oriented change. Within the GNSS, utilities were identified as one of the main field forces opposing change (i.e. the acceptance of the SOL approach). One reason why German utilities opposed the methodology is that they had developed their own approach for a more systematic appreciation of human factors and were in the process of implementing it on a trial basis. However, it was deliberately confined to human factors involved in the immediate human–machine interface. Therefore, any more comprehensive approach (such as the one proposed by the FSS, which includes organisational, management, and extra-organisational aspects) was bound to be seen by utilities as a rival that would disturb an ongoing, more limited, programme.

Besides, the German utilities have certain, not totally unjustified, reservations about adopting an approach that transcends the human–machine interface, since they fear that regulatory bodies might try to extend their control to organisational processes within nuclear power plants. Moreover, the hierarchical structure within nuclear power plants or utilities presents yet another hindrance to introducing a learning system that goes beyond the human–machine interface to cover possible latent failures in management and organisations as well. It seems unrealistic to expect subordinates to call attention to errors and wrong decisions by superiors within the framework of a formalised reporting system. These factors seem to be intrinsic psychological barriers to change, and they pose formidable methodological hurdles.

Regulatory bodies are a second force resisting change. Given the total systems approach of the FSS, regulatory bodies ought to be included as components in SOL. The implication is that regulatory agencies, too, must routinely question their own regulatory actions. However, given that this regulation is determined by laws and administrative rules, it seems difficult to expect such self-critical distance from public servants. To put it somewhat provocatively, people learn from mistakes; public servants do not make mistakes.

Consulting organisations, which are routinely called upon by regulators to evaluate licensee proposals for changing administrative or technical arrangements in nuclear power plants, serve as important intermediaries. Their role has mainly been that of offering technical expertise. Considerations of human factors have only recently become an aspect of legitimising review activities. Hence the methodological challenge of how to upgrade the human-factors competence of a large and regionally dispersed group of consultants.

By virtue of the field-force role that utilities, regulatory bodies, and consulting organisations have, they may be considered critical gatekeepers in Lewin's terms (Lewin, 1951) and can be compared to the nodes in the network of organisations within the GNSS. In order to implement sustained change, we deemed it necessary to weaken the resistant forces and strengthen change-oriented forces in all three groups. We therefore reverted to classical methods of traditional organisational development and conceptualised training modules and seminars for utilities (including the personnel of nuclear power plants), consulting organisations, and regulators.

9.4 PRACTICAL PROBLEMS

At the practical level we were faced with a bedevilling set of obstacles to the implementation of our proposals. They were based on psychological factors (Baram, 1997), hierarchical structures (Becker, 1997), the intrinsic contradictions between learning and regulation (Wilpert *et al.*, 1994), and the problems due to the political conflicts facing the nuclear industry in Germany.

The psychological barrier experienced by employees of nuclear power plants when it comes to reporting incidents, near-misses, or organisational weaknesses results from concern for co-workers, fear of possible sanctions against themselves, and fear of being made liable. To have employees report erroneous actions by their superiors would require a level of trust within an organisation rarely found in industry. It is precisely in this area that German legislation governing industrial relations (the Works Constitution Act of 1952–72) grants employee representatives an important say in how personnel evaluations are to take place. Union officials were afraid that employees submitting reports might be sanctioned or otherwise abused, since their anonymity and impunity were not regarded as guaranteed.

Conflict between regulation and organisational learning seems to be intrinsic in the contradictory goals of these two functions. Regulation, in the German context of the nuclear industry, is inspired by strategies for ensuring relatively close control over compliance and a frequent search for scapegoats. Both of these control elements hinder the emergence of trust and uninhibited reporting in the interests of organisational learning.

Lastly, the political scene in Germany is characterised by an articulate public minority opposed to nuclear energy production. As mentioned above, this political stance has found its way into the stated policies of *Länder* governed by the SPD and the Green Party, both of which seek to halt nuclear power production as soon as possible. Utilities in this context are fearful that identification of some organisational or managerial shortcomings might be interpreted by *Land* regulatory bodies as managerial inability to ensure the safe operation of nuclear power plants. And Germany's Atomic Safety Act of 1985–97 empowers *Land* authorities to shut down nuclear power plants for which such incapacity is proven. It is no wonder that utilities resist our group's proposals to expand the reporting system.

A planned change in the GNSS is, nevertheless, possible. The first step should be to diagnose the organisational behaviour of relevant participants. Subsequent intervention should make the participants aware of themselves as part of an interdependent system, of eroded trust and its consequences, and of their power to improve system-wide organisational learning.

9.5 INTERVENTION STRATEGIES FOR IMPLEMENTING SOL

In response to the problems at different levels, a planned change in the GNSS was conducted along the lines just presented. In part, the diagnosis step was a review of the laws, regulations, and literature relevant to the formulation of intervention strategies for implementing SOL in the German nuclear industry. We also interviewed managers of every nuclear power plant and every regulatory body in Germany, in order to ascertain their understanding of tasks involved in the reporting and analysis of incidents and to familiarise ourselves with their understanding and practice of regulatory measures. The results of this step showed what the actual performance of the relevant actors was and what each actor perceived the performance of the others to be. We identified weaknesses, such as overregulation by regulatory bodies, sanctions, as well as inadequate reporting and under-reporting by personnel of nuclear power plants.

These weaknesses were taken as the first point of attack in the subsequent step – intervention. Being a small research group with little power, and operating under various constraints on our efforts to institutionalise our proposals, we settled on an overall strategy of kindling small fires in critical parts of the inter-organisational field, hoping to spark a conflagration. Three categories of actors were prime targets: the aforementioned gatekeepers or nodes of the network of organisations within the GNSS: regulatory bodies, utilities, and private consulting agencies. We believed it necessary to commence intervention with them, but to have it accompanied by measures at various additional levels.

We therefore designed training workshops for regulatory bodies, for managers of nuclear power plants, and for private consulting agencies (TÜV). Complementary efforts were directed to the Reactor Safety Commission (RSK) and the VGB. Last, but not least, we established cooperative links with international bodies such as the Electricité de France (EDF, France's state-run electricity-supply monopoly) and the Institute of Nuclear Safety Systems (INSS) in Japan, in order to widen our competence and credibility through outside expertise and evaluation.

Intervention for the regulatory bodies was pursued in meetings with all the different regulatory bodies of the *Länder*, during which we presented and discussed our research design, findings, and intervention targets. These meetings were followed by a seminar for regulatory personnel. The goals were to introduce SOL in more detail, to increase sensitivity to the consequences of different control strategies, and to initiate a change of regulatory practice through group decision-making (Lewin, 1951).

A similar strategy was chosen for a seminar for managers of nuclear power plants. After the introduction of SOL, we discussed existing methods for event analysis and its consequences for reporting behaviour. The next step was to have personnel of nuclear power plants evaluate our method of event analysis, and to follow up with the practical testing of SOL in one German nuclear power plant. Representatives of a Swiss nuclear power plant cooperated in a special seminar on evaluating the feasibility of SOL.

Two seminars with union representatives addressed the intellectual reservations that union members have about possible abuse of voluntary reporting by employees. The seminars resulted in an acceptance of the proposed SOL approach.

The intervention strategy for the private consulting agencies was similar to that for the other target groups. After SOL was introduced, a seminar was held to help participants identify safety-related human factors, such as organisational ones, when visiting and examining nuclear power plants.

The RSK, alerted by comparable research ventures in other European countries and the rising concern with safety culture outside Germany, agreed to hold a two-day international workshop on Man–Technology–Organisation: Safety Culture, an event that led to heightened awareness of human factors within this governmental advisory body. We also shared our ideas and proposals with the human-factors task force of the VGB, making that group, too, a target of our intervention, by holding several meetings during which the participants voiced their opinions or apprehensions about SOL. Cooperation with the Japanese INSS led to two international conferences in Berlin on human-factor research in nuclear power operations (ICNPO), in November 1994, and November 1996. It also led to the previously mentioned evaluation of SOL.

Because the loss of consensus in the GNSS had intensified conflict between some of its relevant actors, one of the main intervention goals of our approach was to stimulate trust and cooperation between all organisations involved. Another important goal was to ensure that all relevant actors developed a new conception of the GNSS, a shared concern for a systemic approach to safety and human factors. Every actor was supposed to realise how important it is to learn from experience and how to promote the requisite learning processes. In other words, the intention was to establish a new culture of organisational learning. To this end, our research developed proposals for changing the existing incident-reporting system, and we suggested the implementation of a new event-analysis methodology. Both ideas were aimed at maximising what people can learn from events and hence from experience.

Planned change in the GNSS as an inter-organisational field has only just begun, and we realise that substantial change may take years to achieve. With the initiatives described in this chapter, however, we hope to have taken at least the first steps on that long journey.

References

ALDRICH, H. and WHETTEN, D. A. (1981) Organisation-sets, action-sets, and networks: Making the most of simplicity. In NYSTROM, P. and STARBUCK, W. (eds), *Handbook of organisational design* (Vol. 1, pp. 385–408). London: Oxford University Press.

ASTLEY, W. G. and FOMBRUN, C. J. (1987) Organisational communities: An ecological perspective. *Research in the Sociology of Organisations*, **5**, pp. 163–85.

BARAM, M. (1997) Safety management: organisational learning disabilities. In HALE, A., FREITAG, M. and WILPERT, B. (eds), *After the event – From accident to organisational learning* (pp. 163–77). Oxford: Elsevier Science.

BECKER, G. (1997) Event analysis and regulation: Problems of organisational factors. In HALE, A., FREITAG, M. and WILPERT, B. (eds), *After the event – From accident to organisational learning* (pp. 197–214). Oxford: Elsevier Science.

BREHM, J. W. A. (1966) *Theory of psychological reactance.* New York: Academic Press.

CUMMINGS, T. G. (1984) Transorganisational development. In STAW, B. M. and CUMMINGS, L. L. (eds), *Research in organisational behaviour* (Vol. 6, pp. 367–422). Greenwich, CT: JAI.

D'AUNNO, T. A. and ZUCKERMANN, H. S. (1987) A life-cycle model of organisational federations: The case of hospitals. *Academy of Management Review*, **12**, 534–45.

DiMAGGIO, P. J. and POWELL, W. W. (1983) The iron cage revisited: Institutional isomorphism and collective rationality in organisational fields. *American Sociological Review*, **48**, pp. 147–60.

EVAN, W. M. (1965) Toward a theory of inter-organisational relations. *Management Science*, **11**, B 217–31.

LEWIN, K. (1951) *Field theory in the social sciences*. New York: Harper and Bros.

METCALFE, J. L. (1979) Organisational strategies and inter-organisational networks. *Human Relations*, **29**, 327–43.

PROVAN, K. G. (1983) The federation as an interorganisational linkage network. *Academy of Management Review*, **8**, pp. 79–89.

REASON, J. (1993) Managing the management risk: New approaches to organisational safety. In WILPERT, B. and QUALE, T. (eds), *Reliability and safety in hazardous work systems* (pp. 7–22). Hove: Lawrence Erlbaum.

SYDOW, J. (1992) *Strategische Netzwerke* [Strategic networks]. Wiesbaden: Gabler.

TRIST, E. (1983) Referent organisation and the development of inter-organisational domains. *Human Relations*, **36**, pp. 269–84.

VAN DE VEN, A. H. (1976) On the nature, formation, and maintenance of relations among organisations. *Academy of Management Review*, pp. 25–34.

WARREN, R. L. (1967) The inter-organisational field as a focus of investigation. *Administrative Science Quarterly*, **12**, pp. 123–38.

WHETTEN, D. A. (1987) Inter-organisational relations. In LORSCH, J. W. (ed.), *Handbook of organisational behaviour* (pp. 238–54). Englewood Cliffs, NJ: Prentice Hall.

WILPERT, B., FANK, M., FAHLBRUCH, B., FREITAG, M., GIESA, G., MILLER, R. and BECKER, G. (1994) *Weiterentwicklung der Erfassung und Auswertung von meldepflichtigen Vorkommnissen und sonstigen registrierten Ereignissen beim Betrieb von Kernkraftwerken hinsichtlich menschlichen Fehlverhaltens* [Improvement of reporting and analysis of significant incidents and other registered events in nuclear power plants in terms of human errors] (BMU-1996–457). Bonn: Bundesminister für Umwelt, Naturschutz und Reaktorsicherheit.

WILPERT, B., BECKER, G., MAIMER, H., MILLER, R., FAHLBRUCH, B., BAGGEN, R., GANS, A., LEIBER, I. and SZAMEITAT, S. (1997) *Umsetzung und Erprobung von Vorschlägen zur Einbeziehung von Human Factors (HF) bei der Meldung und Ursachenanalyse in Kernkraftwerken* [Implementation and testing of concepts for including human factors in event reporting and analysis in nuclear power plants] (Endbericht SR 2039/8). Salzgitter: Bundesamt für Strahlenschutz.

PART TWO

Nuclear power operations: organisational aspects

Introduction

This section contains all the contributions that focus on the subsystem organisation as a subsystem and on its interfaces with technology and individual as subsystems.

Apostolakis develops a method for integrating organisational aspects into human reliability analysis and probabilistic safety analysis. An integrative approach to safety management is demonstrated by Hale, Kirwan, and Guldenmund. Misumi and Yoshida give an impressive example of the interaction of the organisation and the individual as subsystems, showing the massive influence that leadership style has on accident rates.

Reason, in his contribution, raises a fundamental problem of safety-related work questions. Does an ever-expanding investigation of organisational factors that contribute to incidents in complex systems really help the safety practitioner propose and implement safety-related measures?

CHAPTER TEN

Organisational factors and nuclear power plant safety

G. E. APOSTOLAKIS

Department of Nuclear Engineering
Massachusetts Institute of Technology, Cambridge, MA

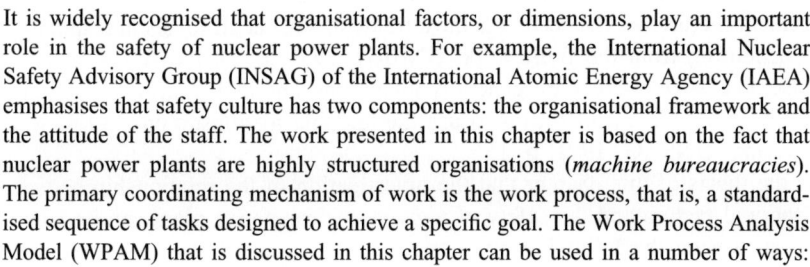

It is widely recognised that organisational factors, or dimensions, play an important role in the safety of nuclear power plants. For example, the International Nuclear Safety Advisory Group (INSAG) of the International Atomic Energy Agency (IAEA) emphasises that safety culture has two components: the organisational framework and the attitude of the staff. The work presented in this chapter is based on the fact that nuclear power plants are highly structured organisations (*machine bureaucracies*). The primary coordinating mechanism of work is the work process, that is, a standardised sequence of tasks designed to achieve a specific goal. The Work Process Analysis Model (WPAM) that is discussed in this chapter can be used in a number of ways:

- to contribute to the assessment of the safety culture at the plant;
- to expand root-cause analysis, so that organisational factors contributing to the incident can be identified;
- to include organisational factors in Probabilistic Safety Assessments (PSA).

The WPAM consists of two parts: the WPAM I is a mostly qualitative analysis of a work process that includes an assessment of the importance of organisational dimensions in the overall quality and efficiency of that work process; the WPAM II provides the link between organisational factors and PSA parameters. The WPAM I utilises a three-step procedure to investigate systematically the types of failure that may occur within a work process, the potential breaches of defences, the organisational dimensions that may bring about these failures, and the relative importance of these dimensions. The three steps are:

- the performance of a task analysis of the work process;
- the development of the organisational factors matrix;

- the determination of the relative importance of the dimensions for each task, a step for which the Analytic Hierarchy Process is used.

The primary purpose of the WPAM II is to address dependencies that might be introduced by organisational factors. It concentrates primarily on capturing the common-cause effects of organisational factors on PSA basic-event probabilities and, thus, on minimal-cut-set frequencies.

10.1 INTRODUCTION

It is well recognised that organisational factors and management policies are major elements of the context in which nuclear power plant personnel function. Reason's type-tokens model of accident causation describes the manner in which the attitudes and practices of different (hierarchical) groups within the organisation can help to facilitate and/or proliferate the occurrence of human errors (Reason, 1990). In this model *types* are defined as general classes of organisational failures; *tokens*, as the more specific failures related to individual situations. These failure types contribute, for example, to the error-forcing context that plays a significant role in second-generation human reliability analysis (HRA) models (Cooper *et al.*, 1996).

Similar issues are discussed by the International Nuclear Safety Advisory Group (INSAG), which has introduced the concept of *safety culture* (INSAG, 1991). This body has emphasised that 'safety culture has two general components. The first is the necessary framework within an organisation and is the responsibility of the management hierarchy. The second is the attitude of staff at all levels in responding to and benefiting from the framework' (INSAG, 1991, p. 5). INSAG went on to recognise that 'such matters are generally intangible' and that 'a principal requirement is the development of means to use the tangible manifestations to test what is underlying' (INSAG, 1991, p. 1).

It is the thesis of this chapter that the investigation of the relationship between organisational factors and the safety of a given nuclear power plant can be facilitated by answering the following questions:

1. *How does the organisation of the nuclear power plant work?* The answer to this question will place the investigation in the right context. The unique features of nuclear power plants should define the boundary conditions for the analysis. Insights from organisational theory and experience from other organisations, such as hospitals, must first be scrutinised for their applicability to nuclear power plants before any recommendations are developed. For example, the organisation of nuclear power plants functions very differently before and after the occurrence of an accident initiator, a fact that should be recognised explicitly. The methodology presented in this chapter deals with pre-accident conditions.

2. *How well does the organisation of the nuclear power plant work?* The answer to this question will rate the organisation with respect to the best and worst industry practices. The development of proper measurement instruments and indicators will be required to answer this question.

3. *What are the safety implications?* The answer to this question will identify the safety implications of the findings on questions 1 and 2. It can be a qualitative

analysis, although a probabilistic analysis is preferable, because the plant's probabilistic risk assessment (PRA) is a more complete description of its potential for harm.

4. *How can the organisation of the nuclear power plant be improved?* This question is, of course, the reason for undertaking safety studies. The answer shall be considered in the broader context of risk management in which all of these questions are properly addressed.

10.2 HOW DOES THE ORGANISATION OF THE NUCLEAR POWER PLANT WORK?

Two basic elements determine the structure of an organisation: the division of labour and the coordination of effort (Galbraith, 1973). The division of labour allows an overall task to be decomposed into subtasks that can be performed by highly specialised individuals (or groups). The coordination of effort integrates these subtasks into a single effort so that the overall goal is achieved.

Nuclear power plants, like many other industrial facilities, exhibit in many respects the characteristics of what is known in organisational theory as a *machine bureaucracy*, namely

> highly specialised, routine operating tasks, very formalised procedures in the operating core, a proliferation of rules, regulations, and formalised communication throughout the organisation, large-scale units at the operating level, reliance on the functional basis for grouping tasks, relatively centralised power for decision making; and an elaborate administrative structure with a sharp distinction between line and staff.
>
> (Mintzberg, 1979, p. 315)

The prime coordinating mechanism of machine bureaucracies is the standardisation of work processes, where a *work process* is defined as a standardised sequence of tasks designed within the operational environment of an organisation to achieve a specific goal.

Figure 10.1 shows the flow diagram of the corrective maintenance work process. After a maintenance work request is issued, it is reviewed by several individuals to determine its significance and the required priority. After these preliminary reviews, the work request is forwarded to the department responsible for planning. The main function of planning is to assemble a work package that addresses all issues related to the proper execution of the job. Scheduling and coordination of maintenance activities ensure that maintenance is accomplished in a timely manner. Concerns about the work are discussed during scheduled daily meetings. Once the work is scheduled, a field work package is assembled, which includes all the work permits, procedures, drawings, and instructions. The execution phase begins when the control room operators tag out the necessary components and the technicians start the field work. Depending on the situation, the work may be executed in the presence of health physics and/or quality control personnel. Upon completion of the job, the person in charge describes the work in the work-tracking form, which, in turn, is reviewed by the department supervisor, who forwards it to the work control centre (WWC). The WCC supervisor reviews the work package and determines

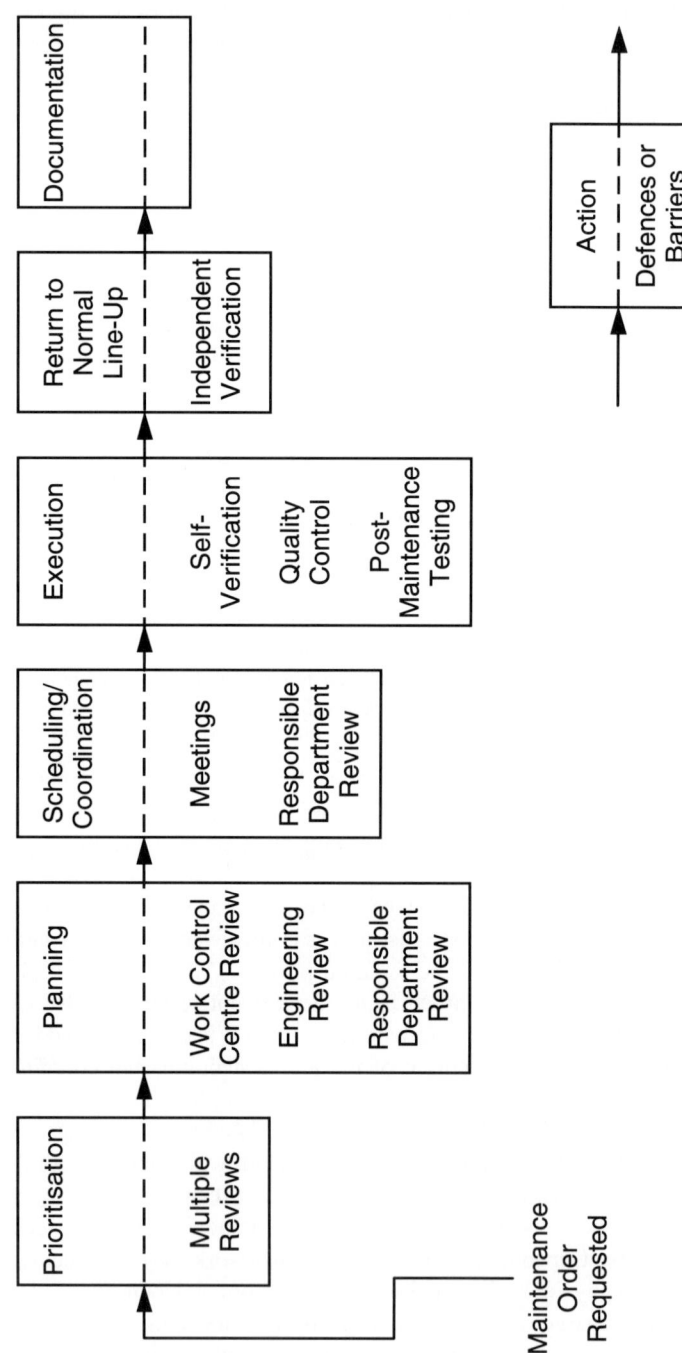

Figure 10.1 The flow diagram for the corrective maintenance work process.

whether post-maintenance testing is required. After any necessary tests are performed satisfactorily, the return to normal line-up takes place, followed by documentation.

This very brief description of the work process (many of the actual details vary from plant to plant and have been omitted) helps illustrate the highly standardised nature of the work processes in nuclear power plants. In addition to maintenance, there are work processes related to operations (plant start-up, power operations, shutdown, and system line-ups), engineering (technical and engineering support of plant activities), and plant support (radiological controls, fire protection, and so forth). It is evident, then, that the work processes must play a prominent role in any investigation of what INSAG has called the structural characteristics of nuclear power plant organisations. However, this description must also include the organisational dimensions in order to be complete.

10.2.1 Organisational dimensions

Jacobs and Haber (1994) have proposed five general categories of organisational dimensions (factors):

- administrative knowledge (e.g. organisational knowledge, formalisation);
- communications (e.g. external, interdepartmental, and intradepartmental);
- decision-making (e.g. goal prioritisation, organisational learning);
- human resource allocation, (e.g. performance evaluation, technical knowledge, training);
- culture, (e.g. ownership, safety culture, time urgency).

As an example, consider the definition of *organisational learning*: 'the extent to which plant personnel and the organisation use knowledge gained from past experiences to improve future performance' (p. 77). *Goal prioritisation* is defined as 'the extent to which plant personnel understand, accept and agree with the purpose and relevance of goals' (p. 77).

The important point is that the work processes and the organisational dimensions provide the basis for answering the question 'How does the organisation work?' The search for 'tangible characteristics' (INSAG, 1991, p. 2) of a sound safety culture must be conducted in the context of these two defining elements of the nuclear power plant's organisational structure. The next question is 'How well does the organisation work?' which, of course, is another way of referring to the question of tangible evidence.

10.3 HOW WELL DOES THE ORGANISATION OF THE NUCLEAR POWER PLANT WORK?

10.3.1 The Work Process Analysis Model I (WPAM I)

The predictable nature of the work processes suggests that a systematic analysis can be conducted to identify the desirable design for a given process and to develop performance measures with respect to the strengths and weaknesses in that process.

Table 10.1   Cross-reference table for the corrective maintenance work process

Task	Action/Barriers	Departments	Personnel
Prioritisation	Prioritisation reviews	WCC WCC/Operations	WCC supervisor Shift supervisor
Planning	Planning WCC review Engineering reviews Responsible departmental review	Maintenance/I&C WCC Engineering Maintenance/I&C	Planner WCC supervisor Systems engineer Mech./elec. I&C engineer
Scheduling/coordination	Scheduling/coordination Meetings Responsible departmental review	Planning Various Various (Operations for system tagging)	Scheduler Variable Operations for work authorisation
Execution	Execution Self-verification QC Post-maintenance testing	Maintenance/I&C Maintenance/I&C QC Maintenance/I&C/WCC	Mech./elec./I&C Mech./elec./I&C QC/QA engineer Mech./elec./I&C
Return to normal line-up	Return to normal line- up 2nd verification	Operations Operations	Control room operator Control room operator
Documentation	Documentation	WCC (nuclear plant reliability data)	Clerk

Note: I&C = Instrumentation and control; QC = Quality control; QA = Quality assurance;
WCC = Work control centre.

The WPAM I (Davoudian, Wu, and Apostolakis, 1994a) consists of a mostly qualitative analysis of a work process, including an assessment of the importance of organisational dimensions in the overall quality and efficiency of that work process.

The WPAM I utilises a three-step procedure to investigate systematically the types of failure that may occur within a work process, the potential breaches of defences, the organisational dimensions (aspects) that may bring about these failures, and the relative importance of these dimensions. The three steps are (a) the performance of a task analysis of the work process; (b) the development of the organisational factors matrix, and (c) the determination of the relative importance of the dimensions for each task.

Task analysis

The task analysis is focused on understanding the following three elements of the work process:

- the tasks (e.g. planning) that are involved;
- the actions in each task and their failure modes (e.g. identifying incorrect procedures during planning);
- the defences or barriers relevant to each task (e.g. the various reviews).

The products of the task analysis are a flow diagram for the process, a cross-reference table, and a design/implementation checklist. The flow diagram depicted in Figure 10.1 shows the tasks involved. The cross-reference table (Table 10.1)

Table 10.2 A sample design/implementation checklist for scheduling and coordination of maintenance work

---

Design checklist
1 Is there a central organisation (planning department and work control centre) in charge of coordinating planned work?
2 What is the organisational relationship of the planning department and work control centre to other departments in the working core (i.e. operations, maintenance, and instrumentation and control departments)?
3 What are the barriers built into the system to correct errors made in the scheduling or coordination of maintenance work? What are their functions?
4 What is the standard process for the scheduling and coordination of maintenance work?

Implementation checklist (normative)
1 Where is the work control centre located?
2 What is the information transfer mechanism used in the work-tracking system?

Implementation checklist (behavioural)
1 How many items are on the maintenance backlog?
2 How many items on the maintenance backlog are of priority 1?
3 How many equipment deficiency tags are on the main control panels?
4 Does the plan-of-the-day schedule reflect the ongoing work?
5 Does scheduling reflect prioritisation of work with respect to safety impact?
6 What is the management philosophy in scheduling – that is, which type of work is usually assigned higher priority?
7 Does the shift superintendent need to review most of the work packages?
8 Are technicians often allowed to negotiate with the shift superintendent regarding assigned work?
9 Does the planning department communicate well with other departments?
10 Does the daily meeting on scheduling and coordination show conflicts among departments regarding scheduled work?

---

identifies the personnel, departments, and actions involved in each task. Lastly, the checklist contains a series of questions aimed at comparing the design and implementation of the tasks (Table 10.2).

Organisational factors matrix

This is the second step of the WPAM I. Its purpose is to show the organisational factors or dimensions that may have an impact on the safe performance of each task. As expected, one organisational failure may cause several unsafe acts. Communication, for example, affects all tasks. This matrix helps in localising the problem areas and guiding the direction of the analysis.

Relative importance of organisational factors

The organisational factors matrix tells which dimensions have an impact on the various tasks of the work process, but it does not prioritise them. Priorities can be set with the Analytic Hierarchy Process (AHP), which allows one to utilise expert judgment. The most important dimensions for the task of setting priorities for the maintenance work process can be sequenced, as is shown in one example given by Davoudian, *et al.* (1994a): formalisation, technical knowledge and time urgency, and interdepartmental communication.

10.3.2   A first set of tangible manifestations of a good safety culture

The process flow diagram, the cross-reference table, and the design/implementation checklist provide basic information about the operation of the organisation, that is, how the organisation is *supposed* to operate. In particular, the design/implementation checklist can provide basic indicators of the quality and efficiency of task performance. For example, from the ergonomic point of view the physical location of the work control centre with respect to the control room may be of interest. Similarly, the number of items on the maintenance backlog is an indication of the quality of plant safety culture.

Although this analysis gives an indication of the organisational structure at the plant, it does not tell how well this structure is implemented. To check that aspect, one needs to continue work with the organisational dimensions. It is at this point that rating scales come in.

10.3.3   Rating scales

As stated earlier, INSAG (1991) identified the attitude of the staff at all levels as the second component of a safety culture (the other component being organisational structure). To understand better what is meant by 'attitude of the staff,' I turn to Pidgeon (1991), who proposed 'a working definition of culture as the collection of beliefs, norms, attitudes, roles, and practices shared within a given social grouping or population' (p. 134). It is important to bear in mind that the actual beliefs, norms, and attitudes are not necessarily as prescribed in management directives and procedures. Reporting the results of the analysis of data collected on full-scale simulators by *Electricité de France* (EDF), Montmayeul, Mosneron-Dupin, and Llory (1994) state that 'the operations performed do not always correspond exactly to what is indicated in the instructions, not that deviations from them are necessarily errors' (p. 69). Their observation confirms once again what has been known for a long time, namely, that an informal structure, in addition to the formal one, exists in organisations; that members of an organisation constantly adapt to their environment, and that unofficial relationships may develop within a group.

To answer the question 'How well does the organisation work?' one clearly needs to think in terms of the dimensions that have been mentioned earlier, and to devise scales that would enable one to measure them. These constructed scales are, in fact, utilised in one form or another by the authors of the cited references. Furthermore, methods need to be devised to capture the actual culture (the mix of formal and informal cultures). These methods include interviews and behavioural checklists. A great deal of subjectivity will be involved in these assessments, but this is unavoidable.

Of particular interest are the Behaviourally Anchored Rating Scales (BARS) (Jacobs and Haber, 1994). These scales, which provide examples of poor, average, and superior behaviours for the assessed dimension, are developed by expert groups that include plant staff familiar with that dimension and with industry practices. Okrent, Abbott, Leonard, and Xiong (1993) have applied these scales to the dimension of deep technical knowledge.

10.4   WHAT ARE THE SAFETY IMPLICATIONS?

Given the work processes and the organisational dimensions, one can proceed in a qualitative way and identify areas where safety-related improvements in the organisation of nuclear power plants may be needed. Such an evaluation is generally not rigorous and relies on the judgment of experienced people. Importance to safety is one of the arguments usually invoked when such improvements are requested, although an explicit demonstration of the link is usually lacking.

A more systematic, but also more time-consuming, way of identifying safety implications of changes to the organisation of nuclear power plants, is to utilise the WPAM II (Davoudian, Wu, and Apostolakis, 1994b). This part of the WPAM provides the link between organisational factors and PRA parameters. The quantitative impact that the dimensions of a work process have on the core damage frequency could be used to identify the important dimensions. The quantitative part of the WPAM II is still in the development phase, but some qualitative insights provided by the WPAM II may be immediately useful.

Typically, PRA results include a set of major (dominant) accident sequences presented in logical combinations of minimal cut sets (MCSs). MCSs contain basic events, such as hardware failures, human errors, and common-cause failures, as well as component/system unavailabilities due to testing and maintenance. Each of these basic events is represented in the PRA by its corresponding parameter (e.g. failure and repair rates, and human error rates). What is important, however, is that the treatment in the PRA does not include the dependencies among the parameters that are introduced by organisational factors. To address this issue, the WPAM II has been designed to modify MCS frequencies to include organisational dependencies among the PRA parameters.

The primary purpose of the WPAM II is to address dependencies that might be introduced by organisational factors. A point of concern has been whether the existing plant-specific data, such as failure rates, already contain the influence of

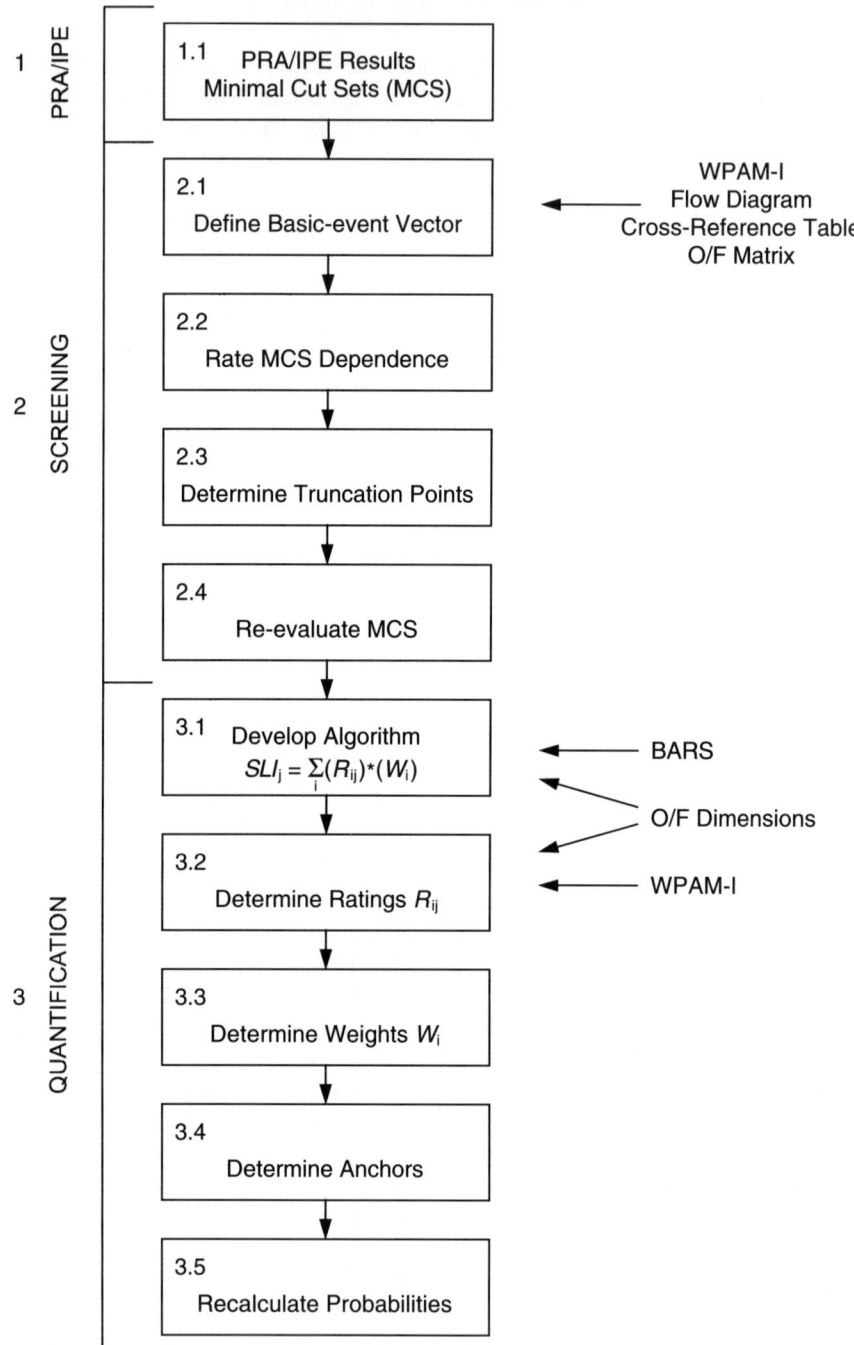

Figure 10.2 Work process analysis model II (WPAM II) flow diagram.
(PRA = Probabilistic Risk Assessment; IPE = Individual Plant Examinations;
MCS = Minimal Cut Set; SLI = Success Likelihood Index; BARS = Behaviourally
Anchored Rating Scales)

Table 10.3 Results of the WPAM II showing the impact of organisational factors on results of probabilistic risk assessment (PRA)

	Minimal-cut-set frequency
PRA results	6.7×10^{-9}
Base case	5.5×10^{-7}
Inter-departmental communication (2)	
Intra-departmental communication (2)	
Coordination of work (2)	
Formalisation (1)	
Training (2)	

Scale: 1 (*worst*) to 5 (*best*)

Note: The minimal-cut-set frequency of 6.7×10^{-9} per year without organisational dependencies is increased to 5.5×10^{-7} per year when the organisational factors shown in the table are included. The parentheses show the rating for each organisational factor as measured by the Behaviourally Anchored Rating Scales.

organisational factors. Assume, for example, that the PRA for a plant has reported that experts determining the uncertainty in the probability of station battery depletion at a given time subsequent to the loss of AC power were asked to consider the condition of batteries at the time of the event as one of the factors. It is, of course, reasonable to claim that the distribution of battery depletion times has accounted for organisational factor(s) in a generic manner. However, the specific organisational factors that can result in reduced battery lifetime (e.g. ineffective maintenance procedures and deficient training) may also have an impact on the availability and/or success of other equipment in the plant. Therefore, while the WPAM II makes it possible to recalculate basic events in each MCS, it concentrated primarily on capturing the common-cause effects of organisational factors on basic-event probabilities and, thus, on MCS frequencies. In effect, the WPAM II is analogous to the common-cause failure (CCF) analysis of hardware, where an additional term is introduced to account for single-cause failure of redundant equipment. However, in contrast to current CCF treatment, the WPAM II goes one step further in that organisational common-cause failures of *dissimilar* systems and/or components are considered.

The WPAM II is composed of two basic steps (see Figure 10.2): minimal-cut-set screening and quantification. The former reduces the list of MCS for each dominant accident sequence by highlighting only those whose basic-event parameters or, simply, parameters that show strong organisational dependence. This highlighting is achieved through an in-depth analysis of the basic events. Having this revised list, the analyst following the quantification process, then reassesses the MCS frequencies by using an approach similar to that of the Success Likelihood Index Methodology (SLIM) (Embrey, Humphreys, Rose, Kirwan, and Rea, 1984; for details on the above steps, see Davoudian *et al.*, 1994b).

The results of the WPAM II can be demonstrated in terms of a simple hypothetical example. Table 10.3 shows the PRA estimate of the frequency of an MCS

$(6.7 \times 10^{-9}$ per year) without organisational dependencies. It also shows that this frequency increases by a factor of about 80 (to 5.5×10^{-7} per year) when such dependencies are included through the WPAM II. Hence, the impact of organisational dependencies could be significant. The significant organisational factors and the rating of the plant with respect to them are also shown.

10.5 HOW CAN THE ORGANISATION OF NUCLEAR POWER PLANTS BE IMPROVED?

10.5.1 The use of WPAM II results

Results such as those shown in Table 10.3 can be used to identify potential improvements in the organisation of nuclear power plants. First, one carries out a sensitivity analysis by increasing the plant ratings with respect to the identified organisational factors. The purpose of performing sensitivity analyses on the ratings is to rank the organisational factors in terms of their effect on plant risk. Once this ranking is done, the results can be used to guide the direction of organisational improvements and the allocation of resources.

Table 10.4 contains the results of a sensitivity analysis on the ratings. In this example improvements leading to a rating of 3 in formalisation and training seem to be more advantageous in terms of risk reduction than improvements in communications. Of course, in a realistic application, the frequency of a given MCS would not be used as a measure of risk. It is more likely that the core damage frequency would play such a role. For the purpose of this chapter, however,

Table 10.4 Results of a sensitivity analysis for risk management: changes in minimal-cut-set frequencies when the original ratings of organisational factors as measured by the behaviourally anchored rating scale are raised

	Minimal-cut-set frequency
PRA results	6.7×10^{-9}
Base case – original O/F ratings	5.5×10^{-7}
Inter-departmental communication (3)	5.3×10^{-7}
All ratings raised to '3'	1.24×10^{-7}
Intra-departmental communication (3)	4.5×10^{-7}
Coordination of work (3)	3.8×10^{-7}
Formalisation (3)	3.5×10^{-7}
Training (3)	3.6×10^{-7}
Inter- and intra-departmental communication (3)	4.3×10^{-7}
Formalisation and training (3)	2.3×10^{-7}

the example in Table 10.4 sufficiently demonstrates the usefulness of WPAM II results.

10.5.2 Root-cause analysis

An important way to reduce risk is to learn from actual experience and take necessary action. Root-cause analysis is a methodology used at nuclear power plants and other industrial facilities to help identify and isolate an incident's key contributing factors. When the occurrence involves many factors, including human performance and/or management decisions, identifying the root cause of the event may be very difficult, and in many cases may involve the underlying safety culture of the organisation. Researchers using traditional methods of root-cause analysis focus primarily on material deficiency and human error, but stop short of looking more deeply into the many work processes and organisational factors affecting everyday operation and support of the plant.

The methodology of the International Atomic Energy Agency (IAEA), which runs the Assessment of Safety Significant Event Team (ASSET), is designed to address nuclear power plants' latent weaknesses that have resulted in an incident or accident. ASSET analyses significant events by preparing a descriptive narrative, establishing a chronological sequence of events, and preparing the logic tree of occurrences that led to the event. Significant occurrences in the logic tree are then investigated in detail and summarised in an Event Root Cause Analysis Form (ERCAF) (Reisch, 1994).

The qualitative approach of the WPAM I can be used to expand the ERCAF to include a section that specifically addresses possible latent weaknesses in the key work process(es). In turn, resulting improvements in the work process(es), combined with appropriate preventive measures, can significantly enhance the effectiveness of the organisation's corrective actions.

The ERCAF is divided into three sections (see Table 10.5). The first section describes the incident by stating specifically what failed to perform as expected, including the nature of the occurrence (i.e. an equipment, personnel, or procedure failure). The second section addresses the direct cause of the incident by focusing on why the event occurred. The analyst looks into possible latent weaknesses of the component that has failed. The third section is directed toward the root cause through examination of reasons why the event was not prevented or, more specifically, of the failure to eliminate contributing latent weaknesses in a timely manner.

Modifying the ERCAF to include analysis of possible latent weaknesses in the organisation's work processes and organisational factors makes it possible to link the organisation directly or indirectly to the specific incident. As the example in Table 10.5 shows, a WPAM I analysis can be used to identify the key work processes and organisational factors that contributed to the incident. Detailed applications to chemical process facilities and nuclear power plants can be found in Tuli and Apostolakis (1996) and Tuli, Apostolakis, and Wu (1996).

Table 10.5 Expanded event root-cause analysis form

Event title:	Oconne: Loss of offsite power and Unit trip	19 Nov. 1992
Occurrence	What failed to perform as expected?	
Occurrence title: Nature:	Breaker failure relays failed to withstand excessive voltages. Equipment failure	
Direct cause	Why did it happen?	Corrective action
Latent weakness Contributor to existence of the latent weakness	Zener diodes in BF relays passed spurious voltage signal causing ACBs to trip open DC power system was being operated with the battery isolated from the bus with the battery charger acting as the only source of voltage.	Breaker relays were modified per vendor instructions. Modification procedure was revised to maintain Busses tied together.
Root cause	Why was it not prevented?	Corrective action
Failure to eliminate the latent weaknesses in time Contributor to the existence of the deficiency	There was inadequate detection of possible problems when operating battery charger without battery in circuit. Management deficiency resulted from inadequate corrective action to perform required maintenance.	Other Oconne procedures were revised and precautions added where appropriate OEP was revised to improve both programme and periodic assessments of programme effectiveness.
Expanded root cause	Which work process(es) and organisational factors played signficant roles in the incident?	Corrective action
Latent weaknesses in the organisation leading to the incident Contributors to the existence of the deficiency	There were various deficiencies in corrective maintenance work process. Technical knowledge and organisational learning were lacking within the task of planning. Problem identification was lacking in various department reviews prior to issuance of the work order.	Key organisational factors within each task were assessed. The OEP was expanded to include improvements in organisational learning Schemes were introduced to assess and upgrade technical knowledge at the plant (e.g. use of behavioural checklists.

References

COOPER, S. E., RAMEY-SMITH, A. M., WREATHALL, J., PARRY, G. W., BLEY, D. C., LUCKAS, W. J., TAYLOR, J. H. and BARRIERE, M. T. (1996) *A Technique for human error analysis (ATHEANA)*. (Report NUREG/CR-6350). Washington, DC: US Nuclear Regulatory Commission.

DAVOUDIAN, K., WU, J.-S. and APOSTOLAKIS, G. (1994a) Incorporating organisational factors into risk assessment through the analysis of work processes. *Reliability Engineering and System Safety*, **45**, pp. 85–105.

DAVOUDIAN, K., WU, J.-S. and APOSTOLAKIS, G. (1994b) The work process analysis model (WPAM). *Reliability Engineering and System Safety*, **45**, pp. 107–25.

EMBREY, D. E., HUMPHREYS, P. C., ROSE, E. A., KIRWAN, B., and REA, K. (1984) *SLIM-MAUD: An approach to assessing human error probabilities using structured expert judgment* (NUREG/CR-3518). Washington, DC: US Nuclear Regulatory Commission.

GALBRAITH, J. R. (1973) *Designing complex organisations*. New York: Addison Wesley.

INSAG (International Nuclear Safety Advisory Group) (1991) *Safety Culture*. (Safety Series No. 75-INSAG-4). Vienna: International Atomic Energy Agency.

JACOBS, R. and S. HABER (1994) Organisational processes and nuclear power plant safety. *Reliability Engineering and System Safety*, **45**, pp. 75–83.

MINTZBERG, H. (1979) *The structuring of organisations*. Englewood Cliffs, NJ: Prentice-Hall.

MONTMAYEUL, R., MOSNERON-DUPIN, F. and LLORY, M. (1994) The managerial dilemma between the prescribed task and the real activity of operators: Some trends for research on human factors. *Reliability Engineering and System Safety*, **45**, pp. 67–73.

OKRENT, D., ABBOTT, E. D., LEONARD, J. D., XIONG, Y. (1993, January) Use of BARS for deep technical knowledge. Paper presented at the Proceedings of the American Nuclear Society, La Grange Park, IL.

PIDGEON, N. F. (1991) Safety culture and risk management in organisations. *Journal of Cross-Cultural Psychology*, **22**, pp. 129–40.

REASON, J. (1990) *Human Error*. Cambridge, England: Cambridge University Press.

REISCH, F. (1994) The IAEA-ASSET approach to avoiding accidents is to recognise the precursors to prevent incidents. *Nuclear Safety*, **35**, pp. 25–35.

TULI, R. W. and APOSTOLAKIS, G. E. (1996) Incorporating organisational issues into root-cause analysis. *Transactions of the Institution of Chemical Engineers (Process Safety and Environmental Protection)*, Vol. 74, Part B, pp. 1–16.

TULI, R. W., APOSTOLAKIS, G. E. and WU, J.-S. (1996) Identifying organisational deficiencies through root-cause analysis. *Nuclear Technology*, **116**, pp. 334–59.

CHAPTER ELEVEN

Capturing the river: multilevel modelling of safety management[1]

R. HALE,[1] B. KIRWAN[2] AND F. GULDENMUND[1]

[1] *Safety Science Group, Delft University of Technology, Netherlands*
[2] *ATMDC, National Air Traffic Service, Bournemouth, England*

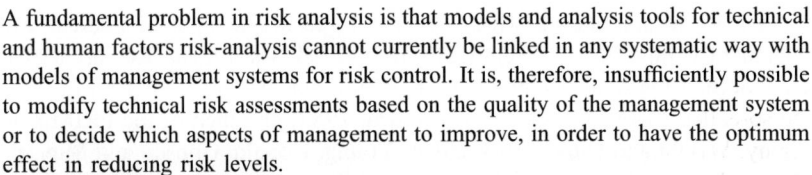

A fundamental problem in risk analysis is that models and analysis tools for technical and human factors risk-analysis cannot currently be linked in any systematic way with models of management systems for risk control. It is, therefore, insufficiently possible to modify technical risk assessments based on the quality of the management system or to decide which aspects of management to improve, in order to have the optimum effect in reducing risk levels.

This chapter combines the safety management framework developed in Delft, using the PRIMA audit system developed by Four Elements to come up with a modelling technique which is potentially capable of making this integration, and of capturing the dynamic nature of safety management. It uses the notation of the Structured Analysis and Design Technique to model safety management as a series of activities designed to identify and control risk at three levels, governed by three feedback loops to enable the system to learn and improve. The critical factors for the quality of the management system are the integrity of these loops, and the quality of the controls and resources provided to carry out the management activities. Advantages and remaining problems with the approach are also discussed.

The analysis of safety management systems (SMS) controlling high-technology organisational systems currently entails a detailed investigation or audit of a host of parameters linked to the likely successful performance of the system. Such audits create a sophisticated and insightful picture of the adequacy of the safety management structure and its operation. However, whereas such analyses give insights into the potential root causes of safety system failures and certain vulnerabilities in those systems (with notable exceptions to which we return below), they do not actually model the linkage between safety management activities and large-scale failures. Traditional commercial audits, even if in-depth and highly detailed, pale, in terms of time investment and detail of modelling, when compared to the risk assessments that are frequently carried out in high-risk industries, such as chemical, offshore, and nuclear power systems. Such risk assessments are intended to represent detailed, thorough, comprehensive, and quantitative invest-igations of these systems, predicting risk levels well beyond the expected opera-tional life cycle of the systems themselves. Currently, however, such risk assessments largely ignore what it is that controls the safety of the system. There is, therefore, a paradoxical gap in the way in which the safety of potentially high-risk instal-lations is assessed: audits assess safety management, but cannot link inadequacies causally to their accidental consequences; and risk analyses cannot establish a causal link between the accident consequences and their immediate causes and their more distal safety management origins. This gap, itself a vulnerability in current assessment approaches, is, therefore, between accidents and risk analysis frame-works on the one hand and safety management policies, procedures, and operations on the other. This chapter builds on the work of two groups which have worked to bridge this gap: the so-called PRIMA audit of Four Elements; and the Delft frame-work of safety management. It outlines a prototypical approach that attempts to link the two halves of the risk equation.

11.1 PREVIOUS APPROACHES

There are some approaches which represent recent attempts to address safety management within, or as an adjunct to, formal risk assessment. Researchers have either modified the final results of the risk assessments, based first on the results of the so-called MANAGER audit (Technica, 1988) and then of the PRIMA audit (Bellamy, Wright and Hurst, 1993; Hurst, Young, Donald, Gobson and Muyselaar, 1996), or have attempted to change probabilities lower down within the risk assessment framework (Davoudian, Wu and Apostolakis, 1994a, b; see also the contribution by Apostolakis in this volume). In both these approaches, assessors extrapolate from observations on management functions and performance to hard-ware and 'liveware' failure events. This is quite an extrapolation, which must be justified with evidence. Only in the case of the work originating from studies at Four Elements (Bellamy et al., 1993; Hurst, Bellamy, Geyer and Astley, 1991) has such evidence been presented in the form of detailed analysis of data on several hundred incidents related to loss of containment from pipes, vessels, and hoses and the relationship of those data with human and organisational factors. The generalisability of such data for other types of incident and accident in other types of industry remains to be proven.

11.2 THE PROPOSED APPROACH

An alternative approach is to attempt to be more mechanistic and deterministic. This perspective entails defining the relationships between management decisions

and actual failures (hardware faults, human errors, system failures, accidental outcomes), and then seeing how such management activities can contribute directly to
these hardware and liveware failures, and hence to the accidents themselves. Such
an approach is one of detailed modelling and inevitably will be probabilistic in
nature, because it is predictive in its aims. But there will be more determinism in
the system, because the safety management effects on actual operations and hardware are modelled explicitly, via actual links within the system boundaries. Such
a system is highly specifiable. Our initial studies (Hale *et al.*, 1997) with the
method show that the overall structure can be specified functionally and hence quite
generically. The detailed operationalisation and, hence, the evaluation of the SMS
are relatively specific to a particular organisation and its culture (here, meaning the
specific way things are done in that company compared to others operating the
same technology). One advantage of a specifiable model is that its predictions can
be compared to experience. This means that it can be proved wrong – that is, it is
falsifiable. This also means that, as a model and modelling approach, it can learn
(i.e. can be adapted to experience), and one can learn from it. The philosophy of
the PRIMA audit approach and of the approach described in this chapter is embodied
in Figure 11.1, where it is compared to the two current alternative approaches that
have hitherto dominated the literature.

The labels along the top of Figure 11.1 represent a generalised causal progression from root causes to accidental outcomes. Thus, the structure and culture of an
SMS underpins the safety plans and procedures, which determine how the safety
measures are executed in work practices and system configurations, which in turn
afford certain errors and failures, which lead to accidents.

As Figure 11.1 shows, safety audits are strong in addressing safety management
and some culture issues, but they either do not state how safety management and
organisational culture affect risk levels, or they achieve such a link solely via expert
judgment or global modification of quantified risk estimates. Risk analyses do not
usually address safety culture and safety management issues, although the assessors
will implicitly carry in their minds a projection of what the safety culture and safety
management processes will be like.

It is proposed to merge the two approaches. Clearly, however, what is missing
at present, and what is suggested by the identified gap, is an appropriate *interface*
between the two approaches. This missing interface is not simply a means of taking
qualitative audit information and rendering it as quantitatively usable products,
since this has already been tried in the audit and the risk-analysis approaches.
Rather, the interface requires a meeting-point of the two approaches, which are
philosophically different. The important difference is not so much qualitative versus
quantitative. Rather, the crucial difference is one of mechanism versus holism. Risk
analysis is a search for causal chains, failure pathways, and failure combinations.
It is failure-driven, working back from a limited number of scenarios defined as top
events in fault trees (or forward from these in limited event trees). This mechanistic
philosophy would be pure determinism if it were not for the necessary probabilistic
and stochastic nature of failures and, hence, of risk analysis itself. Audits, on the
other hand, are more holistic in nature. They start from management system elements

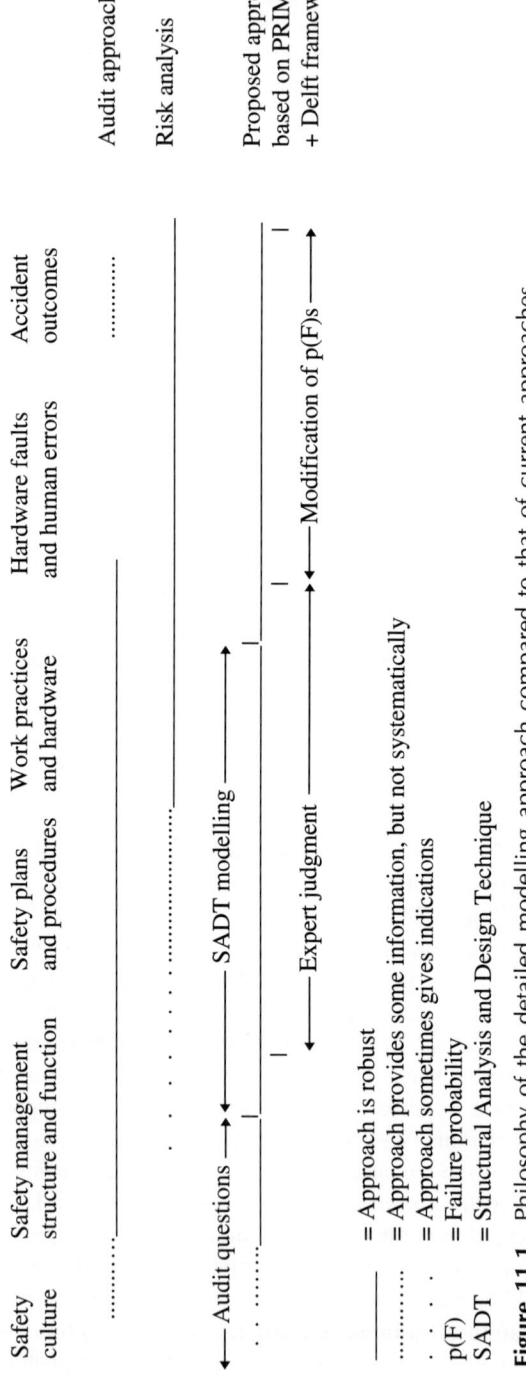

Figure 11.1 Philosophy of the detailed modelling approach compared to that of current approaches.

that should be present. In most audits those elements are presented as checklists, often arbitrarily grouped under headings (Hale and Hovden, 1998). There is much assessor judgment necessary to interpret the combinatorial effects of the answers to many audit questions. There is usually no attempt to test whether the elements used are independent of each other in their working and influence on risk, and inspection of the audit questions indicates that there can be no question of such independence (see also Chaplin and Hale, 1996). The sum will therefore always be less than its parts if there is significant dependency. This lack of independence may be appropriate and valuable when dealing with constructs like safety management and culture, but it is inappropriate when dealing with accident progression analysis.

However, the two approaches do appear to offer an interface that could enable one to analyse failure mechanisms in the safety management itself. Such detailed modelling of safety management activities could then be linked to hardware and software failures, but at the same time the performance of the safety management activities could be driven by both the outputs and considerations of an audit tool. This interface would provide more insight into safety management and hence into its failure potential, while maintaining the appropriate and arguably necessary holistic inputs from audit tools. The system or approach required is therefore one that maintains the holistic inputs at the safety culture end of the progression in Figure 11.1 and translates them ultimately into the more specific and concrete mechanistic projections at the accident prediction end of Figure 11.1. The benefits are potentially large:

■ Safety culture and management influences will be linked to the risk equation: the contribution of safety culture can be ascertained or evaluated quantitatively in a more theoretically acceptable and empirically verifiable way.

■ If risk is high, then detailed recommendations can be made at several levels – hardware changes, human factors changes, changes in the safety management operation, and ultimately changes to the safety culture. The most effective targets for change can be predicted via sensitivity analysis of the risk assessment and underlying modelling.

■ Such a modelling tool could be used to explore the ways in which risk is affected by changes in safety management, or the impact of events in the environment. It would therefore be a potential tool for evaluating and investigating organisational change.

The approaches cited above and below the bottom line in Figure 11.1 are the tools that we suggest will enable the desired interface to be constructed. They are believed to be more useful at this time than alternatives that could be considered for this purpose. In the following sections, we will discuss the rationales of these approaches, and how they fit together to constitute the desired approach.

An additional objective of this method is indicated by this chapter's title, which is drawn from an old parable. It tells of a blind man from the desert who has never encountered a river before. He is travelling and comes at last to a river. He asks a youth he meets to bring him a sample of the river so that he can experience it. The youth goes to the fast-flowing river, brings back a cup of water, and presents it to the old man. 'Ah', says the old man, holding the cup and dipping his finger into the still water, 'so that is what a river is like!' Our contention is that most analysis of SMSs has been far too static and fails to capture any of an SMS's constituent dynamic flows, without which it withers and dies. The analogy of the river links also to our contention that an SMS has a number of levels by which the shop-floor reality of the business process and its barriers and controls directly influencing safety are connected with the decisions made in the company board-room (or by regulators outside the company). These levels may be the hierarchical levels of the management, but they represent also the abstract levels connecting company goals to daily work practice and hardware. Just as a river may show laminar flow, with bodies of water flowing over others without mixing, so the levels of the management system may slide over each other with little communication and transmission of effective influence. In other words, they do not always engage and drive each other with the necessary interaction (turbulence in our river analogy) to ensure that what is desired at the top influences what happens at the bottom (and, in the other direction, that the reality of what happens at the bottom is transmitted to the top to influence decisions there). Any satisfactory modelling method must be capable of reflecting such dynamic aspects and the potential they represent for the corrosion of safety management controls (Bellamy, personal communication, 1993) and for improvement of the system through learning (e.g. Senge, 1990).

The most advanced modelling technique in this area has been PRIMA (Hurst et al., 1996). It was developed to incorporate an aggregate management factor into the risk assessment of accidents due to loss of containment in the chemical industry in the context of the Seveso 2 directive. The central concept of PRIMA is the control and monitoring loop, which depicts the development and implementation of controls and standards. The loop passes through five levels, extending from the external influences that partly drive the company's safety programme, to their effect on the system's reliability and safety, and to the monitoring and review of performance, then back up through these five levels to permit learning and improvement. The five levels, which are largely equivalent to the steps in Figure 11.1, are:

- system climate;
- organisation and standards;
- communication, control, and feedback;
- human reliability;
- containment reliability.

The loop is applied to eight areas of the SMS that accident analysis showed to be particularly crucial for accidents due to loss of containment. The loops are very similar to the problem-solving cycles of the Delft framework described below. The insights derived from PRIMA have been incorporated into the approach described below.

11.3 IC(H)OR

The modelling technique developed is known as IC(H)OR.[2] Its elements and development history are described in detail in Hale et al. (1997). The basic framework can be summarised in four points:

1   Safety management is envisaged as a problem-solving activity in which potential and actual deviations from the desired safety standard must be recognised,
analysed, and prioritised. Solutions must be generated, selected, and implemented, and the results must be monitored to check whether the desired control
has been established (or re-established).

2   This process must go on in all life-cycle phases of a technology, project, or
organisation (conception, design, construction, operation, maintenance, decom-
missioning, and disposal), and both feed-forward and feedback loops must be
present between all the phases to enable control and learning. This arrangement
emphasises that safety management concerns both the safety of the process of
conducting a given phase (e.g. the safety of designers at their drawing boards
or of the maintenance personnel carrying out repairs and modifications) and
the safety of the product of that phase, which will partly determine the safety
of subsequent phases (e.g. the ability of the plant as designed, and of the maintained plant, to function safely when commencing or resuming operation).

3   Safety management is seen as having three levels:

(a)   The actual execution (*E-level*) of the primary business process. At this
level the issue is the direct control of the hazards of the technology.

(b)   The formulation of plans and procedures (*P-level*) to guide the E-level
actions, prepare them for coping with new hazards and criteria, and capture the learning at the E-level in the form of new plans and procedures.
The issues at this level are new hazards or demands and the observation
that the functioning of the E-level is not adequate in relation to specific
hazards or activities.

(c)   The conception and improvement of the system structure and culture
(*S-level*), within which the P-level works. The issues at this level are
observations that the SMS is non-existent, outmoded, or deficient in many
areas, that a step change in performance is required and that the SMS
is not consistent internally, or does not match the desired company culture
(as in a take-over of new subsidiaries).

4   The activities at all levels are modelled as functional management tasks,
with the generic steps of the problem-solving cycle at each level being broken
down into the constituent tasks required. The modelling technique used is the
Structured Analysis and Design Technique (SADT), which is borrowed from
knowledge engineering and which models all activities (e.g. steps in the problem-
solving cycles, life-cycle phases, and primary processes in the business process)
as transformations turning Inputs into Outputs through the use of Resources
and under the influence of Controls or Criteria (see Hale *et al.*, 1997 for a more
detailed explanation).

The basic notation of SADT is illustrated in Figures 11.2, 11.3, and 11.4, which
depict a generic part of the SMS concerned with the design and implementation of
safety, health, and environment (SHE) procedures. The three figures demonstrate how
the activity can be successively broken down into subordinate tasks, in order to detail

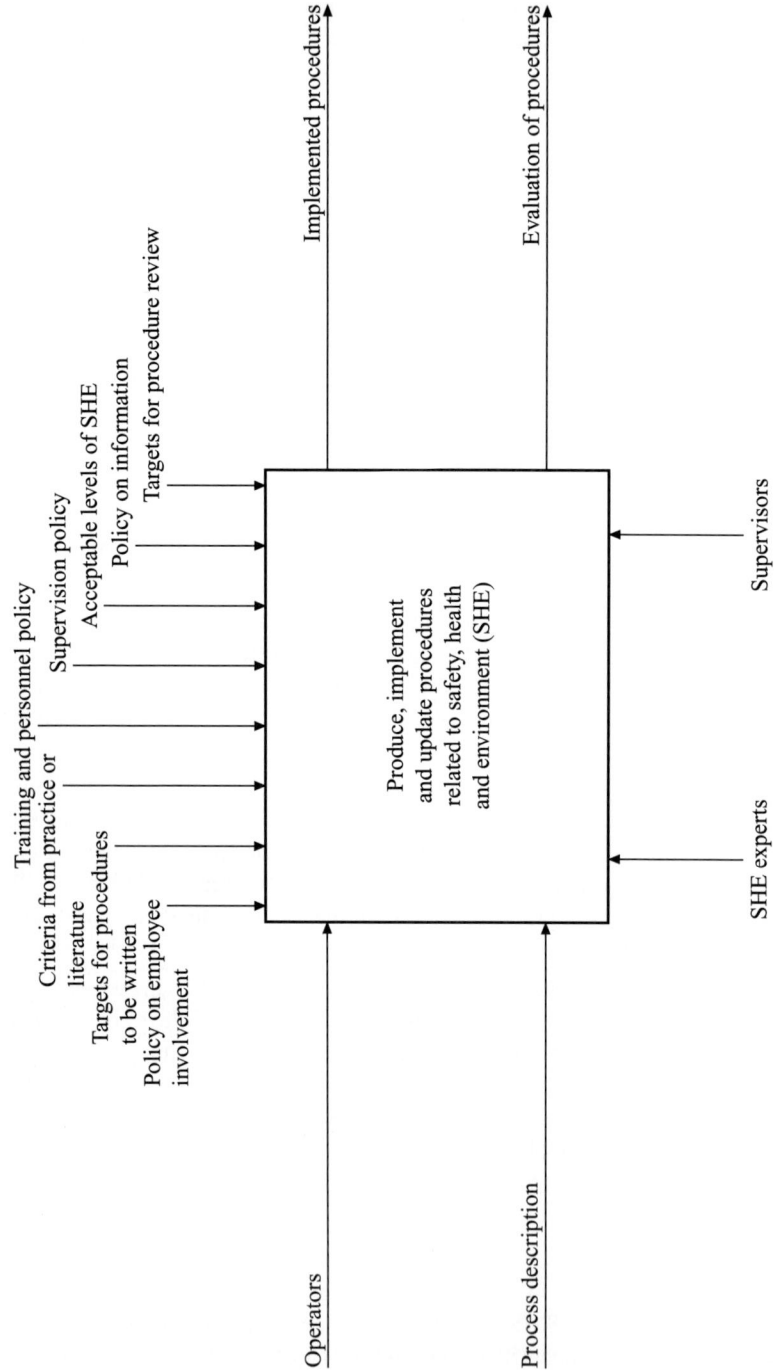

Figure 11.2 Generic Structured Analysis and Design Technique for the design and implementation of safety, health and environment procedures.

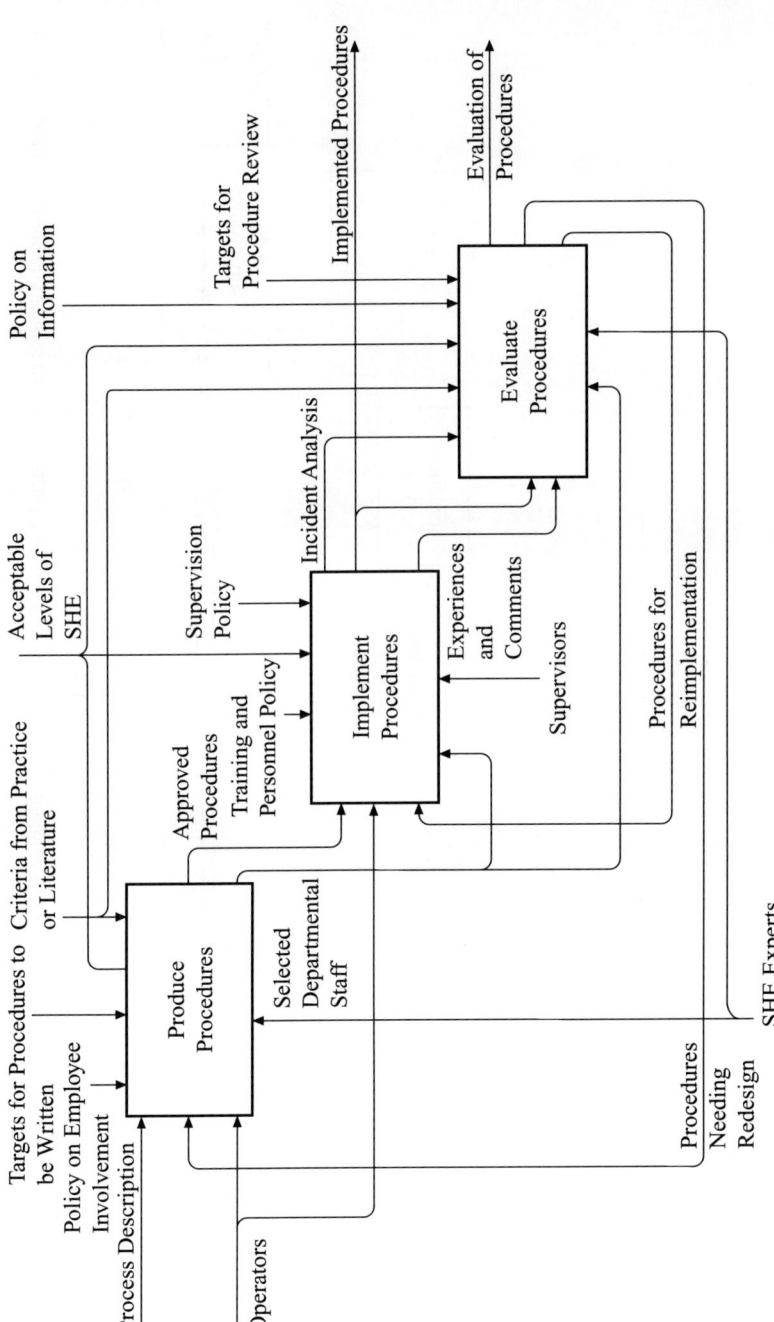

Figure 11.3 Second-level Structured Analysis and Design Technique: detail of producing, implementing and updating procedures related to safety, health and environment.

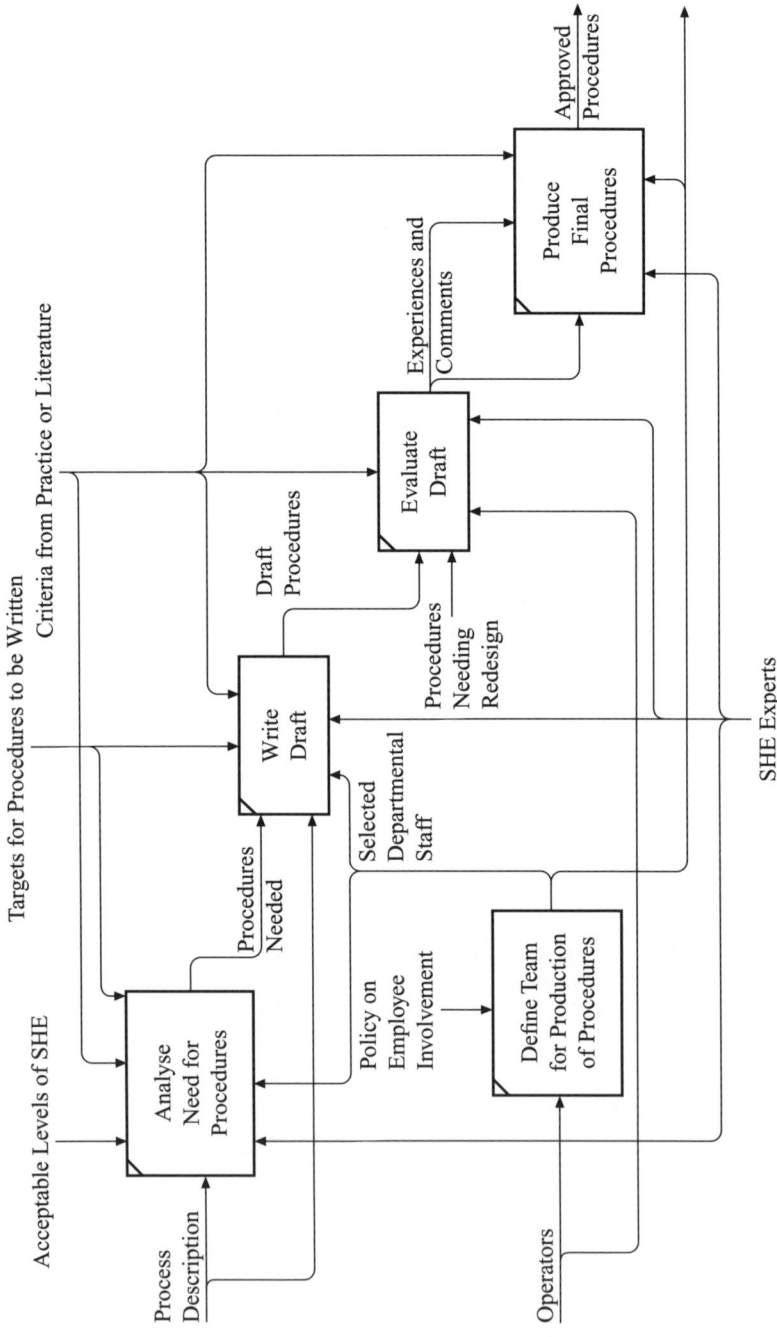

Figure 11.4 Third-level structured Analysis and Design Technique: detail of producing procedures related to safety, health and environment.

the manner in which it is carried out. Figure 11.3 is an 'unpacking' of the first aggregate task description. Figure 11.4 is an unpacking of the first box in Figure 11.3.

This process of unpacking can be carried out to whatever level of detail is required. Indeed the top-level box shown in Figure 11.2 can be considered a part of the unpacking of a still higher level SADT, which would represent the total SMS for the operations phase of a company life-cycle. The unpacking of the SADT boxes is analogous to breaking down task objectives in Hierarchical Task Analysis (see Kirwan and Ainsworth, 1992), but has the advantage that it preserves the sequential relations between the subordinate tasks throughout the analysis of resources and controls, thereby helping to ensure that the activities are carried out according to suitable safety criteria. This breakdown of the SMS into safety management tasks provides the basis for defining the types of failures and mechanisms in these tasks and in the flows between them. Such failures arise because the appropriate controls and resources are not made available, or are not of the appropriate quality, or because constituent tasks are not assigned as a responsibility, not carried out with the proper quality criteria in mind, or not performed on time.

Of the modelling techniques applied to the SMS thus far, SADT has the unique advantage that it indicates flows and has the potential for showing the dynamic nature of management (i.e. the potential for capturing the river). This characteristic of SADT comes from its origins as a tool for knowledge engineering in software development. It was designed to facilitate decision support and make it possible to write expert system programs that could be implemented and run in a computer. Inherent, therefore, in the use of SADT is the idea that it can become a simulation of the SMS, which one should in the future be able to operationalise as a computer model.

The modelling potential of IC(H)OR is being explored in a project (I-Risk), financed by the European Commission under the ENVIRONMENT programme, and by the Dutch Ministry of Social Affairs. The rest of this chapter is confined to insights gained in the project and to their potential for future development.

11.3.1 Links between levels: laminar flow v driving force

Essential to an understanding of the SMS is an understanding of how different levels interrelate and drive each other. The IC(H)OR model identifies three levels in the SMS, the lowest of which, the execution level (E-level), directly controls the hazards in the primary business process of the organisation, whether that be producing nuclear energy, controlling chemical processes, manufacturing automobiles, or healing patients in a hospital.

An analysis of the risks in a technology, and of the barriers and controls (hardware and work practices) that the organisation has put in place to control them, can be derived from an SADT analysis of the primary business process (and of its subsidiary elements, such as the provision of power and utilities). Many other techniques, such as MORT (Management Oversight and Risk Tree) (Elsea and Conger, 1983), exist to carry out such an analysis (see also Visser, 1996), but the advantage of using SADT is that the analysis is congruent with the subsequent steps to be described below. The link with the E-level of IC(H)OR is that the operators and hardware specified as safety-relevant resources, and the rules, pro-cedures, and methods specified as control criteria in this analysis, must function as required. This level and its links can be illustrated as in Figures 11.5 and 11.6, which are simplified SADTs applied to the factors relevant to the functioning of a chlorine filling-pipe for loading tankers in a plant studied by one of the authors. It is focused on the factors determining pipe strength, and so it is important for the scenario of pipe rupture. Only the personnel resources are modelled and not the hardware.

The links between the management tasks and the specific hardware failures allow one to establish the link postulated in Figure 11.1 between management and technical risks. Note that the C and R items identified as relevant for the functioning of the various steps provide the link with the next level of IC(H)OR, the P-level, where such criteria are developed and decided upon, and where the necessary resources are planned and provided.

It seems possible to include all controls and resources under a limited number of headings:

- Procedures/rules, methods, and techniques, whether defined outside the organisa-tion (e.g. laws and contractual agreements) or within it (e.g. company rules).
- Output criteria/performance indicators, whether defined outside the organisation (e.g. laws and contractual agreements) or within it (e.g. risk perception).
- Motivation/commitment to carry out tasks.
- Resolution of conflicting pressures/criteria.
- Competence and availability of the personnel (whether the company's own or the contractor's).
- Hardware (quality and availability), both operational hardware and its replace-ments during maintenance (spares) and the operating interface.

The success of an SMS depends on the way in which these controls and re-sources are planned and delivered, by the P-level to the E-level, for each task that is critical for safety. It becomes clear in using these generic categories that they are not independent influences. There is almost always some dependency between controls and resources, because they are physically linked. For example, a piece of hardware used as a resource (say, a relief valve) comes with a limited range of set-points, which are built-in criteria for its functioning. A competent person who, as a resource, is carrying out a task, comes with a certain commitment and risk perception, which are only partially modifiable. The competence and risk-perception may be inversely correlated (e.g. see Reason, Manstead, Stradling, Baxter and Campbell, 1989, who showed that persons who rated their competence as high were more likely than those who rated it as low to violate rules on the grounds that such action was not a risk for them).

It is also clear that different combinations of the influences may be alternatives for achieving reliable control of an activity. Competence, rules, and supervision form such a combination (with a minor link to commitment). It is possible to envisage an activity that can be controlled equally well by highly competent people with few imposed rules or supervision, or by people with little knowledge, but

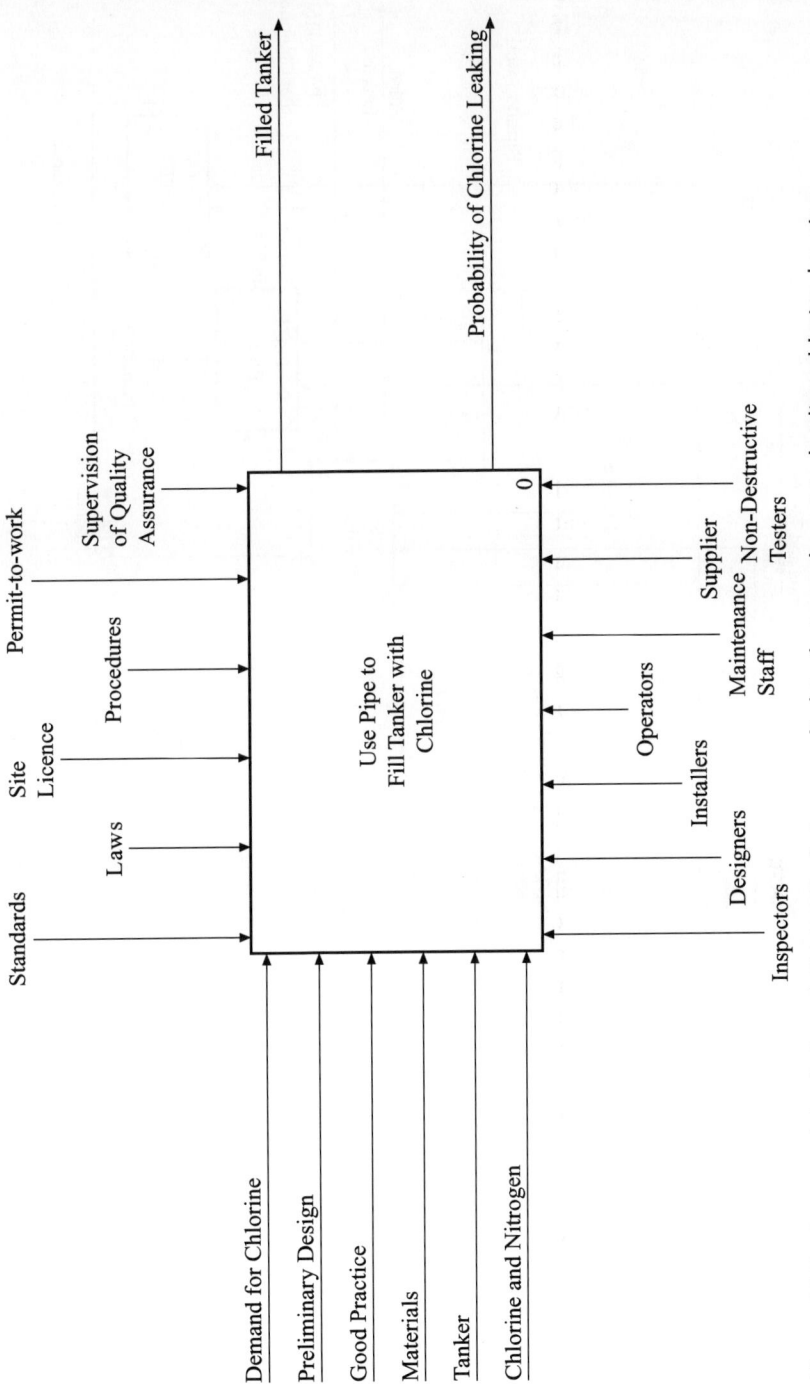

Figure 11.5 Structured Analysis and Design Technique applied to factors relevant to loading chlorine by pipe.

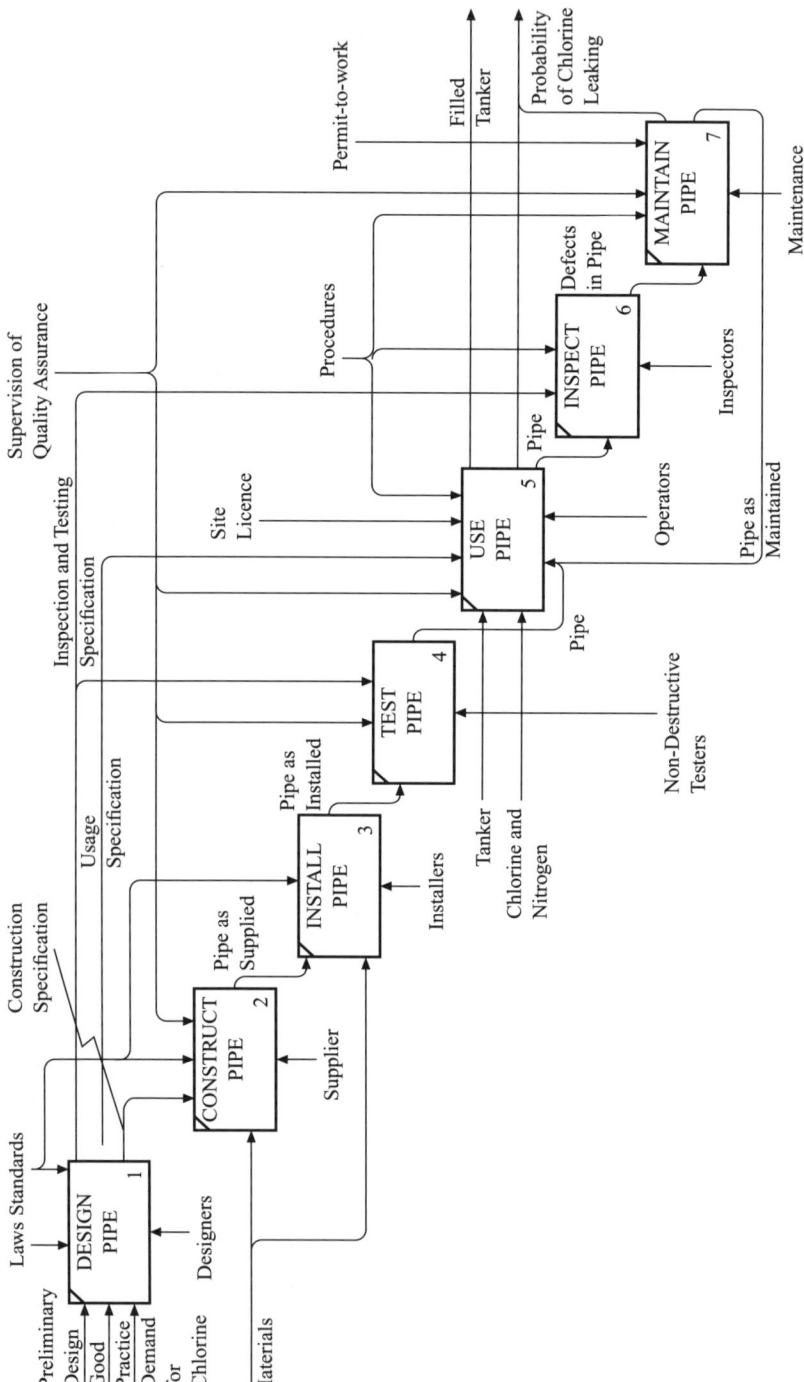

Figure 11.6 Breakdown of Figure 11.5 into life-cycle of the pipe.

clear, detailed rules, close supervision and great commitment to stick to the rules. Considering the generic influences and their interaction forces the modeller to come to terms with these factors.

Each SADT box potentially has feedback loops attached to it. They monitor the outputs of the box in order to learn whether changes need to be made in its Cs, Rs, or Is so that the box will work more safely next time. These feedback loops can be envisaged as the 'genes' of the company system (Bellamy, personal communication) that govern its immune system and adaptation to external change and threat. In other words, they are the means for preventing the organisation's SMS from 'corroding' and for adapting it constantly to new and changing circumstances. Some of the feedback loops stay at the same level in the SMS; they are direct check activities that pick up deviations and correct them. Other loops circle up to a higher level (E to P) and are responsible for the company's redefining, or re-emphasising the safety controls (C) on an activity, or reconsidering, refreshing, or renewing the resources (R) needed to carry it out. Still other loops circle from the P-level boxes up to the structure (S) level and trigger major redesign of the P-level tasks.

One can postulate that the absence or weakness of these loops allows the quality or preparedness of the Cs and Rs to decline over time, a change that will, in turn, mean that they are less effective at ensuring that the activity or transformation in the box is carried out to a high quality, and hence that the generic failure rate being influenced by the box stays low. If one wishes to use such a gene in quantitative modelling as an input to the failure-probability monitoring, it will be necessary to make assumptions (or elicit expert judgments) about the frequency and quality with which these loops must operate in order to prevent corrosion (or, indeed, to produce improvement in the transformation).

This example demonstrates a generic aspect of SADT modelling – namely, that, if the I – >O links are followed, one stays on one level of the SMS model. If, on the other hand, one pursues a C or R arrow (at right angles) out of a transformation box and inquires how that control or resource is decided upon, provided, and kept functional, one moves up a level in the SMS, from the E- to the P-, or from the P- to the S-level. This means one drives deeper into the management system and the characteristic operationalisation of its controls and resources that constitute the safety culture of the organisation. It is in following these links that one must expect to be able to capture the common mode failures represented by a less-than-adequate safety culture. The quality of these links between abstraction levels can indicate whether the levels interact and drive each other successfully, or simply flow over each other like lamina, in which case only pseudo-control will occur, characterised by such phenomena as well-kept safety manuals that bear no relationship to reality, frequent violations of rules that are impracticable, and false impressions at the senior level that problems are under control. The interrelationships of elements in the IC(H)OR model are shown in Figure 11.7. This functional notation also lends itself to use as a way of depicting flows suitable for a management-system equivalent of a Hazard and Operability (HAZOP) technique, where the principles of the HAZOP (e.g. key words used to uncover failures in flows) are used to detect mechanisms of failure in safety management (Kennedy and Kirwan, 1996).

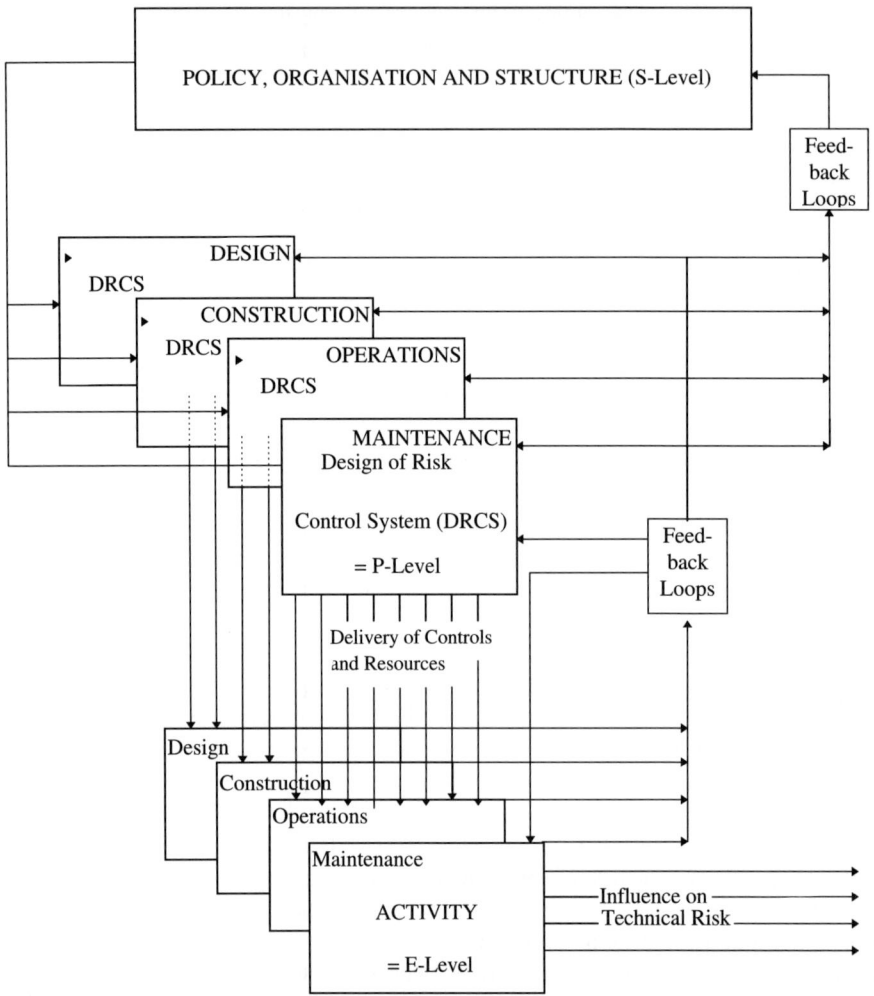

Figure 11.7 IC(H)OR Safety Management System.

11.3.2 Quantification and sensitivity analysis

The connection between IC(H)OR and quantitative risk analysis can be made by indicating that a principal throughput of the SADT boxes at the E-level is the failure probability of the hardware or work-practice system element to which the SADT analysis relates. If the unmodified technical failure rate is fed into any SADT box as input, the effect that the management and other factors represented by the I, C, and R arrows have on a box can be represented as a combined influence on that failure rate. The rate can be increased or decreased according to the quality of the organisational and management factors. Such factors could, in principle, be assessed with techniques of influence modelling (Phillips, Humphreys and Embrey,

1983) or other techniques that use expert judgment in a systematic way to assess the effect of different factors and the interaction of influences on an activity (see also Goossens and Cooke, 1997). Considerable developments in such a technique are necessary to make it suitable and efficient for SMS modelling and quantification, but it offers a potential solution to the problem of linking qualitative with quantitative modelling in this area. The expert weightings for the influence of the different controls and resources can provide the basis for a sensitivity analysis, which, in turn, can indicate which aspects of the SMS are likely to have a major (common mode) effect on a number of tasks critical to technical or human failures that may have been regarded as independent in the fault tree analysis.

11.3.3   Generic *v* specific modelling of influences

The division of influences into categories of I,C,O, and R already indicates that SADT can operate at a generic level. Experimentation with the technique suggests that it provides an interesting and challenging way of linking the generic to the specific, a benefit manifested in several ways that are of value to SMS modelling. The first has already been indicated: the ability to model generic delivery systems for the criteria and resources, which can be filled-in in different ways depending on the company.

A second aspect of the generic-*v*-specific nature of the modelling is found when either the SADT boxes are unpacked or the various influences (Cs and Rs) are worked out (i.e. traced back in the management system to their origins and the parts of the system that ensure their presence and quality). At a high degree of aggregation, such as the SADT diagrams shown in Figures 11.2 to 11.4, the diagrams apply to any organisation. However, as the boxes are unpacked further, it becomes increasingly difficult to remain at a generic level, and the analyst soon has to know how the company is organised, which department is responsible for which activity, and which specific technology and hazards are present. In other words, the analysis becomes more industry-and-site-specific (see also Hale and Swuste, 1993). The same phenomenon is familiar to task analysts using hierarchical task analysis. This suggests some useful ideas and distinctions. It may help forge consensus on defining safety culture, as the company-specific unpacking and specification of the generic elements of an SMS. In particular it suggests that the specification of aspects that are characteristic of the technology, and the hazards present in it should be distinguished from that which reflects the choices made in companies operating with the same technology, but with different organisational ways of achieving the same generic results. In other words, SADT offers a much clearer conceptual framework and vocabulary with which to talk about safety culture than has been the case up to now (Hale and Hovden, 1998). One can locate the cultural differences that determine such aspects as: which balance of different controls and resources is used in which parts of the SMS; which default values there are for risk-perception and commitment; who is involved or consulted in what safety activity; which parts of the system are managed within the company; and which parts are contracted out.

Implicit in the above discussion of culture is the assumption that the alternatives chosen by different companies can offer the same overall degree of control of the risk being managed. The literature is currently poor at making that assumption explicit. Indeed, some authors use the term *safety culture* as a synonym for a culture that promotes and values safety, a definition that completely confounds two ideas: the means of achieving the result and the result achieved. As a modelling technique SADT offers a clear distinction between the two by distinguishing a generic ideal system from a site-specific unpacking and specification of how (and how far) a given company implements that ideal, i.e. achieves the result.

The final aspect of the generic-*v*-specific debate dealt with in this chapter is the consideration of the link between the SMS and the failure scenarios identified in classical analyses of technical risk. One of the ideas driving the development of the original PRIMA model, and now of IC(H)OR, is the idea that site-specific modification of the quantitative risk estimates derived from generic data on failure is essential if all companies in a given industry (in particular the good companies among them) are not to be tarred by the failures experienced only by some of its poorer ones. The MANAGER technique (Technica, 1988) modified the quantitative risk from the technical model with one factor derived for the whole SMS. PRIMA, its successor, derived factors weighted in three different ways to link to failure rates of vessels, pipes, and hoses. These factors were derived from assessments of eight key management areas. The objective of the current project is to make a much more fine-grained analysis, which makes the links at least at the level of a moderately large number of different failure scenarios, if not 'deeper into the river'. However, the links must be of a manageable number, or the modelling burden will be far too great. This condition implies that the failure scenarios and the elements in the technical model must be categorised into a limited number of classes that one can hope to be able to model with limited slices through the SMS. For example, some groups of scenarios may be more sensitive to good maintenance, others to careful operation, still others to well-planned installation.

11.4 SHORT-TERM *V* LONG-TERM DEVELOPMENTS

Underlying the discussions in this paper has been a modelling dilemma that presents a problem in reconciling short- and long-term goals. It was foreshadowed in the introduction in the contrast drawn between risk analysis as failure-driven, and auditing and management analysis as holistic and system-driven. These are, in fact, two fundamentally incompatible paradigms that researchers have been attempting for years to join on to each other. Work on the SADT modelling makes this incompatibility clearer than it has been, and suggests that, ultimately, progress in this area may depend on a paradigm shift in risk analysis.

Risk analysts start from a small number of carefully defined failure events and regard as relevant only those factors feeding those events. The success of the analysis depends upon ruthlessly simplifying it by eliminating any aspects that do not have a significant effect on the probability or the size of the consequences. A basic tenet of the techniques of risk analysis is that events in its trees must be independent of each other. Common-mode failure, whatever its origins, has always presented risk analysis with problems. Coping with common mode has resulted in compromises and fudging, one reason why risk analysis has always had problems with human reliability. When considering management factors, one moves into an area that almost totally contradicts the assumption of independence between factors in the fault tree. Management systems are designed to introduce dependence in this sense of reliability engineering. Their very purpose is to bundle specific aspects of the company's activities into coherent policies on maintenance, personnel selection and training, incentives, performance improvement, and company culture. By doing so, management systems ensure that, by and large, the decisions made about any one area will be consistent with each other. Hence, good or poor decision-making about allocation of maintenance resources for one piece of equipment that is critical for safety will tend to go together with good or poor decisions related to all other such equipment. Because of the complexity of management in large and technologically advanced companies, the dependence is not complete: certain aspects of management may be weaker than others (i.e. independent in their influence on risk). But the degree of common mode failure will be large. In terms of modelling, it is interesting to ask: 'How does one recognise which parts of an SMS will be highly consistent with others, and so represent a large common mode failure?' To the extent that parts are independent, they need to be modelled as separate influences; to the extent that they are common mode, they may be treated as one management factor (as MANAGER and PRIMA have done thus far).

The IC(H)OR model is one that logically starts from the opposite end of the spectrum set out in Figure 11.1 (i.e. the company management system for safety in its entirety). Once unpacked to the suitable level of detail, it would contain the management and behavioural elements that influence all the failure scenarios one wishes to consider. It would also contain all those that risk analysts currently ignore as too minor, but that still strongly influence the company's total performance on conventional safety, health, and absenteeism. However, the factors relevant to a given failure scenario would appear at very diverse parts of the model, with some being closely linked (high common mode) and others being more remote (lower common mode).

Building and validating such a complete model of the SMS either generically or for a given specific industry or site requires a great deal of time and expertise. Once these tasks have been accomplished, the model's potential is huge, since it could be seen as a complex SMS simulation that could be used, for example, to conduct experiments and to see the effects of potential changes. Building and validating a complete model of the SMS is the long-term challenge. If it were to be met, it could turn risk analysis on its head by making the management analysis drive the technical analysis, and not vice versa.

In the meantime the dominant paradigm is that of risk analysis, reasoning backwards from failure into the management system. It calls for management factors as probabilities to be fed into fault trees (leading either to top events or to the steps in event trees). At present our holistic models, or parts of them, have to be hung under these diverging fault trees in order to show how the links between the branches are made by common-mode management failure. It remains to be seen what compromises make this approach possible, pending the long-term development of the IC(H)OR model as a full simulation.

Notes

1   Work on this paper was supported partly by funding through the project I-Risk from the Commission of the European Union and the Dutch Ministry of Social Affairs and Employment. The views expressed are entirely the responsibility of the authors.
2   The divine blood that flowed in the veins of the gods in ancient Greece. In the acronym the letters ICOR are derived from the four principal elements appearing in the notation used in the model: Inputs, Controls, Outputs, Resources. The name is chosen to emphasise the idea of flows as the lifeblood of the SMS.

References

BELLAMY, J. L., WRIGHT, S. M. and HURST, W. (1993, September) History and development of a safety management system audit for incorporation into quantitative risk assessment. Paper presented at the International Process Safety Management Conference and Workshop, Part II, San Francisco.

CHAPLIN, R. P. E. and HALE, A. R. (1996, June) An evaluation of the use of the International Safety Rating System (ISRS) as intervention to improve the organisation of safety. Paper presented at the NeTWork Workshop on Safety Management and Organisational Change, Bad Homburg.

DAVOUDIAN, K., WU, J.-S. and APOSTOLAKIS, G. (1994a) Incorporating organisational factors into risk assessment through the analysis of work processes. *Reliability Engineering and System Safety*, **45**, pp. 85–105.

DAVOUDIAN, K., WU, J.-S. and APOSTOLAKIS, G. (1994b) The work process analysis model (WPAM). *Reliability Engineering and System Safety*, **45**, pp. 107–125.

ELSEA, K. J. and CONGER, D. S. (1983) Management Oversight and Risk Tree. *The Risk Report*, (Vol. 6, No. 2). Idaho: International Risk Management Institute.

GOOSSENS, L. H. J. and COOKE, R. M. (1997) Applications of some risk assessment techniques: Formal expert judgment and accident sequence precursors. *Safety Science*, **26**, pp. 35–48.

HALE, A. R. and SWUSTE, S. (1993, May) Safety rules: Procedural freedom or action constraint? Paper presented at the NeTWork Workshop on The Use of Rules to Achieve Safety, Bad Homburg.

HALE, A. R. and HOVDEN, J. (1998) Management and culture: The third age of safety. A review of approaches to organisational aspects of safety, health and environment. In FEYER, A. M. and WILLIAMSON, A. *Occupational Injury: Risk, prevention and intervention*. London: Taylor & Francis.

HALE, A. R., HEMING, B., CARTHEY, J. and KIRWAN, B. (1997) Modelling of safety management systems. *Safety Science*, **26**, pp. 121–40.

HURST, N. W., BELLAMY, L. J., GEYER, T. A. W. and ASTLEY, J. A. (1991) A classification scheme for pipework failures to include human and sociotechnical errors and their contribution to pipework failure frequencies. *Journal of Hazardous Materials*, **26**, pp. 159–86.

HURST, N. W., YOUNG, S., DONALD, I., GOBSON, H. and MUYSELAAR, A. (1996) Measures of safety management performance and attitudes to safety at major hazard sites. *Journal of Loss Prevention in the Process Industry*, **9**, pp. 161–72.

KENNEDY, R. and KIRWAN, B. (1996) The Safety Culture HAZOP: An inductive and group-based approach to identifying and assessing safety culture vulnerabilities. In CACCIABUE, P. C. and PAPAZOGLOU, I. A. (eds), *Probabilistic Safety Assessment and Management* (pp. 910–15). Berlin: Springer.

KIRWAN, B. and AINSWORTH, L. K. (eds) (1992) *A guide to task analysis*. London: Taylor & Francis.

PHILLIPS, L. D., HUMPHREYS, P. and EMBREY, D. E. (1983) *A socio-technical approach to assessing human reliability*. (Technical Report No. 83–4). London: London School of Economics, Decision Analysis Unit.

REASON, J. T., MANSTEAD, A. S. R., STRADLING, S. G., BAXTER, J. S. and CAMPBELL, K. A. (1989, May 26–28) Errors and violations on the roads: a real distinction? Paper presented at Commission of European Communities Workshop on Errors in the operation of transport systems, Cambridge.

SENGE, P. M. (1990) *The fifth discipline: The art and practice of the learning organisation*. New York: Doubleday.

TECHNICA (1988) *The MANAGER technique. Management safety systems assessment guidelines in the evaluation of risk*. London: Technica.

VISSER, J. P. (1996, June) Developments in HSE management in oil and gas exploration and production. Paper presented at the NeTWork Workshop on Safety Management and Organisational Change, Bad Homburg.

CHAPTER TWELVE

The effects of leadership and group decision on accident prevention

JYUJI MISUMI, LITT.D.

Director, Institute for Social Research, Institute for Nuclear Safety System, Inc., Kyoto
Director, Institute for Group Dynamics, Fukuoka

MICHIO YOSHIDA, M.ED

Associate Professor, Kumamoto University, Kumamoto

Inspired by Lewin's research on leadership and group decision, we launched a series of studies in Japan. With regard to leadership, we constructed a model based on the assumption that leadership consists of two behavioural dimensions – (P) performance and (M) maintenance – and conducted studies to verify the validity of this PM model. It was demonstrated that leadership has influence on many factors, such as subordinates' morale and the incidence of accidents at workplaces in industries, businesses, and other organisations. In addition to the study of leadership, we conducted action research by introducing group decision-making to promote safety and prevent accidents in workplaces. The action research resulted in reduced incidence of accidents caused by drivers in a bus company and employees at a shipyard. Furthermore, it was proved that the incidence of accidents attributable to employees vary in frequency according to their superiors' leadership. Drawing on this experience, we developed and implemented leadership training designed to improve their leadership in preventing accidents in nuclear power plants.

Studies of accidents have mainly focused on personality factors. In Japan, Yuzaburo Uchida developed the Uchida-Kraepelin test by using a test by E. Kraepelin, a German psychomedical scientist, and it has been widely utilised in industrial circles (Misumi, Shirakashi, Ando, and Kurokawa, 1961). Also, Japanese industries have since devised various kinds of accident prevention methods and have translated them into practical applications. One concrete example is the system developed by the Labour Research Institute of the National Railroad Corporation. It is called 'pointing a finger and calling out', whereby a railroad employee points a forefinger at a thing to confirm its safety, while saying aloud such phrases as, 'The switch is all right' or 'The speed is all right'. These words are self-directed only in order to cause a state of self-arousal, which, in turn, contributes to action prevention.

We have been engaged in a series of studies on accident prevention, combining the experiments on leadership and group decision-making that were originally initiated by Kurt Lewin. This research focuses on group dynamics for accident prevention. In some cases, accidents involving only individuals occur, but in many cases accidents happen in the course of group activities or in connection with organisational life. Accidents at nuclear power plants occur in the form of incidents within an organisation. It was in those studies conducted by Kurt Lewin and his disciples that human behaviour was proved to be closely related to an organisation's social climate. It was in that context that Kurt Lewin came to develop his point of view of group dynamics.

No repetition of accidents like the one at Chernobyl can be permitted. Therefore, no research can be conducted in anticipation of any repeat of nuclear power plant accidents. Once an accident has occurred, research is launched to find the causes of the accident. If the causes are identified and removed, some people might think that accidents will no longer occur. However, if the causes are related to some human factors, it is hardly possible to remove only the specific ones involved.

When human factors are to blame for accidents, what must be removed is not part of the human factor, but the human factor as a whole. Here, each part equals the whole. The part and the whole are indivisible. Consequently, everybody involved at the scene of an accident must be dealt with and not just some of them.

This leads us on to another question. Suppose some human factors were to blame for the accident at Chernobyl. If the same accident were to be repeated even once more, it could easily become a subject for scientific research. Needless to say, such an accident shall not occur. What can be done is to push ahead with 'comparative studies' under relatively analogous conditions. Just imagine how many guinea pigs have been used for the advancement of medical science in the past. By the same token, it is probably through comparative studies under a large number of variously analogous conditions that ways and means of preventing accidents at nuclear power plants might gradually be identified. It is true that human factors vary from culture to culture, from place to place, and from organisation to organisation and that there are other situational and environmental variations. But there are some common, universal features, and the important thing is to identify them.

In this chapter we discuss the studies concerning leadership and group decision-making that we have carried out on accident prevention. In doing so, we take into consideration the results of our on-site research at nuclear power plants as well as other studies on accident prevention. That work, much of which is still in progress, builds on two major US studies that Kurt Lewin conducted on leadership and group decision (Lewin, 1947; Lippit and White, 1943). After World War II, we began our studies on leadership by carrying out an American and Japanese comparative study that recreated, as closely as possible in Japan, the conditions obtaining in Lippitt and White's research done under Lewin's guidance.

12.1 AN INTERCULTURAL APPROACH TO UNDERSTANDING LEADERSHIP

Soon after the war, the General Headquarters of The Supreme Commander for The Allied Powers (SCAP) carried out a series of programmes at universities in major cities in Japan under the title of Institute for Educational Leaders (IFEL). It was through the IFEL programme that we became familiar with the work of Kurt Lewin (1939) and the procedures and results of Lippitt and White (1943). After having participated in the programme, we had doubts as to whether the results obtained in the experiments conducted in the United States could be duplicated in Japan, so we decided to undertake a follow-up study in Fukuoka.

The children in our study (Misumi and Haraoka, 1958; Misumi, Nakano and Ueno, 1958) were 10- and 11-year old boys, just as in Lippitt and White's study. We tried to match as closely as possible that study's experimental conditions of the three leadership types: democratic, autocratic, and laissez-faire. However, our working hypothesis was that group dynamics in Japan would differ from those in American society because of the difference in sociohistorical conditions between the two countries. We were surprised, however, to find almost the same group dynamics as Lippitt and White. The visible behaviour toward the leader differed, whereas the principle of behaviour exhibited by the children turned out to be exactly the same in both countries.

Although differences might be found in the concrete reactions of the two groups of children, Misumi and Haraoka (1958) and Misumi and Nakano (1960) obtained results very similar to Lippitt and White's (1943). The Japanese children, who had spent their important early formative years under militaristic totalitarianism, did not, on that account, prefer the autocratic leadership style. Under democratic leadership they showed relatively similar behavioural patterns to those of the American children, who were brought up in a democratic country.

In our second study (Misumi and Nakano, 1960), related to Lippitt and White, the children were given two different problem situations. In the first one they were to draw a picture together; in the second, they were asked to construct a model of their school. The finished products were evaluated by experts on arts and crafts. The results showed that the effects of leadership differed according to the conditions and nature of the task. That is, while doing relatively easy tasks such as drawing a picture, the democratic groups were found to be more effective than the autocratic and laissez-faire groups; in relatively difficult tasks, however, such as constructing a model of a school, the autocratic groups were the most effective.

Since no researcher in the field of leadership was advocating a contingency model at that time, our research can be considered a starting-point of the contingency approach. Fiedler's contingency theory (Fiedler, 1967) dates only from 1964. The tasks chosen for Lippitt's study were recreational, whereas ours were related to classroom curriculum. Consequently, Lippitt's tasks were interesting for the children, and conducive to voluntary participation, so they probably increased receptivity to democratic leadership. The effectiveness of leadership was found to differ in relation to the problem situation. Thus, leadership style itself should be considered contingent upon the task situation. The behavioural classification of the leadership types employed by Lippitt and White is commonly used. However, classification by common usage has limitations. First, it is unidimensional (democratic or autocratic, conservative or progressive, liberal or authoritarian, hawkish or dovish, employee-centred or production-centred) rather than multidimensional. Secondly, the terms used in this classification have multiple meanings in common usage, which makes them difficult to operationalise. Thirdly, these terms are heavily value-laden. Fourthly, the categories are used as historical and functional concepts.

12.2 DEVELOPMENT OF THE PM LEADERSHIP CONCEPT

As a remedy, we developed the concept that leadership is based on performance (P) and maintenance (M). This leadership PM concept

- allows multidimensional analysis;
- can be operationally defined;
- is itself value-neutral;
- makes experimental research and statistical studies possible.

Measuring leadership, which is very much a group phenomenon, requires a group-functional concept like the PM concept (Misumi, 1985).

In the concept of PM, P represents the kind of leadership that is oriented to achievement of the group's goals and to problem-solving. M stands for the kind of leadership that is oriented to the group's self-preservation, or maintenance, and to the strengthening of the group process itself. These two conceptual elements (P and M) are similar to Bales' (1953) *task leader* and *emotional leader*.

The concept of PM is a constructive concept to classifying and organising the factors obtained from leadership at different levels. It is not merely a descriptive concept for the factors obtained from factor analysis; it is at a higher level of abstraction. Because of this high level of abstractness, the PM concept applies to industrial organisations as well as to many other social groups. P does not concern production only, but also more general group goals or problem-solving tasks. This characteristic is what principally distinguishes it from Blake and Mouton's (1964) model.

In the case of PM concept, we consider P and M to be two axes on which the level of each type can be measured (high or low), thus obtaining four distinct types of leadership (see Figure 12.1). The validity of these four PM types was proved, using correspondence analysis, which was first developed by Guttman (1950) and later by Hayashi (1956).

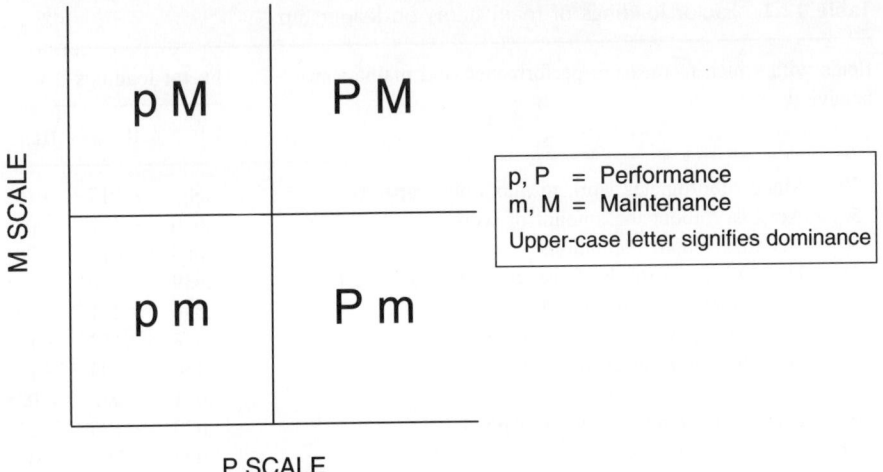

Figure 12.1 Conceptual representation of four patterns of performance-maintenance (PM) leadership behaviour.

Our research on the PM model consisted of both field surveys in different kinds of organisations and laboratory studies. Regarding measurement in the field, we found that subordinates' evaluation of their superiors was more valid than evaluation by superiors, peers, or self-evaluation. We therefore had subordinates evaluate the leadership of their superiors on the P and M dimensions.

To determine the level of P and M leadership for each subject, we first calculated the mean score of all subjects on each item of the two dimensions (P and M). As discussed by Misumi (1985), these P and M items, represented in Table 12.1, are the results of factor analysis. The first and third factors in Table 12.1 are the items with which to measure P behaviour, and the second factors are the items with which to measure M behaviour. A leader whose score in P and M is, for example, higher than the mean is thought to provide a leadership of the PM-type. A leader whose score is higher than the mean only in P dimension is classified as providing a P-type (or Pm-type) leadership. When a leader's score is higher than the mean only in M dimension, he is referred to as an M-type (pM-type). When a leader receives a score lower than the mean in both dimensions, he is thought to provide leadership of the pm-type. This scoring results in our final four-type classification: PM, P, M, and pm.

To test the validity and reliability of these leadership categories in industrial organisations, we examined their relation with a variety of objective and cognitive variables, such as productivity, accident rate, rate of turnover, job satisfaction, satisfaction with compensation, sense of belongingness to company and labour union, team work, meetings quality, mental hygiene, and performance norms. More than 300,000 subjects were surveyed. As indicated in Table 12.2, the PM-type was found to provide the best results of the four types, and the pm-type the worst. In the long run the M-type ranks second, and in the short run, the P-type ranks second.

Table 12.1 Factor loadings of main items on leadership

Items with which to measure performance and maintenance behaviour	Factor loadings		
	I	II	III
59. Make subordinates work to maximum capacity	.687	−.017	−.203
57. Act fussy about the amount of work	.670	−.172	.029
50. Act fussy about regulations	.664	−.072	.001
58. Demand that a job be finished within the time limit	.639	.070	.065
51. Give orders and instructions	.546	.207	.198
60. Blame the poor job on the employee	.528	.113	−.121
74. Demand reporting on the progress of work	.466	.303	.175
86. Support subordinates	.071	780	.085
96. Understand subordinates' viewpoint	.079	.775	.229
92. Trust subordinates	.024	.753	−.003
109. Favour subordinates	.067	.742	−.050
82. Subordinates talk to the superior without any hesitation	−.026	.722	.059
101. Concerned about subordinates' promotion, pay-raise, etc.	.147	.713	.134
88. Show consideration for subordinates' personal problems	.132	.705	.150
94. Express appreciation for job well done	.058	.651	.129
104. Impartial to everyone in work group	−.143	.644	.164
95. Ask subordinates' opinion of how on-the-job problems should be solved	.049	.643	.121
85. Make efforts to fill subordinates' request when they request improvement of facilities	.110	.606	.333
81. Try to resolve unpleasant atmosphere	.233	.538	.338
87. Give subordinates jobs after considering their feelings	−.276	.478	.457
76. Work out detailed plans for accomplishment of goals	.229	.212	.635
75. Waste no time because of inadequate planning and processing	.038	.333	.614
70. Inform subordinates of plans and contents of the work for the day	.254	.278	.607
52. Set time limit for the completion of the work	.319	.299	.554
53. Indicate new method of solving the problem	.251	.489	.479
56. Show how to obtain knowledge necessary for the work	.295	.492	.472
61. Take proper steps for an emergency	.360	.451	.305
69. Know everything about the machinery and equipment subordinates are in charge of	.255	.304	.458

This order of effectiveness is not limited to businesses only: it is the same for teachers (Misumi, Yoshizaki and Shinohara, 1977), government offices (Misumi, Shinohara and Sugiman, 1977), athletics coaches (Misumi, 1985), and religious groups (Kaneko, 1986).

What we have just explained is the general introduction to the Leadership PM Theory.

Table 12.2 Summary of comparison of the effectiveness of four patterns of PM leadership behaviour on various kinds of factors of work group

Factor	Pattern of leadership behaviour (ranking of effectiveness in each factor)			
	PM[a]	M[b]	P[c]	pm[d]
Productivity				
Long-term	1	2	3	4
Short-term[e]	1	3	2	4
Accidents[f]				
Long-term	1	2	3	4
Short-term[e]	1	3	2	4
Turnover	1	2	3	4
Group norm for high performance				
Long-term	1	2	3	4
Short-term[e]	1	3	2	4
Job satisfaction (a narrow sense)	1	2	3	4
Satisfaction with salaries	1	2	3	4
Team work	1	2	3	4
Evaluation of work group meeting	1	2	3	4
Loyalty (belongingness) to				
Company	1	2	3	4
Labour union	1	2	3	4
Communication	1	2	3	4
Mental hygiene (excessive tension and anxiety)[g]	1	2	3	4
Hostility to supervisor[h]	1	2	3	4

[a] Above-average score in both performance and maintenance. [b] Above-average score in maintenance. [c] Above-average score in performance. [d] Below average score in both performance and maintenance. [e] The data obtained by laboratory studies are included. [f] Smaller figures indicate lower rate of accidents or turn over. [g] Smaller figures indicate less tension and anxiety. [h] Smaller figures indicate less hostility to supervisor.

12.3 GROUP DECISION AND ACCIDENT PREVENTION

We turn to our action research on accident prevention and explain our leadership research. The hypothesis was that accidents will not necessarily always happen, even if one engages in unsafe activities in manufacturing factories or drives unsafely in the course of day-to-day work. These unsafe activities thereby become habitual. In fact, these habitual, unsafe activities often lead up to accidents. Like eating habits, such unsafe activities are difficult to change.

To change these habitual unsafe activities, we used the idea of group decision-making that Lewin introduced into his action research. The first action research study was conducted with the Nishi-Nippon (Western Japanese) Railroad Company (hereafter referred to as Nishitetsu), which operates electric train and bus services and has its headquarters in Fukuoka City in southwestern Japan. Our research, which was conducted in 1963, involved only 45 of the company's bus drivers who had caused accidents during the past three years. They were assembled in one place, ostensibly for the purpose of training, and were divided into groups of five or six drivers each. Our research design called for the group decision procedure to consist of six steps:

- a small meeting of the drivers for the purpose of warming-up;
- a meeting to discuss their workplace problems;
- a meeting in which each group made a report on the results of its discussion;
- a second meeting to re-discuss workplace problems;
- a second meeting in which each group made a report on the results of its second discussion;
- a session of post-discussion self-decision.

At the final session each participant was required to determine, on his own, a concrete action goal aimed at expressing his intention before all other members of the group to which he belonged. Figure 12.2 puts these changes into a broader context so that the overall impact of the group decision procedure can be appreciated.

The Nishitetsu project inspired the Nagasaki Shipyard in Nagasaki City, which is not far from Fukuoka, to start a programme designed to allow employees to participate in decision-making concerning their weekly task assignments. The Nagasaki Shipyard programme proved to be successful, and it, in turn, affected the Nishitetsu programme in a positive way. Thus, the two programmes became mutually supportive.

As indicated in Figure 12.2, the group decision programme aimed at accident prevention was conducted at a time when Japan was just beginning to achieve high economic growth, a period during which both the supply and the demand for bus services were increasing rapidly. The number of buses in operation and the distance people travelled by bus increased more during that period than at any time in the past. Accidents also increased accordingly. Thirty years have elapsed since then. In the ten years following 1968, when the accident rate decreased sharply, the annual accident rate was lower than at any time during the ten years preceding 1968. Furthermore, Nishitetsu's accident rate during that period was lower than that of any other bus company in Japan.

In 1995 leadership evaluation was again conducted, this time with 4110 bus drivers at Nishitetsu. They evaluated the leadership of their supervisors (business office directors and their second-in-command managers in the company's 47 business offices) and we measured the drivers' attitude (e.g. morale and satisfaction with the job) dividing the business offices into those that had had 'no accident' and

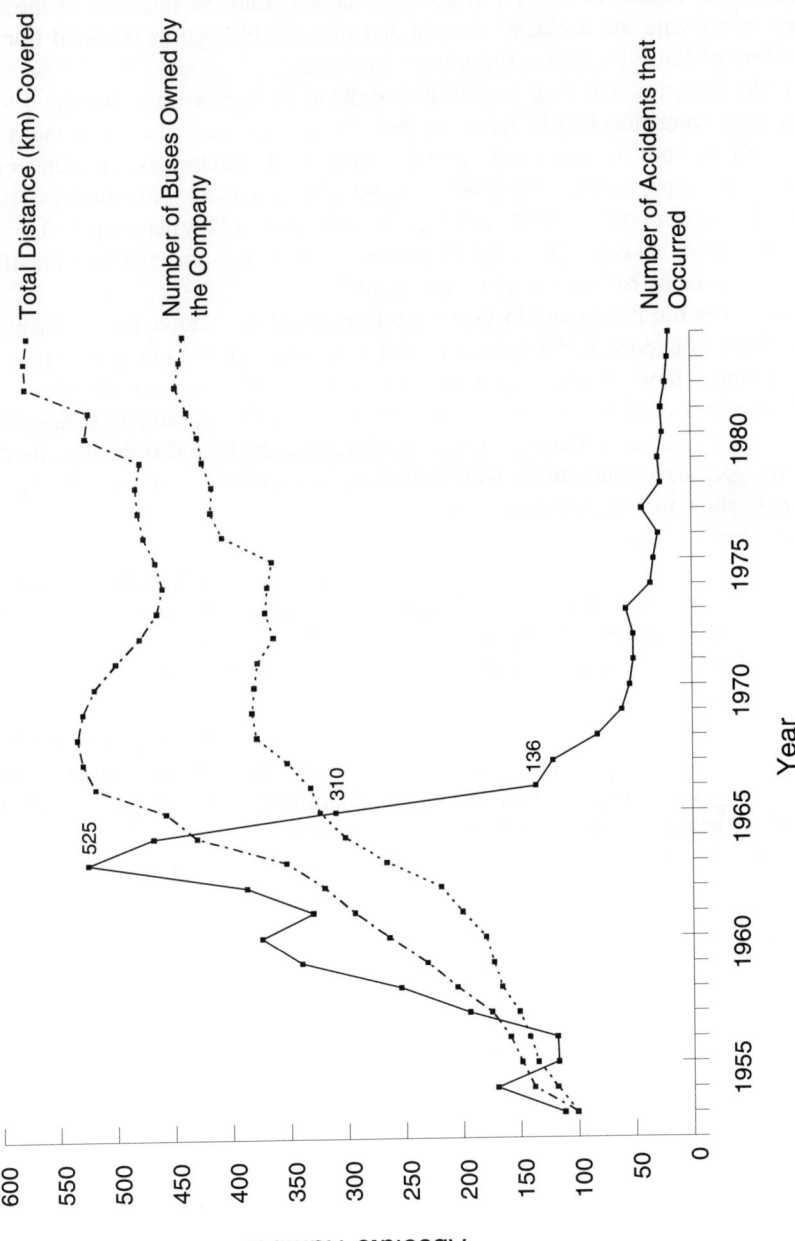

Figure 12.2 Fluctuation of road accidents attributable to bus drivers of a Japanese transport company (indices, using 1952 as 100).

those that had had one or more 'accidents' during the one and a half years prior to the survey. The respective leadership PM scores were then compared.

As shown in Figure 12.3, the PM scores obtained from the directors of those business offices with 'no accident' were higher than the PM scores obtained from the directors of those business offices with 'accidents'.

Next, the directors and their second-in-command managers were examined to see how they were matched in terms of their leadership types. In 19 of the 47 business offices both directors and second-in-command managers were either of PM type or pm type, that is, 'PM-PM' or 'pm-pm' respectively. Nine of these 19 business offices were of 'PM-PM' and 2 of these 9 offices (22.2 per cent) had one or more accidents, whereas 10 of the 19 offices were of 'pm-pm' and 8 of the 10 offices (80 per cent) had one or more accidents.

In this survey the P-type and M-type were few, and it was shown that accidents are less likely to happen under leaders of PM-type, and that the pm-type is more accident-prone. These results support the results obtained by Misumi (1984).

Nishitetsu buses are operated by one person and their drivers have little day-to-day, personal contact with the directors. It is, therefore, unlikely that the directors' leadership types have much to do with their drivers' accidents. However, the data of our study show that the top leadership in the workplace has a considerable effect on the incidence of accidents.

After the action research at Nishitetsu, the same method was introduced to conduct large-scale action research at the Nagasaki Shipyard (8000 employees) of the Mitsubishi Heavy Industries Company. As shown in Figure 12.4, the 172 accidents at the outset of our action research drastically decreased to 4 over the 10-year period involved.

Leadership surveys were conducted in 1995 at three nuclear power plants with the cooperation of the Kansai Electric Power Company. Drawing on the data obtained from the evaluations of leadership by subordinates, we carried out leadership training, examining the effects of group decision in practice. With the data thus obtained, training, consisting of various steps of group decision-making, was implemented.

As indicated in Figure 12.5, the leadership training began with the first survey, and continued over a period of six months. In order to avoid any excessive burden on the participants during the programme, the training programme was made up of steps 2, 5, and 8, which together required a total of four full days. At the final stage of the three training sessions, the participants made self-decisions about action goals for each of them to put into practice at their workplaces. The first two days of training were devoted to basic training, during which the participants learned about leadership, group theory, and technique for data analysis and analysed their own confidential data feedback. At the end of the two-day programme, the participants made a group decision. Subsequently, leadership surveys were conducted twice at an interval of about three months. On both occasions a one-day collective training session was held to analyse changes in leadership scores, and upon conclusion of the programme, the participants made a group decision. The results are shown in Figure 12.6, which clearly indicates changes in leadership behaviour.

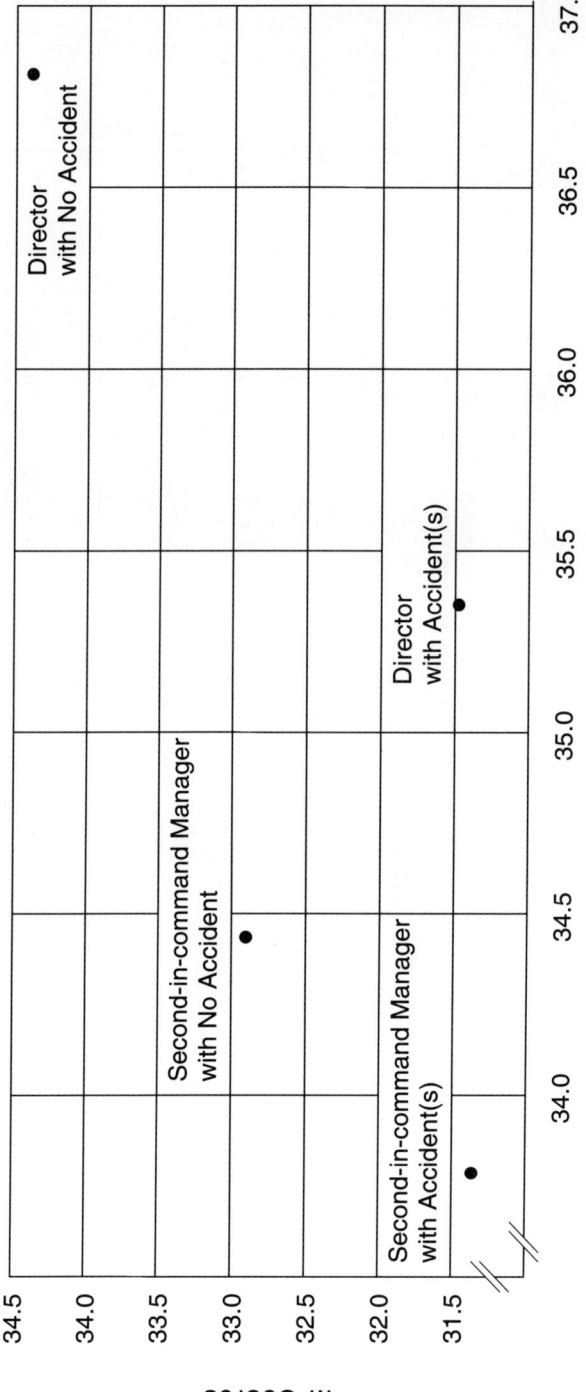

Figure 12.3 Performance (P) scores and Maintenance (M) scores of directors and second-in-command managers with accidents and those with no accidents.

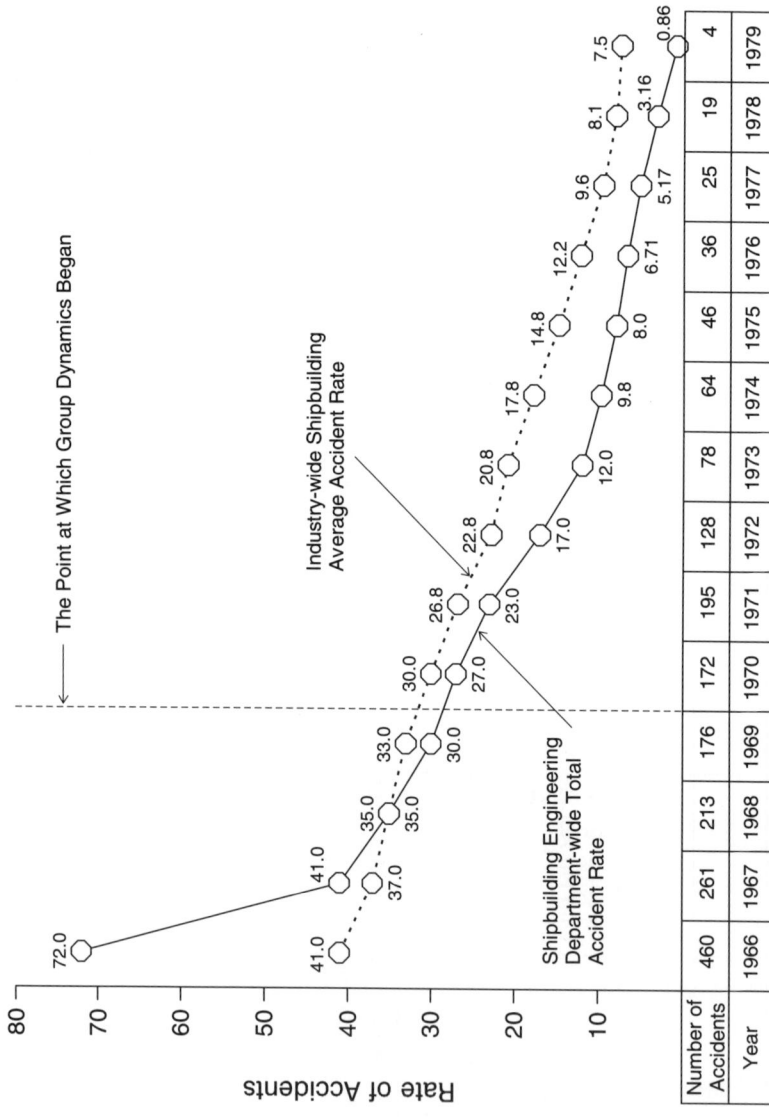

The Point at Which Group Dynamics Began

Industry-wide Shipbuilding Average Accident Rate

Shipbuilding Engineering Department-wide Total Accident Rate

Rate of Accidents

80 — 70 — 60 — 50 — 40 — 30 — 20 — 10 —

| Number of Accidents | 460 | 261 | 213 | 176 | 172 | 195 | 128 | 78 | 64 | 46 | 36 | 25 | 19 | 4 |
| Year | 1966 | 1967 | 1968 | 1969 | 1970 | 1971 | 1972 | 1973 | 1974 | 1975 | 1976 | 1977 | 1978 | 1979 |

Figure 12.4 Shipbuilding department-wide accident rate by year.

(The shipbuilding industry-wide accident rate is based on the statistics released by the Japanese Shipbuilding Industry Association, which encompasses 30 major shipbuilders.)

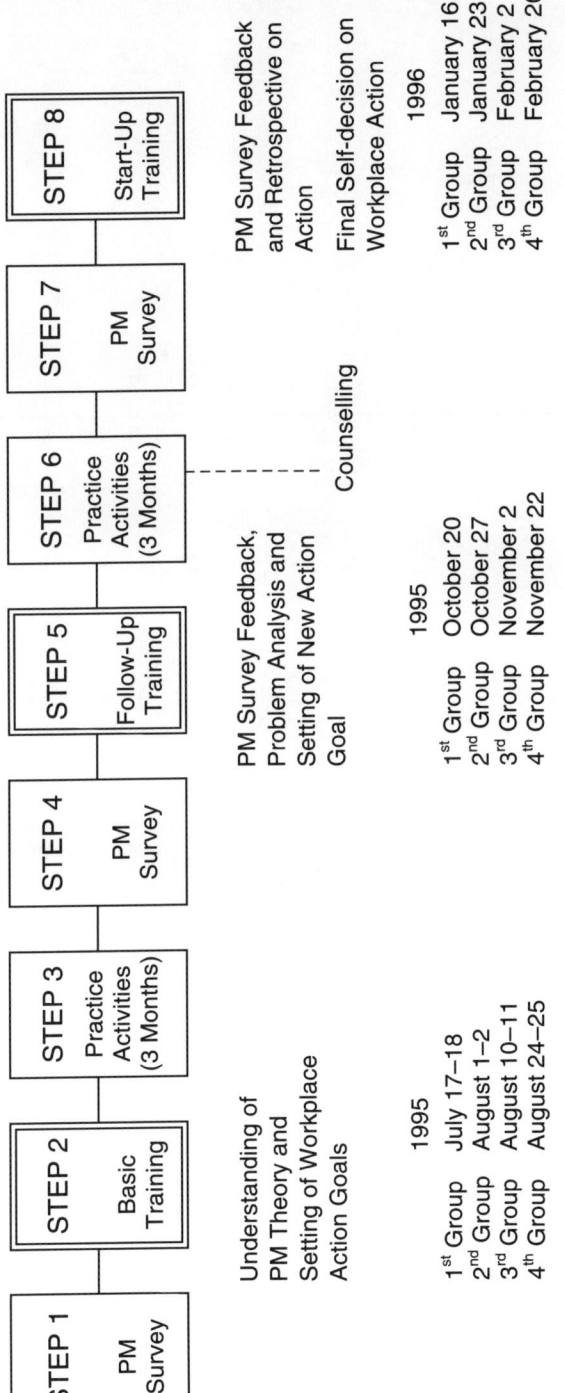

Figure 12.5 Overall 'flow' (composition) of the leadership training programme.

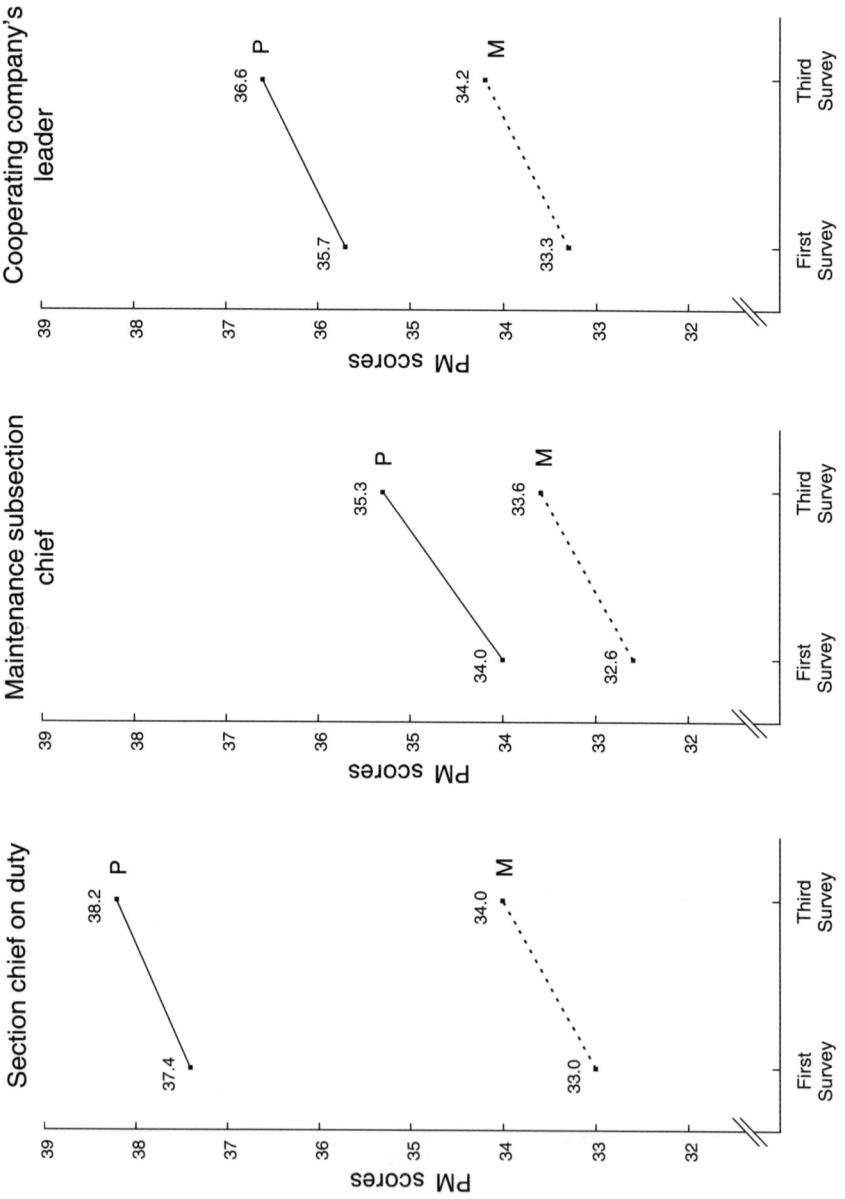

Figure 12.6 Changes in Performance (P) scores and Maintenance (M) scores during leadership training.

12.4 CONCLUSION

In this chapter, we discussed the action research on leadership and group decision that was launched after World War II. As regards leadership, Misumi and his colleagues constructed the PM theory, on which an examination of the relationship between leadership types and various factors in organisations was based. As a result, it was proved that leadership has influence on many factors in workplace situations, such as subordinates' morale and the incidence of accidents in workplaces. We simultaneously developed action research to prevent workplace accidents. In so doing, group decision-making and leadership factors were introduced. These efforts resulted in a drastic decrease in accidents attributable to one transport company's bus drivers, and in a similar reduction in employees' accidents in a shipyard. In the process it became clear that the incidence of accidents caused by employees varies with the leadership types represented by the superiors of the employees involved. Thus, leadership training at nuclear power plants was developed so as to improve leadership in preventing accidents. Although it is extremely difficult to eliminate accidents completely, it is incumbent upon everyone to continue to grapple with the task of accident prevention in keeping with the spirit of scientific research into human behaviour.

References

BALES, R. F. (1953) The equilibrium problem in small groups. In PARSONS, T., BALES, R. F. and SHILS, E. A. (eds), *Working papers in the theory of action* (pp. 111–61). Glencoe: Free Press.

BLAKE, R. R. and MOUTON, J. S. (1964) *The managerial grid*. Houston: Gulf.

FIEDLER, F. E. (1967) *A theory of leadership effectiveness*. New York: McGraw-Hill.

GUTTMAN, L. (1950) The principal components of scale analysis. In STUFFER, S. (ed.), *Measurement and prediction* (pp. 312–61). West Sussex, UK: John Wiley.

HAYASHI, C. (1956) Theory and examples of quantification (II). *Proceedings of the Institute of Statistics and Mathematics*, (Vol. 4, pp. 19–30). Tokyo: Institute of Statistics and Mathematics, Japanese Ministry of Education.

KANEKO, S. (1986) Religious consciousness and behaviour of religious adherents' representatives. (Report No. 65–86). Kyoto Survey Research Center of Jodo-shin-syu.

LEWIN, K. (1939) Experiments in social space. *Harvard Educational Review*, **9**; pp. 21–32.

LEWIN, K. (1947) Group decision and social change. In NEWCOMB, T. M. and HARTLEY, E. L. (eds), *Readings in social psychology* (3rd edn, pp. 197–211). New York: Holt, Rinehart and Winston.

LIPPITT, R. and WHITE, R. K. (1943) The social climate of children's groups. In BARKER, R. G., KORNIN, J. S. and WRIGHT, H. F. (eds), *Child behaviour and development* (pp. 485–508). New York: McGraw-Hill.

MISUMI, J. (1984) *The behavioural science of leadership* (2nd edn). Tokyo: Yuhikaku.

MISUMI, J. (1985) *The behavioural science of leadership*. Ann Arbor, MI: University of Michigan Press.

MISUMI, J. and HARAOKA, K. (1958) Experimental Study of Group Decision-II. *Research Bulletin Kyushyu University, School of Education*, **5**, pp. 61–81.

MISUMI, J. and NAKANO, S. (1960) A cross-cultural study of the effect of democratic, authoritarian, and laissez-faire atmospheres in children's groups (II). *The Japanese Journal of Educational and Social Psychology*, **1**, pp. 10–22.

MISUMI, J., NAKANO, S. and UENO, Y. (1958) A study of group dynamics concerning class atmosphere (2): A cross-cultural study concerning effects of autocratic and laissez-faire leadership types. *Research Bulletin of Kyushu University*, **5**, pp. 41–57.

MISUMI, J., SHINOHARA, H. and SUGIMAN, T. (1977) Measurement of leadership behaviour of administrative managers and supervisors in local government offices. *The Japanese Journal of Educational and Social Psychology*, **16** (2), pp. 77–98.

MISUMI, J., SHIRAKASHI, S., ANDO, N. and KUROKAWA, M. (1961) A study of the validity of the Uchida-Krapelin mental test. *The Japanese Journal of Educational and Social Psychology*, **2**, pp. 1–21.

MISUMI, J., YOSHIZAKI, S. and SHINOHARA, S. (1977) The study on teacher's leadership: Its measurement and validity. *The Japanese Journal of Educational and Social Psychology*, **25**, pp. 157–66.

CHAPTER THIRTEEN

Are we casting the net too widely in our search for the factors contributing to errors and accidents?

JAMES REASON

Department of Psychology
University of Manchester

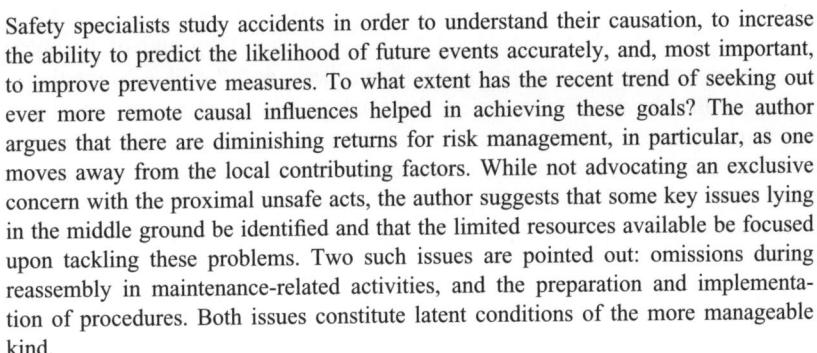

Safety specialists study accidents in order to understand their causation, to increase the ability to predict the likelihood of future events accurately, and, most important, to improve preventive measures. To what extent has the recent trend of seeking out ever more remote causal influences helped in achieving these goals? The author argues that there are diminishing returns for risk management, in particular, as one moves away from the local contributing factors. While not advocating an exclusive concern with the proximal unsafe acts, the author suggests that some key issues lying in the middle ground be identified and that the limited resources available be focused upon tackling these problems. Two such issues are pointed out: omissions during reassembly in maintenance-related activities, and the preparation and implementation of procedures. Both issues constitute latent conditions of the more manageable kind.

This chapter is concerned with establishing a workable balance between local factors and more remote influences in the search for the human causes of accidents involving complex well-defended systems, such as nuclear power plants and modern commercial aircraft. This topic is prompted by a suspicion that the pendulum may be swinging too far in present attempts to track down possible error and accident contributions that are widely separated in both time and place from the events themselves. The relative worth of the various causal categories is evaluated by reference to three questions central to the pursuit of system safety: to what extent does the consideration of individual, contextual, organisational, systemic, and societal factors add value to:

- an understanding of the causes of accidents and events?

- the ability to predict the likelihood of future accidents and events?

- most important, the remedial efforts that are made to reduce their future occurrence?

Although it is clear that the present situation represents a major advance over knee-jerk human error attributions, some concerns need to be expressed about the theoretical and the practical utility of this ever-spreading quest for contributing factors. We safety specialists seem to have reached, or even exceeded, the point of diminishing returns, particularly when it comes to safety management. We need to find some workable middle ground that acknowledges both the cognitive and the contextual influences on human performance, as well as the interactions between active failures and the latent conditions that serve, on rare occasions, to breach the system's defences.

This discussion draws heavily on aviation experiences. It is my contention that aviation and nuclear power generation have, from a systems safety point of view, more common features than they have differences. As such, everyone can only benefit from the richer data provided by these inquiries into both domains.

13.1 THE SEARCH FOR MORE REMOTE CAUSES

Twenty years ago the search for the human causes of accidents usually ended with the identification of errors committed by those at the sharp end of the system: control room operators, pilots, air traffic controllers, and the like. Since then, we safety specialists have come to realise that such proximal errors are consequences rather than causes. Their identification represents the beginning rather than the end of the search for the explanations of accidents. We have also come to appreciate that the mental states immediately preceding an error (e.g. momentary inattention, forgetfulness, preoccupation, and distraction) are the last and least manageable parts of a causal history that reaches back to the nature of the task, the local conditions within the workplace, and the upstream systemic factors that shaped those conditions.

These developments have been summarised in the distinction between active failures and latent conditions. Active failures are the local unsafe acts (errors and violations) committed by those at the human-system interface. They can have an immediate impact on safety. Latent conditions, on the other hand, arise from top-level decisions and organisational processes. They are present in all systems, and manifest themselves in a variety of ways: poor design, undue time pressure, unworkable procedures, inadequate training, conflicting goals, clumsy automation, unsuitable equipment, and so on. These potentially adverse conditions may lie dormant for long periods of time, becoming critical only when they combine with

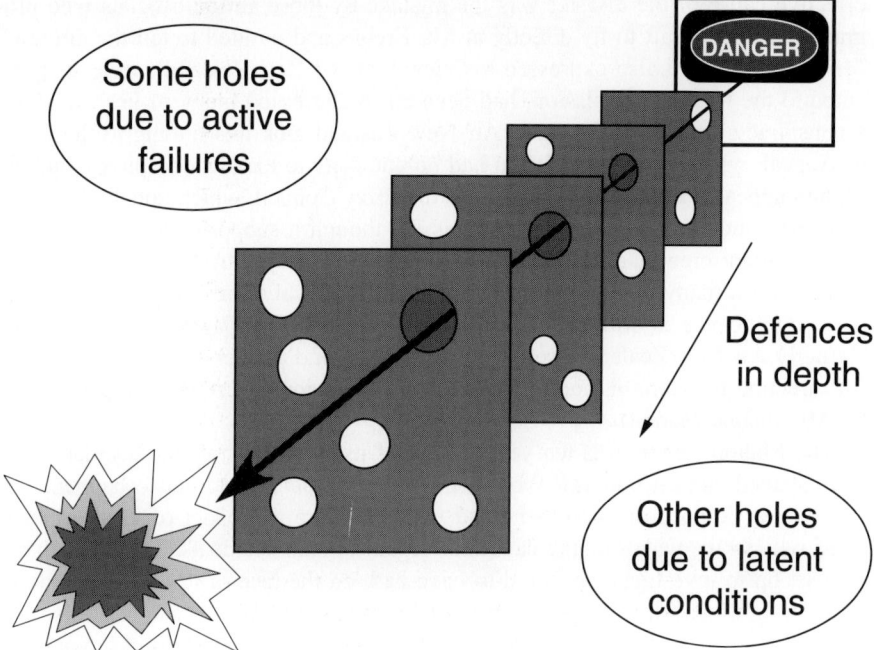

Figure 13.1 Showing how both active and latent failures create holes in a system's multiple defences, barriers and safeguards, allowing hazards to come into damaging contact with victims, assets and the environment.

active failures and local triggers to penetrate the aviation system's many defences, barriers, and safeguards (see Figure 13.1).

   Although these broader systemic issues have come to dominate the concerns of safety specialists only in recent years, there is little new in the ideas themselves. Far more important in creating the present interest in organisational, cultural, and systemic issues were the accident inquiries into the Three Mile Island, Challenger, Bhopal, Chernobyl, Zeebrugge, Clapham Junction, King's Cross Underground, Exxon Valdez, and Pipe Alpha disasters, to name but a few. Within aviation, two of the most influential investigations have been Mr. Justice Mahon's Royal Commission Report into the Air New Zealand DC10 crash at Mount Erebus on 18 November 1979 (Mahon, 1981) and Mr. Justice Moshansky's Commission of Inquiry into the Air Ontario F-28 crash at Dryden, Ontario, on 10 March 1989 (Moshansky, 1992). All these accidents have been discussed at length elsewhere. For the present purposes of this chapter, it is sufficient to recall some of the main conclusions.

   One of the most significant features of the Mahon Report is that it completely rejected the findings of an earlier investigation carried out by the chief investigator of New Zealand's Office of Air Accident Investigations (1980), who concluded that the crash had been due almost entirely to flight crew errors. The Mahon Report, on the contrary, found no fault with the pilots. It concluded: 'The single dominant and effective cause of the disaster was the mistake by those airline officials who pro-grammed the aircraft to fly directly at Mt. Erebus and omitted to tell the aircrew'. Mr. Justice Mahon also expressed his view that Air New Zealand's case, as pres-ented to the Royal Commission, had been an 'orchestrated litany of lies'; in short, a conspiracy to conceal the truth. Air New Zealand took this finding to the Court of Appeal, which ruled that Mahon had no authority to make a conspiracy finding. Mahon appealed against this decision to the Privy Council, which upheld the Court of Appeal judgment on the conspiracy issue, though it supported Mahon's finding that the flight crew had not been at fault. Unfortunately, Mr. Justice Mahon died in 1986 with many of these organisational and political issues unresolved, though their significance is still being pursued energetically by Captain Gordon Vette, formerly Air New Zealand's senior training pilot, and the person most responsible for directing the Commission to its eventual conclusions (as readily acknowledged by Mr. Justice Mahon).

The Mahon Report was ten years ahead of its time. Most of the accidents that have shaped current thinking about organisational factors had yet to happen. The Moshansky Report, on the other hand, was far more a product of its time in its sensitivity to the wide-ranging causes of organisational accidents. The crash that it was set up to investigate appeared to have had, on the face of it, all the hallmarks of a straightforward pilot error accident. A commuter F-28 took off in snowy conditions without being de-iced, and crashed just over half a mile beyond the runway due to the accumulation of ice on its wings. In another age, the flight crew's flawed decision to take off under such conditions could have been the end of the causal story. But Mr. Justice Moshansky interpreted his mandate very widely. In order to make the required recommendations in the interest of air safety, he felt it necessary 'to conduct a critical analysis of the aircraft crew, of Air Ontario [the carrier], of Transport Canada [the regulator] and of the environment in which these elements interacted' (p. xxv). After considering the evidence provided by 166 wit-nesses, the 1700-page final report implicated the entire Canadian aviation system:

> Had the system operated effectively, each of the [various causal] factors might have been identified and corrected before it took on significance. It will be shown that this accident was the result of a failure in the air transportation system.
>
> (Moshansky, 1992, pp. 5–6)

The need for air accident investigators to take account of the wider managerial and systemic issues was formally endorsed by the International Civil Aviation Organ-ization (ICAO) in February 1992. The minutes of the ICAO Accident Investigation Divisional Meeting reported the following discussion item:

> Traditionally, investigations of accidents and incidents had been limited to the persons directly involved in the event. Current accident prevention views supported the notion that additional prevention measures could be derived from investigations if manage-ment policies and organisational factors were also investigated.

The meeting concluded with a recommendation that a paragraph relating to man-agerial and organisational factors be inserted in the next edition of ICAO Annex 13 – the standards and recommended practices for air accident investigators throughout the world. The following paragraph appeared in the Eighth Edition of Annex 13 (Aircraft Accident and Incident Investigation, 1994):

> 1.17 Management information. Accident reports should include pertinent information concerning the organisations and their management involved in influencing the operation of the aircraft. The organisations include, for example, the operator; the air traffic services, airway, aerodrome and weather service agencies; and the regulatory authority. The information could include, but not be limited to, organisational structure and functions, resources, economic status, management policies and practices, and regulatory framework.   (p.18)

A short while after the publication of the Eighth Edition of Annex 13, an accident investigation adopting this particular recommendation played a significant role in producing a radical change in the Australian civil aviation system. The Bureau of Air Safety Investigation cited regulatory surveillance failures by the Australian Civil Aviation Authority as having contributed to the crash of a small commuter aircraft at Young, New South Wales, in June 1993. As the result of this finding and other factors, the then (Liberal) government disbanded the Australian CAA in 1995, and later reconstituted it as the Civil Aviation Safety Agency.

Perhaps the high water mark of causal attribution was reached in April 1988. Academician Valeri Legasov was the principal investigator of the Chernobyl disaster in April 1986. He was also the Soviet Union's chief spokesman at the international conference on this accident, held in Vienna in September 1986. At that meeting he blamed the disaster upon the errors and violations of the operators. Later, he confided to a friend, 'I told the truth at Vienna, but not the whole truth' (Legasov, 1988). In April 1988, two years to the day after the Chernobyl disaster, he hanged himself from the balustrade of his apartment. Prior to his suicide, he recorded on tape his innermost feelings about the accident. In one tape, he stated: 'After being at Chernobyl, I drew the unequivocal conclusion that the Chernobyl accident was . . . the summit of all the incorrect running of the economy that had been going on in our country for many years'.

13.2 HAS THE PENDULUM SWUNG TOO FAR?

I have traced the ever-widening spread of causal fallout from the time when it was sufficient simply to blame the individuals immediately concerned, through various organisational and system factors, to Legasov's belief that the Chernobyl accident was rooted in the failings of an entire society. The question now is whether this process has gone too far.

There is every reason to believe that Legasov was right in his evaluation. But how does this information help safety specialists? It is impossible to go back to 1917 and replay the years following the Russian Revolution. Although Legasov's verdict was correct, it was unlikely to lead to improvements. Models of accident causation can be judged only by the extent to which their applications enhance the safety of a system. The economic and societal shortcomings, identified by Legasov, are beyond the reach of system managers. From their perspective, such problems

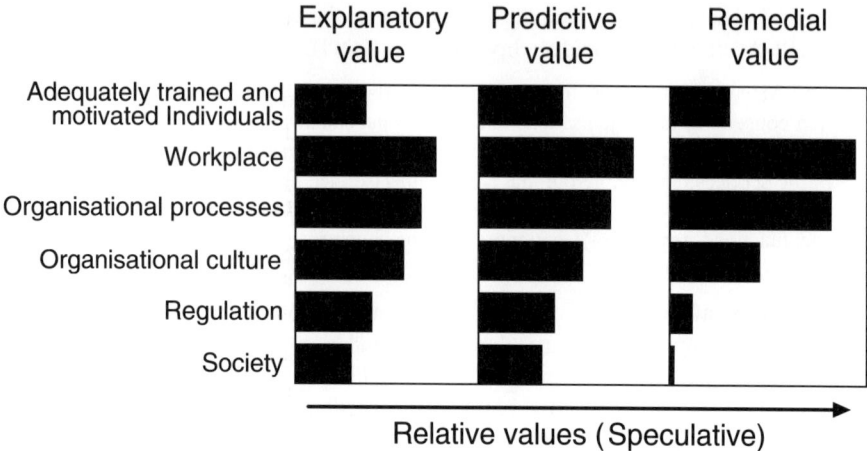

Figure 13.2 Relative values that various causal factors are believed to have for explanation, assessment and management.

are given and immutable. But the main interest of safety specialists has to be what is changeable and controllable. Figure 13.2 summarises the arguments presented so far. In particular, it focuses on the three issues stated at the outset and raises a question: what are the relative values that individual and workplace factors, organisational processes and organisational culture and regulatory and societal issues are judged to have for explanation, assessment, and management. The general conclusions can be summarised as follows:

- Individual factors alone have only a small to moderate value for all three goals.
- Overall, workplace and organisational factors contribute the most added value.
- There are rapidly diminishing returns on pursuing the more remote influences, particularly in regard to countermeasures and risk management.

13.3 THE CONCEPT OF LATENT CONDITION: MINUSES AND PLUSES

The notion of latent conditions, or resident system 'pathogens' is widely used in a number of hazardous domains. Although it is a concept that is hard to avoid in any consideration of accident causation in complex technological systems, one needs to recognise both its minuses and pluses if its full value is to be appreciated. Some of its limitations are listed below:

- Latent conditions are present in all technological systems. Their existence does not discriminate between normal states and accidents. As such, they are more properly conditions rather than causes (see Hart and Honoré, 1985).
- The more exhaustive the accident inquiry, the more latent conditions it will discover.
- Only local events will determine whether there is a bad outcome.

It is the contention of this chapter, however, that these minuses are more than compensated for by the following pluses:

- Latent conditions undoubtedly combine with local triggers to penetrate, on rare occasions, the many defences, barriers, and safeguards characteristic of modern technological systems.

- Potential pathogens within the workplace and the organisation can be identified and removed *before* the event.

- Local triggers and unsafe acts (active failures) are hard to anticipate and almost impossible to defend against completely.

In theory, there is no limit to how far safety specialists can pursue remote causes of organisational accidents and other catastrophic events. Historians, for example, routinely trace the origins of the world wars back through hundreds of years. In practice, however, accident investigators need to establish principled boundaries (or stop rules) to their investigations. The following list provides some guidelines for establishing these stop rules.

- Leave the more distant influences to historians, philosophers, and political scientists.

- Focus on manageable issues. The essence of risk management is to control the controllable, to manage the manageable.

- Stay within the influence range of the system's managers and regulators.

- Concentrate on identifying and reducing the tractable problems.

13.4 HOW CAN THE MANAGEABLE PROBLEMS BE IDENTIFIED?

There appear to be two ways of dealing with this issue. First, one can specify in advance which types of human activity are most likely to attract performance problems of one kind or another. Secondly, one can review the available data linking human performance problems to the various kinds of tasks performed by people within complex systems. Where both the *a priori* predictions and reality correspond, one can be fairly confident of having identified a major problem area upon which to target the limited remedial resources.

It is suggested that there are three key issues in establishing the degree to which any human activity is likely to be associated with less-than-adequate performance.

- Which human activities involve the most hands-on contact with the system?

- Which activities, if performed less than adequately, pose the greatest risks to the integrity of the system?

- How often is any particular activity performed?

In addition, one can identify at least four types of human activity that are universal in the area of hazardous technologies.

Table 13.1 Assessing the degree to which universal sharp-end activities are likely to attract human performance problems

Activity	Hands on	Criticality	Frequency
Normal control	Low	Low	High
Abnormal control	Moderate	High	Low
MCZM	High	High	High

- Controlling the system under normal conditions.

- Controlling and recovering the system under abnormal or emergency conditions.

- Carrying out corrective and preventive maintenance, calibration, testing, and system modification (hereafter, MCTM).

- Preparing and applying procedures, regulations, and administrative controls.

Of these types of human activity, the first three cluster together as sharp-end activities – that is, activities performed by those in direct contact with the system. For analytical reasons and because of the limited scope of the present chapter, it is convenient to treat these sharp-end issues separately from the procedure-related ones. Table 13.1 summarises my beliefs about how each of the first three activities score on the problem-attracting characteristics listed above.

Table 13.1 makes it clear that, on *a priori* grounds, MCTM is the activity most likely to attract human performance problems because it scores high on all three problem-attracting criteria. How well is this prediction supported by the data?

The list below summarises the data from three studies carried out within the nuclear power generation industry, two by the Institute of Nuclear Power Operations in Atlanta (INPO, 1984, 1985) and one by the Central Research Institute of the Electrical Power Industry in Tokyo (K. Tanako, personal communication). In each case, the analyses were performed upon significant event reports. The percentages represent the proportions of the total number of human performance problems associated with each type of activity.

- Maintenance, calibration, testing, and modification (mean: 60 per cent; range: 55–65 per cent)

- Normal plant operations (mean: 16 per cent; range: 8–22 per cent)

- Abnormal and emergency operations (mean: 5 per cent; range: 2–8 per cent)

These data clearly support the prediction that MCTM attracts, on average, around 60 per cent of all the identified performance problems contributing to off-normal states in nuclear power plants. But one can go further to identify which aspect of MCTM disassembly or reassembly is most provocative of unsafe acts. Once again, *a priori* considerations clearly indicate that the reassembling or installation of components is the most vulnerable to human performance problems. For example, imagine that the task is to remove eight marked nuts from a bolt and then to replace them in a predetermined order (a fair model of many MCTM activities). There is really only one way in which the items can be disassembled, but there are at least 40 000 ways (factorial 8) in which they can be reassembled in the wrong order, and this figure takes no account of omissions. The problem is further compounded by the fact that the act of reassembly conceals prior errors, thus reducing their chances of detection and correction.

To what extent do the available data bear out this prediction? Several independent surveys of lapses in aircraft maintenance make it quite evident that the large majority of installation problems are associated with the act of reassembly (Boeing, 1994; Ingham, 1997; Pratt and Whitney, 1992; UKCAA, 1992). These problems comprise incomplete installation, damaged upon installation, and improper installation as well as the original equipment installation. The evidence both from nuclear power generation and aircraft maintenance indicates that omissions during installation, such as the failure to replace necessary components or to remove foreign objects, constitute the largest single category of human error throughout these systems as a whole (INPO, 1984; Rasmussen, 1980; Reason, 1995). As stated by the Institute of Nuclear Power Operations (INPO, 1984):

- 60 per cent of root causes (of significant events) attributable to human performance involved omissions.

- 65 per cent of errors in maintenance-related activities involved omissions.

- 96 per cent of deficient procedures involved omissions.

It is not within the scope of this chapter to discuss the ways in which these omission problems can be managed. They have been discussed at length elsewhere (Reason, 1995; Reason, 1997). It is sufficient for the purposes of this chapter to indicate (a) that the omission-provoking features of task steps are well known to psychologists and can thus be predicted in advance, and (b) most tasks likely to involve errors of omission can be effectively treated by the provision of suitable reminders.

13.5 SUMMARY AND CONCLUSIONS

1   The search for remote causal factors, errors and accidents has probably gone beyond the limits of utility, particularly with regard to safety management.

2   It is better to focus on the tractable problems that have their origins midway between proximal and more distal factors.

3   Two likely candidates for focused attention to human factors are (a) omissions during installation in maintenance-related activities and (b) the preparation and provision of procedures and documentation. This issue was not considered in the present chapter, but it has been discussed in detail elsewhere (Reason, 1997; Reason, Parker, Lawton and Pollock, 1995).

References

BOEING (1994) *Maintenance Error Decision Aid*. Seattle: Boeing Commercial Airplane Group.

HART, H. L. A. and HONORÉ, T. (1985) *Causation and the Law*. Oxford: Oxford University Press.

INGHAM, E. A. (1997) Human errors and their avoidance in maintenance. *Flight Operations/Aircraft Maintenance Standards Departments' Roadshow. April-May.*

INPO (Institute for Nuclear Power Operations) (1984) *An Analysis of Root Causes in 1983 Significant Event Reports* (INPO-84-027). Atlanta, GA: INPO.

INPO (Institute for Nuclear Power Operations) (1985) *An Analysis of Root Causes in 1983 and 1984 Significant Event Reports* (INPO-85-027). Atlanta, GA: INPO.

INTERNATIONAL CIVIL AVIATION ORGANIZATION (1994) *Aircraft Accident and Incident Investigation*. (ICAO Annex 13) 8th Edn. International Civil Aviation Organization, Montreal.

LEGASOV, V. (1988) *The Legasov Tapes*. Washington, DC: Department of Energy.

MAHON, P. (1981) *Report of the Royal Commission to Inquire into the Crash on Mount Erebus, Antarctica, of a DC10 Aircraft Operated by Air New Zealand Limited*. Wellington, New Zealand.

MOSHANSKY, V. (1992) *Commission of Inquiry into the Air Ontario Crash at Dryden, Ontario* (Final Report). Ottawa.

OFFICE OF AIR ACCIDENT INVESTIGATIONS (1980) *Air New Zealand McDonnell-Douglas DC10-30 ZK-NZP, Ross Island, Antarctica, 28 November 1979* (Aircraft Accident Report No. 79–139). Wellington, New Zealand: Office of Air Accidents Investigation.

PRATT AND WHITNEY (1992) *Open Cowl*. March issue.

RASMUSSEN, J. (1980) What can be learned from human error reports? In DUNCAN, K., GRUNEBERG, M. and WALLIS, D. (eds) *Changes in working life*. London: Wiley.

REASON, J. (1995) *Comprehensive Error Management in Aircraft Engineering: A Manager's Guide*. London Heathrow: British Airways Engineering.

REASON, J. (1997) *Managing the Risks of Organisational Accidents*. Aldershot: Ashgate.

REASON, J., PARKER, D., LAWTON, R. and POLLOCK, C. (1995) Organisational Controls and the Varieties of Rule-Related Behaviour. *Proceedings of the Economic and Social Research Council's Meeting on Risk in Organisational Settings*.

UKCAA (United Kingdom Civil Aviation Authority) (1992) Maintenance error. *Asia Pacific Air Safety*. September.

Group and individual performance

Introduction

In focusing on the nuclear power plant as a sociotechnical system, the third part of this volume directs attention to the two subsystems that operate at the 'sharp end' of the production process: the team and the individual. Nishijima collects and describes indicators for human performance in nuclear power plants. Job demands in nuclear power plants are seen as a key condition for the occurrence of human error. Fleishman and Buffardi investigate this connection, with the aim of improving the prediction of human error probabilities at the workplace.

Team aspects in the simulator training of nuclear power plant personnel are considered in the contribution by Antalovits and Izsó. Their computer-aided method helps improve the self-assessment of operators and of the control-room team after a training session. Sasou, Suzuki, and Yoshimura describe an experimental study in which the importance of training for individual and group performance is shown.

CHAPTER FOURTEEN

Human performance indicators

YOSHIMASA NISHIJIMA

Central Research Institute of the Electric Power Industry (CRIEPI)

The contribution addresses the challenging problem of defining human performance indicators for the nuclear industry. Starting from a set of general performance indicators developed by the World Association of Nuclear Operators (WANO), the author embarks on the development of specific human performance indicators that he identifies both on an individual and an organisational level in the three areas of safety, efficiency and welfare. For instance, safety-related human performance indicators are defined for safety-related attitudes and behaviour, the existence of an incident-reporting system, and accident and human error rates. Qualification, productivity and work quality parameters are some of the efficiency indicators. Parameters such as work ethics, the award and promotion system and work climate are specified as welfare indicators. Human error rates, including the difficulties involved in defining human error, are then addressed. Human error rates that appear to be lower in Japanese nuclear plants than in US ones are interpreted as differences between approaches to interpreting and classifying human performance problems as causal factors.

14.1 A MANIFEST NEED

In order to foster safety and reliability in nuclear power plants, human performance in these installations must be improved. Various steps are already being taken – one example being the introduction of the Human Performance Enhancement System (HPES), a system that the Institute of Nuclear Power Operations (INPO) has developed to assist specialists conducting root-cause analysis, which may help mitigate factors that impair human performance. Yet how far have safety specialists, engineers, and operators come toward the shared objective of enhancing the safety and reliability of operations at nuclear installations? If it is true that what cannot be measured cannot be managed, then answering that question makes it essential to identify ways of measuring improvements in safety at individual nuclear power stations and in the nuclear power industry as a whole. At that point it would be possible to monitor and compare progress. Analysts could then evaluate the effectiveness of each measure, and use the results to establish the new standards of human performance that are needed.

Devising a methodology for assessing or improving the knowledge and practice of safety makes it necessary to relate attributes and concepts of human performance to the individual plant's operational realities. A tool for achieving that link would be the development of indicators. World Association of Nuclear Operators (WANO) performance indicators are significant examples of this tool, because they are used by the nuclear power industry to monitor performance and progress and to set new goals for improvement, compare performance from one plant to the next, and detect possible need to adjust priorities and resources accordingly. A WANO performance indicator is based on one or more of the following criteria:

- It provides a quantitative measure of nuclear safety, plant reliability, plant efficiency, or personnel safety.
- It is widely applicable.
- It is objective and fair.
- It lends itself to goal-setting.
- The data on which it is based is available and reliable.
- Its emphasis on increasing the value of the indicator is unlikely to cause undesirable plant behaviour.
- If the indicator primarily monitors plant reliability, it should reflect performance only in areas that can be controlled or influenced by plant management.
- Indicators of a nuclear plant or of personnel safety should reflect overall plant performance, including, in some cases, elements beyond the control of plant management.

There are ten performance indicators in the WANO programme:

- unit capability factor;
- unplanned capability loss factor;
- unplanned automatic scrams per 7000 hours of operation;
- performance of the safety system;
- thermal performance;
- fuel reliability;
- chemical index;
- collective radiation exposure;
- volume of low-level solid radioactive waste;
- rate of industrial accidents.

Since 1990, when data collection for the programme began, WANO has steadily advanced toward its stated goal of assessing or improving the knowledge and practice of safety. These kinds of human performance indicators should also be developed for the nuclear power industry, so that methods can be conceived for monitoring and comparing progress on human performance within and between nuclear power plants throughout the industry.

14.2 AREAS OF HUMAN PERFORMANCE INDICATORS

Historically, the main aim of ergonomics has been to improve productivity, whereas the priority of industrial hygiene, in the sense of occupational safety and health, has been to protect the human being from various industrial hazards. Safety science differs from both these approaches in its emphasis on the effort to ensure safety by improving the working environment. Because human factors are the key elements in ergonomics, industrial hygiene, and safety science, one could say that the afore-mentioned safety, efficiency, and welfare are of equal importance in human performance indicators. Thus, the areas in which to develop indicators for monitoring the improvement of human performance should include aspects of those three areas at both the individual and organisational levels (see Table 14.1).

Table 14.1 Potential human performance indicators

Aspect	Individual	Organisational
Safety	*Safety consciousness*: safety attitude index *Safety behaviour*: injury/accident rate *Safety character*: safety personality diagnosis index	*Safety management*: incident-reporting system, safety officer, industrial accident rate, human error rate *Safety education*: education system, participation records *Safety activities*: award system, demonstration of good practices, kaizen (suggestion box)
Efficiency	*Ability*: qualification, training *Aptitude*: aptitude test	*Productivity*: capacity factor, overtime work, manager ratio *Work quality*: production cost
Welfare	*Work ethics*: absentee *Sense of belonging*: workforce stability, job change	*Award system*: award near-miss event reporting *Welfare system*: resort facility for employee *Promotion system*: merit system *Workplace climate*: communication, leadership *Work environment*: resting place, comfortableness

14.2.1 Safety

At the level of the individual, the awareness, practice, and attitude of the plant personnel with regard to safety may be important. An index of the personnel's attitude toward safety could serve as the human performance indicator of safety awareness, and the plant's rate of injury or incident could fulfill the same function in relation to the practice of safe behaviour. An index for diagnosing personality characteristics that correlate with safety could be the indicator of the safety-related character of plant personnel.

From the organisational standpoint, it is safety management, safety education, and safety activities that are important. Indicators of safety management could include the existence of a system for reporting incidents, the presence of a safety officer, the rate of industrial accidents in the plant, or the rate of human error.

14.2.2 Efficiency

In terms of the individual's efficiency, the ability and aptitude of plant personnel could be considered for a role as indicators of human performance. An employee's degree of qualification and training may serve as an indicator of ability. Results on an aptitude test can be suggested as an indicator of aptitude.

Organisationally speaking, at the very least, productivity and work quality should be considered indicators of organisational performance. In turn, potential indicators of productivity are such factors as capacity – i.e. nuclear plant capacity factor defined as the ratio of the available energy generation over a given period to the reference energy generation over the same period, expressed as a percentage – hours of overtime work, and manager/staff ratio. A potential indicator of work quality would be production cost.

14.2.3 Welfare

Work ethics and the sense of belonging could be taken as indicators of individual welfare. For organisational indicators of welfare, such aspects as an award system, the welfare system, the promotion system, the working climate, and the work environment are considered.

At the level of the individual, the rate of absenteeism could be an indicator of work ethics, and the turnover of the workforce could serve as an indicator of the sense of belonging. Organisationally, a performance indicator of the reward system could be an award for reporting a near-miss event. An indicator of the plant's welfare system could be the availability of a resort facility for employees. A merit system could serve as an indicator of the promotion system. Communication and

Table 14.2 Candidates from investigation items for identifying factors influencing organisational safety

Aspect	Focus of investigation	Content
Safety management	Incident reporting system	Formal or informal, method, scope (near-miss, actual, potential events), feedback, organisation
	Safety supervisor	Structure, number of person (ratio), full-time or part-time, competent Individual system (i.e. STOP in Dupont)
	Other safety programmes	Administrative rules for safety, encouragement of bottom-up activities
Safety education	Education system	Curriculum, materials (media, text), methodology (lecture, field training, and so forth), frequency, term
	Participants	Scope (manager, staff, worker level), frequency, interval
Safety activities	Award system	Object frequency
	Daily activities	TBM (tool box meeting), KY (anticipating possible danger), 4s%5s (tidy up, clean up, discipline, and so on), TQC, small group discussion, touch and call, self-surveillance
	Kaizen system	Methodology, feedback

leadership could be taken as an indicator of the working climate, and comfort or a place to rest could be an indicator of the work environment.

Seeking efficiency is, however, sometimes at odds with preserving safety and promoting welfare. The most desirable solution to this problem is to balance efficiency and safety. Because the fulcrum generally tends to move toward the efficiency side, continuous pressure is needed to push it toward the side of safety. The effort to find the point of balance is a crucial managerial function. With resources often limited, executives must identify what can be done most to improve safety. Hence comes the imperative to develop human performance indicators of organisational factors that affect safety.

The most important safety issue is the human-related incident – industrial accidents involving humans. Future administrative effort should be invested in examining human performance indicators for controllable factors that affect human-related incidents. Table 14.2 specifies several candidates for the focus of a search for human performance indicators of organisational safety. Based on this concept, the project should reveal which indicators can effectively guide the design of administrative and organisational measures intended to reduce the rate of industrial accidents that involve humans.

14.2.4 Human error rate

One typical indicator by which one can monitor some aspects of human performance is the human error rate – that is, the ratio of the abnormal events caused by
human error to the total number of abnormal events. This indicator raises the key
question of how to define human error in quantifiable, and hence measurable, terms.

The definition of human error seems to vary greatly from one country to the
next. In Japan, for example, regulatory authorities and the utilities have agreed to
define human error at nuclear power plants as 'inappropriate human actions devi-
ating from the expected standard in an area such as design, manufacturing, installation, operation, maintenance, or management' (Human Factors Committee, personal
communication). Beyond Japan, one finds definitions such as 'the result of human
performance that fails to meet some acceptable standard or expectation' (Tay-Fuh
Tang, personal communication), 'deviation from the performance intended or
expected' (Tay-Fuh Tang, personal communication), and 'a kind of wrong human
decision and incorrect human action because personnel did not recognise the safety
significance of their actions, violated procedures, were misled by incomplete data
or incorrect, or did not fully understand their plant conditions' (Guo-Shun Chen,
personal communication). When attempting cross-national comparisons of human
error rates, one must be sure that the rates are based on equivalent definitions.

The definition of human error is not the only factor that affects the human error
rate: the application of that definition to incidents has a bearing as well. When it
is applied to Japanese nuclear power plants, the following interpretations are meant:

1 Incidents caused by inappropriate actions at plant sites in such areas as design,
   manufacturing, installation, operation, maintenance, or management are defined
   as *human-error-caused* incidents. Or the cause is attributed to human error in
   cases where

   (a) multiple causes are apparent and the immediate cause is an inappropriate
        action.

   (b) the incident is significant from the viewpoint of human factors, even if the
        immediate cause is not an inappropriate action.

2 Incidents caused by inappropriate actions in factories in such areas as design,
   manufacturing, installation, operation, maintenance, or management are defined
   as human-error-caused incidents errors.

3 Incidents caused by inappropriate actions that were believed appropriate when
   the incident occurred, whether at a plant site or a factory, are not defined as
   human-error-caused incidents.

4 The incident is not defined as human-error-caused in cases where multiple
   causes such as inappropriate actions and equipment failure are apparent, and
   equipment failure was a dominant causal factor.

5 An incident is not defined as human-error-caused when the relationship between
   human factors and the incident's causal factors is obscure (e.g. in some cases
   involving foreign objects or worn parts).

To calculate the difference between the percentages of human performance problems in all causal factors for incidents in US and Japanese nuclear power plants, the Central Research Institute of the Electric Power Industry in Japan (CRIEPI) classified incidents at Japanese plants in the same way that INPO had at US plants. Though the researchers encountered certain difficulties in categorising causal factors of the incidents, it was ultimately possible, with INPO's help, to identify human performance problems whose percentage in the total causal factors was far lower than that at the US plants. The lower figure obviously did not result from superior human performance in Japan, but from the difference between the two countries' interpretation of human performance problems as causal factors (IAEA, 1988).

The same type of discrepancy exists between members of the WANO Tokyo Center with regard to human error rates. Although human error is a delicate problem to discuss among people of different countries, and even different organisations within the same country, it is still a useful potential human performance indicator. Accordingly, an impartial, easily understood definition of human error is greatly needed.

14.3 CONCLUSION

To enhance human performance in the operation and maintenance of nuclear power plants, I have suggested potential indicators of human performance in the domains of safety, efficiency, and welfare. I do not think it possible to quantify all the indicators listed in this chapter, since it seems too difficult to establish a consensus on their definitions in the nuclear utility industry. However, some of the indicators could be defined and quantified by common consent in the industry, and their adoption would be of great importance, because they could then be used to monitor and compare human performance and progress toward the objective of increased safety both within and between nuclear power plants.

Reference

IAEA (International Atomic Energy Agency) (1988) *Man–Machine Interface in the Nuclear Industry* (IAEA-CN-49/13). Vienna: IAEA.

CHAPTER FIFTEEN

Predicting human error probabilities from the ability requirements of jobs in nuclear power plants

EDWIN A. FLEISHMAN AND LOUIS C. BUFFARDI

George Mason University, Fairfax, VA

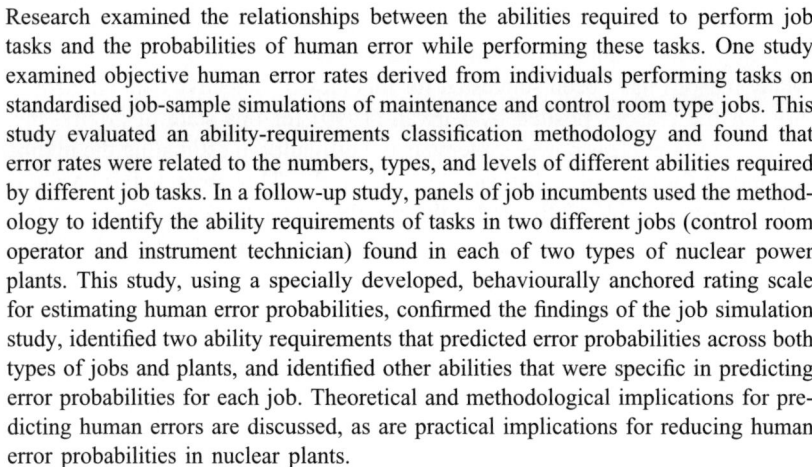

Research examined the relationships between the abilities required to perform job tasks and the probabilities of human error while performing these tasks. One study examined objective human error rates derived from individuals performing tasks on standardised job-sample simulations of maintenance and control room type jobs. This study evaluated an ability-requirements classification methodology and found that error rates were related to the numbers, types, and levels of different abilities required by different job tasks. In a follow-up study, panels of job incumbents used the methodology to identify the ability requirements of tasks in two different jobs (control room operator and instrument technician) found in each of two types of nuclear power plants. This study, using a specially developed, behaviourally anchored rating scale for estimating human error probabilities, confirmed the findings of the job simulation study, identified two ability requirements that predicted error probabilities across both types of jobs and plants, and identified other abilities that were specific in predicting error probabilities for each job. Theoretical and methodological implications for predicting human errors are discussed, as are practical implications for reducing human error probabilities in nuclear plants.

It is estimated that from 20 to 70 per cent of all system failures at nuclear power plants are caused by human error (Arkin and Handler, 1989; Bell and O'Reilly, 1989; Bento, 1990; Dennig and O'Reilly, 1987; Lam and Leeds, 1988; O'Reilly and Plumlee, 1988; Skof, 1990; Wu and Hwang, 1989; Zech, 1988). Despite the widely recognised impact of human error and the various efforts undertaken to reduce its occurrence, human error probabilities (HEPs) in nuclear power plants are not necessarily decreasing (Kameda and Kabetani, 1990). Thus, there is a need to develop improved methods for understanding human errors and the conditions that lead to such errors. Furthermore, it has been estimated that even relatively small improvements in HEPs can result in highly significant improvements in many accident sequences (Samanta, Wong, Higgins, Haber, and Luckas, 1988).

In any work-setting there is a multitude of sources that can contribute to human error rates. For example, in nuclear power settings, plant type (pressurised water reactor versus boiling water reactor) and organisational climate (plant manage-ment's relative concern for safety versus production) are but two factors on which nuclear power plants differ that might have substantial influence on task HEP. Similarly, within the same plant, error rates may vary for different types of jobs (control room operator versus instrument and control technician). At a more 'micro' level of analysis, a critical source of human error may be the abilities required to perform the tasks of the job. For example, cognitive abilities and perceptual-motor abilities have been proposed as important determinants of performance in nuclear power plants. However, not much attention has been paid to establishing empirically whether this view is true, or whether there are particular cognitive or perceptual-motor ability requirements of job tasks or some combination of ability requirements that contribute significantly to producing human error in performing job tasks. The research programme described in this chapter addresses such questions.

One considerable barrier to the investigation of human error is that objective error rates are not readily available in most complex systems. The availability of data for only a limited number of tasks in specific jobs exacerbates the problem of identifying and quantifying human error, and makes it difficult to validate subject-ive estimates of error probability against actual performance data. In fact, it has been argued that this shortage of objective data is the most critical factor impeding the development of human reliability indexes (Dhillon, 1986).

Many reasons have been suggested for this lack of objective data on error prob-ability. Of the reasons posited by Kirwan (1990) for this state of affairs, the one most relevant to nuclear power systems is the difficulty in estimating the number of opportunities for error in realistically complex tasks. This obstacle has been termed the 'denominator problem'. Related to this problem is the tendency for error databases to be developed for specific needs and purposes that do not always address broader needs and applications. For example, insurance companies may collect accident fre-quency data on particular vehicle models, which is very useful information for their purposes. However, human error reliability analysts would like accident data at the task level (e.g. 'failed to obey traffic signal') and would prefer the data to be expressed in error-rate format. This format is difficult to achieve, because it requires collecting denominator information on the number of opportunities the driver had to commit the error (e.g. 'the total number of traffic signals encountered'). This latter informa-tion is less likely to be documented routinely, with the resulting gap in knowledge again requiring subjective estimates to express the data in error-rate format. Given this shortage of objective HEP data, any attempt to investigate human error will require considerable attention to the development of suitable measures of HEP.

15.1 OBJECTIVES

We will now describe a research programme which was designed to address these issues. One of our objectives was to examine the relationships between the ability requirements of different tasks found in nuclear power plant jobs and the human error rates of job incumbents performing these tasks. Another objective was to evaluate the utility of a methodology for classifying tasks in terms of their common ability requirements, and to use this classification as a basis for generalising error probabilities within different classes of job tasks in nuclear power plants.

In our first study, the objectives were to:

- identify and assemble a database of job tasks that included objective error rates;
- develop a method that would allow the classification of these tasks by the abilities they require;
- examine empirical relationships between the ability requirements of different tasks and the objective human error rates of job incumbents performing these tasks.

This initial feasibility study was carried out with tasks in Air Force jobs, where the tasks were similar to those involved in control room and maintenance jobs. The advantage of carrying out this initial feasibility study was the availability of objective error-rate data for these job tasks, information that is not easily found in nuclear settings. In a more extensive follow-up study we used a specially developed criterion of HEP to replicate the model, linking ability requirements to error probabilities of job tasks in nuclear power plants. This more comprehensive study allowed an evaluation of the unique contribution that different ability requirements make to the prediction of error rates of job tasks, and established the generality of these findings across different nuclear plants and jobs.

15.2 STUDY 1: THE RELATIONSHIP BETWEEN TASK ABILITY REQUIREMENTS AND OBJECTIVE HUMAN ERROR RATES

15.2.1 Identification of a task error rate database

Although a number of human error databases have been developed, many of them depend on subjective judgments or fail to report the data in error-rate format. However, it was possible to identify a source of task error data that overcomes both of these deficits. These data are provided by the Job Performance Measurement System (JPMS), developed by the US Air Force to establish objective criteria for the assessment of job proficiency in various Air Force occupational specialities (Lipscomb and Hedge, 1988).

In this system, an in-depth, 'hands-on' work sample proficiency test was developed for each of several occupational specialities. For example, the Jet Engine Mechanic Test is administered at the work site over a seven-hour period, during which the mechanic is required to perform a series of tasks under controlled conditions while being observed and scored by an examiner. Each task is composed of a number of subtasks, with each subtask scored as correct or incorrect. These job-sample tests have been administered to large numbers of job incumbents (usually around 200). The objective error data that exists for each subtask can be expressed in error-rate format – the number of individuals who performed incorrectly on the subtask, divided by the total number of individuals who performed it. Because a large number of job-sample tests have been developed for a variety of occupational specialities, the JPMS provides a rich source of objective human error rate data associated with different tasks. Given our interest in maintenance and control room tasks, we focused on the specialities of Jet Engine Mechanic, Avionics Communication Specialist, Precision Measure Equipment Specialist, and Air Traffic Controller. A total of 1475 subtasks across 94 tasks for these specialities was identified in the JPMS database.

15.2.2   Identification of a method for classifying tasks

In order to assess the comparability of tasks, it was necessary to identify and apply a common 'yardstick' that would allow the classification of many specific tasks into broader categories with common human performance requirements. One approach developed by Fleishman and his colleagues (see e.g. Fleishman, 1972, 1975a, 1982; Fleishman and Quaintance, 1984) describes tasks in terms of the human abilities required for task performance. Given its extensive research history, the ability requirements approach (see e.g. Fleishman, 1975b, 1992; Fleishman and Mumford, 1988, 1989, 1991; Fleishman and Quaintance, 1984; Fleishman and Reilly, 1992a, 1992b) is a methodology potentially useful for classifying tasks into categories having different error rates.

This methodology was developed from a longstanding programme of research involving the identification of ability requirements in human task performance. The general objective of this work has been 'to define the fewest independent ability categories which might be most useful in describing performance in the widest variety of tasks' (Fleishman, 1967, p. 352). The approach distinguishes between abilities and skills, with abilities referring to the individual's more general capacity related to performance in a variety of human tasks. In contrast, a skill is defined as the level of proficiency acquired on a specific task or group of tasks. The development of a given skill on a given task is predicated, in part, on the individual's possession of relevant component basic abilities.

The methodology for identifying human abilities is derived from a whole series of interlocking, experimental-factor analytic studies. The approach includes definitions of more than 50 abilities (e.g. oral expression, problem sensitivity, inductive reasoning, control precision, static strength, and near-vision) spanning the cognitive, perceptual, psychomotor, physical, and sensory domains of human performance. For each ability, a seven-point rating scale has been developed. It includes:

- carefully developed ability definitions;
- distinctions from other abilities;
- definitions of high and low levels of each of the ability requirements;

Written   The ability to read and understand written
Comprehension   sentences and paragraphs

How Written Comprehension is Different From Other Abilities		
Written Comprehension: Involves reading and understanding written words and sentences	versus	Oral Comprehension: Involves listening to and understanding words and sentences spoken by others
		Oral Expression and Written Expression: Involves speaking or writing words and sentences so that others will understand

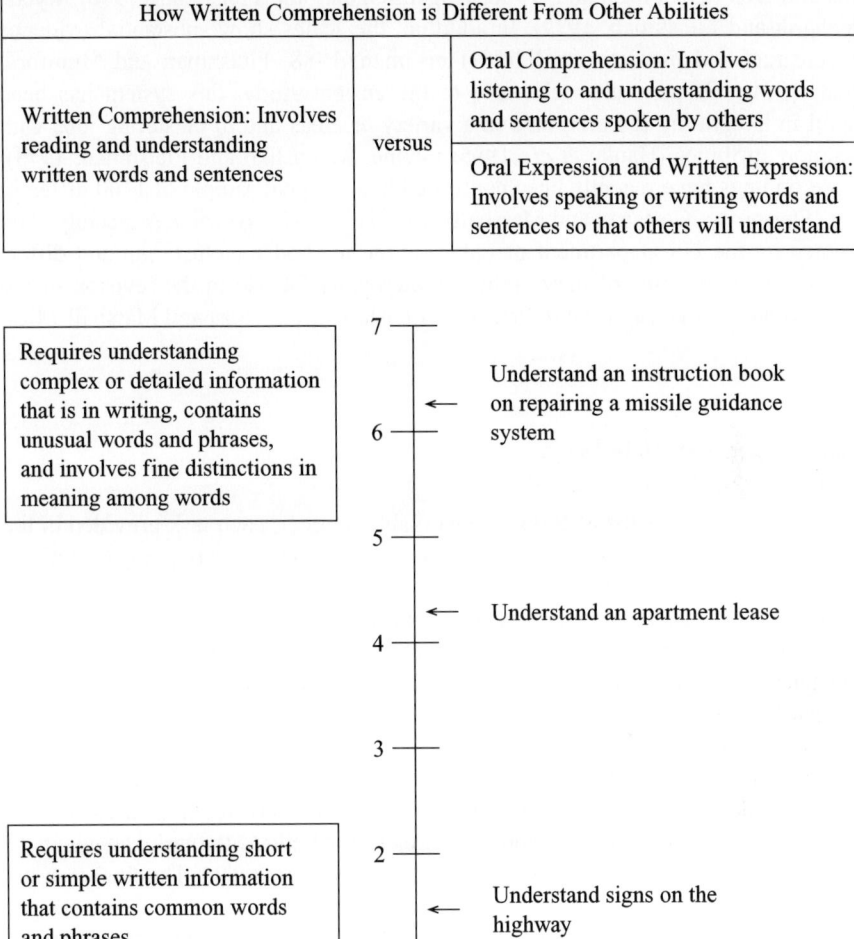

Figure 15.1 An example of an ability requirement scale.
(From Fleishman, E. A. [1992]. *The Fleishman Job Analysis Survey* [F-JAS]. Bethesda, MD: Management Research Institute. Copyright 1975, 1992 Management Research Institute. Reproduced with permission of the publisher.)

■ task anchors to provide raters with examples of everyday tasks that reflect high, moderate, and low levels of each ability.

The scale values of these task anchors have been determined empirically, and they have been selected for their high reliability. These ability requirement scales have been combined into a job analysis system, called the Fleishman Job Analysis Survey (F-JAS) (Fleishman, 1992). Figure 15.1 provides an example of one such scale.

There is an extensive empirical database associated with this approach. Interrater reliability with current versions of these scales, used to describe jobs and tasks in a wide variety of industrial, governmental, and military settings, tends to be in the .80s and .90s (e.g. Fleishman, 1988; Hogan, Ogden and Fleishman, 1978; Myers, Gebhardt and Fleishman, 1979). In addition, the scales show substantial evidence of construct and predictive validity (Fleishman, 1988; Fleishman and Mumford, 1988, 1991). Of particular relevance to the current study, this system has been useful in integrating research data in a variety of areas and in clustering jobs with common attributes (Hauke *et al.*, 1995; Levine, Romashko and Fleishman, 1973). These findings have recently been confirmed in a national sample of 1100 different jobs (Fleishman, Costanza and Marshall-Mies, 1996). The system has recently been adopted by the US Department of Labor as the method for clustering and differentiating jobs in terms of their ability requirements for use in the revision of the *US Dictionary of Occupational Titles* (see Fleishman, Costanza and Marshall-Mies, 1996).

15.2.3   Selection of tasks

In order to reduce the list of tasks to a workable number, each task provided in the JPMS database was categorised into one of nine duty functions (e.g. repair/replace, troubleshooting, and report preparation, etc.) used by Siegel, Federman, and Welsand (1980) in their classification of these tasks. Each task's criticality and frequency of performance contained in a series of technical reports authored by Siegel and his colleagues (e.g. Siegel *et al.*, 1980), were used to narrow the lists further. The tasks retained for the present study were those whose criticality and frequency had been rated in that earlier literature to be highest on their respective duty functions. Indices used to achieve these ratings were frequency × time spent, consequences of inadequate performance, and frequency × consequences. All retained tasks were in the top ten on these indices. In addition, project staff edited the task statements to make them understandable to a wide audience of possible raters. The final task list included 32 tasks across nine duty function areas with no more than four tasks per duty function area. Thus, the list consists of critical and/or frequently performed tasks that represent a wide cross-section of duty functions.

15.2.4   Selection of specific ability dimensions

For each of the nine duty function areas that covered the selected tasks, project staff members selected at least one dimension from the Ability Requirements Taxonomy (see Fleishman, 1975b, 1992; Fleishman and Quaintance, 1984) included in the Fleishman-Job Analysis Survey (F-JAS). This list of dimensions was then compared to the dimensions used by Siegel *et al.* (1980) to rate the requirements of nuclear power plants and Air Force jobs. The following 16 ability dimensions were selected:

1   Oral comprehension.
2   Written comprehension.
3   Oral expression.
4   Written expression.
5   Memorisation.
6   Problem sensitivity.
7   Deductive reasoning.
8   Flexibility of closure.
9   Information ordering.
10   Visualisation.
11   Perceptual speed.
12   Control precision.
13   Multilimb coordination.
14   Manual dexterity.
15   Selective attention.
16   Near vision.

There is a detailed operational definition and rating scale that reflects the level of each ability required to perform particular job tasks (see Fleishman and Quaintance, 1984; Fleishman, 1992).

15.2.5   Administrative procedure

Through the cooperation of the Idaho National Engineering Laboratory, a panel of four subject matter experts (SMEs), knowledgeable in the tasks to be rated, was formed to pretest the procedures. The SMEs possessed from 14 to 20 years of professional experience with maintenance and control-room tasks.

A specially constructed manual for the 16 ability requirements selected from the Fleishman Job Analysis Survey was distributed to the panel along with the task lists and optical scanning forms. After approximately one and a half hours of this training, the panel rated each of the 32 tasks on the level of each ability required to perform them. All tasks were rated on one ability requirement before proceeding to the next ability requirement scale.

15.2.6   Reliability of task ratings of Air Force tasks

Intra-class correlations, a measure of agreement between rates, were calculated for each of the 16 ability dimensions. A mean reliability of .80 for the tasks was found, with only three of these 16 correlations falling below .70. This result indicates that the raters, despite their small number, demonstrated reasonable consensus in their independent judgments of the level of a particular ability required for a given task.

15.2.7   Task similarity and human error probability

Analyses were performed to determine the error rates of various tasks that were similar with respect to their ability requirements. Correlations of the median SME ratings on the 16 dimensions for each possible pair of the 32 tasks were calculated. (Median ratings were used to minimise the effect of a single extreme rating.) A high positive correlation between tasks would indicate that the tasks are similar, that is, that they possess a similar pattern of ability requirements. Each of these 496 correlations (*rs*) between the ability profiles of tasks was transformed to a *z* score according to Fisher's *r* to *z* tables (Thorndike, 1982). The *z* score for each pair of tasks was then correlated with the corresponding absolute discrepancy between the error values of the two tasks, yielding a Pearson $r = 29$, $df = 494$, p < .001. Thus, there was some preliminary indication that tasks requiring a similar pattern of abilities were significantly more likely to have similar error rates (i.e. low discrepancies).

15.2.8   Relation of number of abilities required with human error probability

Another analysis involved an examination of the relation between the number of different abilities having moderate to high ratings (> 3.5 on the 7-point scale) for each of the 32 Air Force tasks and the HEPs of these tasks. The analysis yielded a correlation of .35 ($df = 30$, $p < .05$), indicating a significant relationship between the number of abilities having moderate to high ratings and error rates. Further inspection of the mean HEPs for each number of abilities required (0 to 5) appeared to show a non-linear relationship. Consequently, a one-way analysis of variance (ANOVA) was performed to compare the HEPs for the 24 tasks requiring two or fewer abilities with those eight tasks requiring more than two abilities. The analysis yielded $F(1,30) = 6.75$, $p < .015$, with tasks requiring more than two abilities having a significantly higher error rate (.40 versus .25).

15.2.9   Level of ability required and human error probability

Although there were not sufficient tasks within a given ability dimension to compute a meaningful correlation with HEP, there were 21 separate tasks across ability dimensions that required a moderate to high level of at least one ability. For tasks that required more than one ability, the highest requirement with the highest-ability mean value was chosen. Correlations were calculated between the median 'peak' ability ratings and HEP values associated with each of these tasks. The analysis yielded a correlation of .55 ($df = 19$, $p < .01$). Thus, the higher the peak ability level required (regardless of type of ability), the higher the human error rate.

15.2.10   Type of ability required and human error probability

When the 16 ability dimensions were divided into cognitive and perceptual-motor categories, it was found that the median human error rate was .40 for tasks requiring cognitive abilities and .29 for those requiring perceptual-motor abilities. A sum-of-ranks non-parametric procedure demonstrated that tasks requiring cognitive abilities were characterised by significantly higher error rates than those requiring perceptual-motor abilities ($z = 4.62$, $p < .0001$).

Because some tasks appear on several ability dimensions, the possibility existed that the HEPs for such tasks may unduly influence these data. Consequently, an additional analysis was performed in which a task was categorised according to its primary ability requirements. *Primary ability* was defined as the highest rated ability required by the task. For example, the task 'calibrate scales' requires the abilities of written comprehension (3.5), control precision (4.0), and near vision (3.5). However, because this task requires a higher level of control precision than of any other ability, control precision is the primary ability required for the task. When task HEPs were categorised in this fashion, only four cognitive dimensions (written comprehension, oral expression, problem sensitivity, and visualisation) and three perceptual-motor dimensions (manual dexterity, near vision, and control precision) captured the tasks. Nevertheless, it was found once again that tasks requiring cognitive abilities had significantly higher error rates than those requiring perceptual-motor abilities (.44 versus .24), $z = 3.50$, $p < .001$. Thus, this finding held whether or not tasks were categorised by primary ability.

15.2.11   Conclusions of Study 1

Research during the first phase of this programme (a) identified a large non-nuclear human performance database (JPMS) with objectively derived human error probabilities (HEPs), and (b) evaluated the extent to which classifying tasks in the database, by means of the Fleishman Job Analysis Survey (F-JAS) (a standardised ability requirements taxonomy), could be useful in predicting error rates (HEPs) associated with these task categories. The findings have potentially important implications for generalisation of error rates across tasks and settings. These findings indicate that:

1   The Fleishman Job Analysis Survey (F-JAS) scales successfully captured tasks in Air Force settings and provided a basis for estimating HEP values for these tasks within ability requirement categories.

2   Similar tasks, as determined by the profile of ratings across the 16 ability requirements utilised in this study, tend to have similar HEPs.

3   Tasks with higher peak ability-requirement ratings tend to have higher HEPs.

4   Tasks requiring moderate-to-high levels of more than two abilities have higher HEPs than tasks requiring two or fewer abilities.

5   Tasks requiring cognitive abilities tend to have higher HEPs than those requiring perceptual-motor/sensory abilities.

These results appeared promising for the prediction of human error in job tasks, because the study indicated that tasks can be classified according to their ability requirements into meaningful groups that differ with respect to human-error probability. Furthermore, the F-JAS (ability requirements) methodology, which utilises an ability-requirement measurement system, provides a reliable methodology for classifying job tasks.

15.3 STUDY 2: PREDICTING HUMAN ERROR PROBABILITIES OF TASKS PERFORMED IN NUCLEAR POWER PLANT SETTINGS

Although the results of Study 1 were encouraging, there was a need to demonstrate that this methodology can be successfully applied to the prediction of task HEP in nuclear power plant settings. Similarly, research was needed to determine the *specific* abilities most strongly associated with human error rates as well as their relative contribution *vis-à-vis* the more 'macro' variables of different plants and jobs.

15.3.1 Development of the HEP rating scale method

As mentioned earlier, objective error rates are quite scarce in nuclear power settings. Hence, a major concern was the development of a psychometrically sound measure of HEP. Relevant reports on generating human error reliability estimates using expert judgment were reviewed. This literature generally was supportive of using expert judgment in the development of HEP estimates, a fact that indicates some consistency in these judgments across various techniques. For lack of objective error-frequency data, predictive validity of the techniques has not yet been established, but the use of expert judgments provides substantial practical time and cost advantages in the estimation of task HEPs.

Stillwell, Seaver and Schwartz (1982), in discussing ways of transforming psychological scales into error probability scales, suggest a few tasks with known objective error probabilities be incorporated into the scaling methodology used by the experts. Fortunately for the purposes of this study, Beare, Dorris, Bovell, Crowe, and Kozinsky (1984) reported objective error rates for a limited number of generic nuclear power plant tasks empirically determined through an extensive study of performance in simulators. In order to maximise the use of the little objective HEP data that does exist in nuclear settings, project staff used these generic tasks and their objective error rates as benchmark anchors to be placed at different points, from 9 (high) to 1 (low) likelihood on an error-probability rating scale. This approach provided SMEs with an objectively established guide for judging the relative HEP level of nuclear power plant tasks rated according to a scale format that was similar to the one used with the ability requirement dimensions. In order to accommodate the relatively wide variation in objective HEPs of the task anchors within a 9-point format, a $\log_2$ scale was created (see Figure 15.2). Although a $\log_2$ scale is not common in the literature, it was essentially used in a scale for estimating HEP directly (see Comer, Seaver, Stillwell, and Gaddy, 1984) and was found superior to a paired-comparison rating technique. The Human Error Probability scale developed is presented in Figure 15.2.

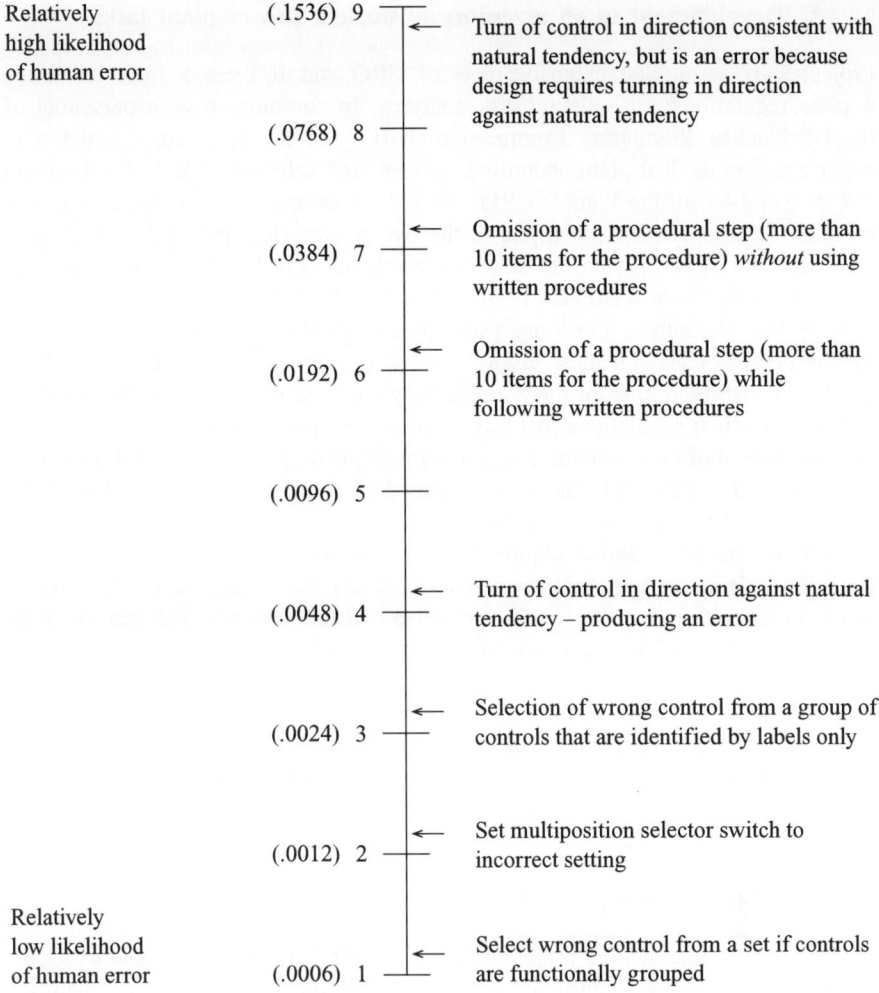

Relatively high likelihood of human error

(.1536) 9 — Turn of control in direction consistent with natural tendency, but is an error because design requires turning in direction against natural tendency

(.0768) 8

(.0384) 7 — Omission of a procedural step (more than 10 items for the procedure) *without* using written procedures

(.0192) 6 — Omission of a procedural step (more than 10 items for the procedure) while following written procedures

(.0096) 5

(.0048) 4 — Turn of control in direction against natural tendency – producing an error

(.0024) 3 — Selection of wrong control from a group of controls that are identified by labels only

(.0012) 2 — Set multiposition selector switch to incorrect setting

Relatively low likelihood of human error

(.0006) 1 — Select wrong control from a set if controls are functionally grouped

Figure 15.2 Scale for measuring Human Error Probability (HEP). (This is a $\log_2$ scale – each scale value represents an error rate twice as high as immediately preceding value. Source of anchors: A simulator-based study of human errors in nuclear power plant control room tasks [NUREG 3309], by A. N. Beare, R. E. Dorris, C. R. Bovell, D. S. Crowe, and E. J. Kozinsky (1984), Albuquerque, NM: Sandia National Laboratories.)

15.3.2 Subjects utilised

A total of 35 experienced nuclear plant job incumbents (the SMEs) participated in this phase of the project. They included 20 instrument and control technicians (ICTs) and 15 control room operators (CROs) drawn from Plant A (a pressurised water reactor plant, PWR), and from plant B (a boiling water reactor plant, BWR). All participants had years of experience in at least eight nuclear power plants.

15.3.3   Development of an inventory of nuclear power plant tasks

Project staff assembled extensive lists of CRO and ICT tasks from published nuclear regulations and existing task analyses. In conjunction with personnel of the US Nuclear Regulatory Commission (NRC) and a project consultant highly experienced in nuclear plant operations, project staff selected 32 tasks for the Plant A CROs and 63 for the Plant B CROs, so as to represent a wide range of responsibilities for control room operators at the two plants. Because of the differences between PWR (Plant A) and BWR (Plant B) plants, only 13 of the tasks presented to the Plant B CROs were also presented to the Plant A CROs.

Fifty-four tasks from a task analysis report conducted by Plant A were selected for the Plant A ICTs. Of these tasks, 50 were also applicable to the Plant B ICTs and thus constituted most of the 55 tasks they received, with the rest coming from published nuclear regulations. All tasks, across both jobs and plants, were selected as being typical of those occurring under normal operating conditions. Before project staff visited the plants, all tasks were screened by plant management and the NRC for clarity and job-representativeness.

From the list of 16 ability requirement dimensions used in Study 1, eight scales from the Fleishman Job Analysis Survey (F-JAS) were selected for the present study, because they had previously demonstrated the strongest independent relations to objective HEP indices on the Air Force tasks. These abilities comprised perceptual-motor abilities (control precision, near vision, manual dexterity) and cognitive abilities (written comprehension, problem sensitivity, oral expression, deductive reasoning, and visualisation). Number facility was added after discussions by project staff suggested that it might be a relevant ability that had not been used in prior research.

15.3.4   Administration procedures

Three project researchers, a project officer from the US Nuclear Regulatory Commission, and the project consultant met with each group of SMEs in each plant for three to four hours to provide instructions to job incumbents SMEs. They were instructed to consider each task as occurring under 'normal operating conditions'. SMEs then rated the HEP of each task listed for their job (CRO or ICT). Afterward, they were instructed on the meaning and use of the ability requirement scales, whereupon they rated each task on the level of each ability required.

15.3.5   Reliabilities of the ability requirements and HEP scales

Reliabilities, determined by intra-class correlations among task ratings, were calculated for the HEP scale and for the nine ability scales for each of the four distinct groups of raters (2 plant types × 2 job types).

Across the nine ability scales, the median reliability coefficients for each of the four groups were:

- .78 (Plant A-ICT);
- .82 (Plant A-CRO);
- .86 (Plant B-ICT);
- .89 (Plant B-CRO).

The reliabilities obtained for the newly developed HEP scale ranged from .63 to .69 across the four rater groups. Although somewhat lower than the reliabilities obtained for the ability requirement scales, these reliabilities are high enough for research purposes. This somewhat lower reliability may have been due to a more restricted range of scores, as SMEs rarely gave ratings higher than 5 on the 9-point HEP scale ($M = 2.37$, $SD = 1.41$).

In addition to these intraclass correlations, separate Pearson correlations for each job type were calculated for mean HEP scores for those tasks that were common across the two types of plants. Significant correlations were found in both cases, with $r = .51$, $p < .01$ for the 50 ICT tasks common to both plants, and $r = .67$, $p < .05$ for the 13 common CRO tasks. Thus, although there may be some differences in the way in which the same task is performed in different types of plant, SMEs provided reasonably consistent HEP ratings of these tasks as they were performed in different plants. The high correlations between HEP indices for common tasks across different types of plant support the validity of the HEP rating scale developed for the project, and suggest its usefulness in identifying, at least on a relative basis, high HEP tasks in either type of plant.

15.3.6 Effects of plant and job on HEP

Separate HEP means were calculated for each task for each of the four SME groups. Two-way ANOVAs were performed to determine the statistical significance of the effects of plant, job, and their interaction on ratings of task HEPs. Significant main effects were found for plant differences ($F(1,200) = 65.69, p < .001$) and for Job differences ($F(1,200) = 107.93, p < .001$), and there was a significant plant $\times$ job interaction ($F(1,200) = 23.19, p < .001$). With respect to the direction of these differences, tasks in the PWR plant were rated higher on HEP than Plant B tasks in the BWR plant ($M = 2.89$ versus 2.00, respectively), and technician (ICT) tasks were rated higher on HEP than tasks performed by CROs ($M = 2.87$ versus 1.81, respectively). The significant plant $\times$ job interaction was due to the particularly low HEP ratings ($M = 1.39$) given to the CRO tasks in the BWR plant.

15.3.6 Relationships between ability requirements and HEP

The number of observations (N) for each ability requirement and for each HEP equalled the number of tasks for a group. We organised data for analysis by correlating the mean rating on each ability with the HEP mean for each task obtained from each SME group.

In general, most of the ability ratings correlated significantly with the measure of HEP, indicating that, the higher the level of a given ability required by a task, the higher the task HEP. Cognitive ability requirement ratings correlated mostly in the .50s and .60s with HEP, with problem sensitivity, written comprehension, and deductive reasoning showing the strongest bivariate correlations (median rs across groups = .68, .65, .65, respectively). Ratings of perceptual-motor ability requirements also correlated significantly, but were lower (mostly in the .30s and .40s). This pattern of cognitive abilities showing stronger correlations with HEP than with perceptual-motor abilities was present across both plants and positions, although it was particularly striking in the PWR groups. The lone exception to this pattern were the low correlations found between oral expression and HEP for the CRO groups in both plants. Apparently, the level of oral comprehension required on job tasks is unrelated to error probabilities reported for those tasks by CROs, although it is related to error probability in ICT tasks.

15.3.7   Multiple regression and replication analyses

Thus far, we have presented relations between individual ability requirements of nuclear power plant tasks and the HEPs for these tasks. We have shown that ratings of these ability requirements are predictive of HEPs. We now turn to an examination of the HEP predictions achieved by *combinations* of different ability ratings.

Multiple regression is an extension of bivariate regression, in which several independent variables (e.g. the nine ability dimensions) are combined to predict a value on a dependent variable (i.e. HEP) for each task. The result of regression is an equation that represents the best prediction of HEP from several optimally weighted ability dimensions. Each of the multiple regression coefficients obtained, ranging from .734 to .848, was highly significant, indicating that, for each group, mean task HEP was strongly predicted from a weighted combination of ability means on these same tasks. The standard errors of estimate associated with the regression equations provided a confidence (or uncertainty) interval around the predicted HEP ratings for each task. For example, if the ability ratings on a given CRO task yielded a *predicted* HEP rating of 3.0 based on the regression equation for that group, then, given a standard error of estimation of .437, its *actual* HEP rating for that task would fall between 2.563 and 3.437 (3.0 ± .437) 68 per cent of the time. Thus, the standard error of estimate is useful in setting 'boundary' values within which the actual HEP rating is likely to fall.

Because it is based solely on the set of data for which the fit was optimised, the multiple Rs (like all multiple Rs) may be statistically biased upward as a measure of how well the regression equation might be expected to fit the data from a *different* sample of tasks or a different sample of SMEs. Thus, a more meaningful and rigorous test of the goodness of this multiple regression prediction model is to see how well the equation holds up when applied to a *different* set of data provided by another, separate SME group. With this in mind, we carried out 20 cross-validation or replication analyses, in which the equation developed on one SME sample (equation source) was applied separately to each of the data sources provided by the other four SME groups.

The results clearly indicated the efficacy of the multiple regression model, with 19 of the 20 analyses reaching statistical significance (median regression coefficient = .653). Although the results appear strongest for predictions made within-job and within-plant, these findings are particularly compelling, considering that most of the analyses involve predicting HEPs of tasks in different jobs (ICT versus CRO) or in different plants (PWR versus BWR) or in different jobs *and* plants! Although the exact weights that should be applied to the various abilities in order to predict the HEP of other nuclear power plant tasks is not yet defined with precision, the model appears to be generalisable across nuclear jobs and plant types for tasks performed under normal operating conditions.

15.3.8 Cross-validated hierarchical multiple regressions

Thus far, we have shown that significant relations exist between HEP and different combinations of ability ratings obtained and that, in general, these relationships generalise to some extent across jobs, plants, and, to a limited extent, across organisations. However, we still need to assess the unique contributions that particular abilities make to the prediction of HEP.

In order to assess whether specific ability requirements of tasks were related to HEP over and above the macro variables of plant and job, three-stage hierarchical multiple regressions were performed. The main effects of plant and job type were entered at Stage 1, plant × job interaction at Stage 2, with the nine ability requirements entered at Stage 3. The entire list of 204 tasks was divided in half (odd versus even), with parallel three-stage regressions performed on each of the two data sets, thus allowing for cross-validation of predictors.

Taking the hierarchical multiple regression weights developed on the odd tasks and applying them to the even data set (and vice versa), we obtained double cross-validation multiple correlations of .84 and .86, respectively. These results indicated that the macro variables of plant and job, in combination with task ability requirements, strongly predicted the HEPs of these tasks. More important, the hierarchical analyses indicate that two specific ability requirements, written comprehension and deductive reasoning, significantly predict task HEP in both sets. Thus, these two abilities contribute to the prediction of HEP for a wide variety of tasks *over and above* the variance accounted for by the main effects of plant and job type and the plant × job interaction. They predict error probability estimates over a wide range of tasks performed in different jobs in different types of plant.

With respect to differential requirements across jobs, it was found for CRO tasks that the requirements for near vision and oral expression significantly predicted HEP over and above the prediction achieved by the requirements of written comprehension and deductive reasoning. For ICT tasks, requirements for the abilities of visualisation and manual dexterity added to the prediction already found for the written comprehension and deductive reasoning requirements.

15.4 DISCUSSION AND CONCLUSIONS

The rather comprehensive study of job tasks in the nuclear power plants confirms the earlier findings with Air Force tasks. The errors in the Air Force study involved real-time recording of errors made by incumbents while performing tasks in job simulations. The error rates in nuclear plants were derived from ratings of error probability on a specially developed, behaviourally anchored, rating scale. In each study, the *levels* of ability requirements of different job tasks were found to be related to the kinds and levels of different abilities required. In each study, tasks requiring higher levels of cognitive abilities had higher error rates than did tasks requiring perceptual-motor abilities. Also, in each study, the *number* of different abilities required was found to be related to error rates. The study in nuclear power plants allowed for greater detail in the regression analysis of the relationships that various *combinations* of ability requirements have to errors in job tasks. It also afforded greater in-depth regression analysis of the *unique* contribution that particular ability requirements make beyond that stemming from such macro variables as type of plant and job. It was possible to identify (a) particular ability requirements that predicted error rates of job tasks and across the two types of plants and two types of jobs, as well as (b) ability requirements that were related to errors only in CRO or ITC jobs.

15.4.1 Some limitations

The present study has a number of limitations. Sampling error is always a concern when the number of raters and tasks is relatively small, as it was for each of our five SME job-incumbent groups. That consideration limits our ability to state unequivocally what weights for each ability should be used in a single multiple regression equation designed to maximise the prediction of HEPs of any set of nuclear power plant tasks. However, it should be noted that virtually all the cross-validation/replication analyses yielded highly significant results, despite concerns about sampling error.

In addition, the question of 'method variance' needs to be addressed. One can argue that these strong correlations and regressions in Study 2 resulted because the same raters provided rating scale values of both the predictors (abilities required) and the criterion Human Error Probability (HEP) scale. Perhaps the raters subconsciously 'justified' a high HEP rating by indicating that the same task also required high levels of the abilities. However, if this were the case, one would expect each of the abilities to correlate to the same degree with error rates, and they plainly do not. Some ability requirements are more related to error rates in these tasks than are others and, for each of the five SME groups, cognitive abilities were always more strongly correlated with error rates than were perceptual-motor ones. Moreover, these results are consistent with the data from the Air Force study, in which the experts who provided the ability ratings were completely unaware that the objectively-derived error data existed. Thus, although method variance may possibly account for some of the magnitude of the ability error relationships, it cannot account for all of it. In order to further minimise such possible effects, future research in this area might well use separate SME groups, one providing the error probability ratings, and the other the ability ratings.

A number of conclusions with respect to the macro variables are limited by possible confusions associated with plant and job types (e.g. organisational climate, task performed, quality and frequency of training, and group safety norms). For example, we would need to increase the number of BWR plants and PWR plants to be able to generalise across these types of plants. The number of plants that would have been necessary for such generalisation was beyond the scope of the present study, which mainly showed the contribution of ability requirements to error over and above the influence of plant differences.

However, because the cross-validated, hierarchical, multiple regressions control the effects of the macro variables and for any rater biases, the results strongly indicate the importance of ability requirements in predicting task-error rates. Both written comprehension and deductive reasoning predict HEPs across both plants and jobs. Thus, these ability requirements generalise across nuclear settings. Additional research is needed to assess the extent to which these particular abilities generalise to maintenance and control room settings in non-nuclear environments.

Another issue is the lack of time pressure associated with the task database chosen. As previously indicated, all tasks were selected to be representative of the activities performed during normal operating conditions. Under such conditions, where time is less of a factor than it is in unusual situations, perceptual-motor tasks are quite straightforward and unlikely to lead to error. However, under emergency conditions, where there may be several alarms competing for attention, those same tasks may be much more error-prone. Clearly, this aspect is an area for further exploration.

15.4.2 Some theoretical considerations

The results showing which of the abilities generalise (written comprehension, deductive reasoning) and which are situation-specific (near vision and manual dexterity) appear to be consistent with Rasmussen's (1981) breakdown of error into two types, which he called slips (execution errors on skill-based tasks) and mistakes (errors of planning or judgment on knowledge-based tasks). However, the complete Rasmussen model, which emerged from studies of error in the nuclear industry (Hale and Glendon, 1987), also contains a rule-based level of functioning that is intermediate to the skill and knowledge levels. This level requires the application of an identifiable procedure and, thus, is at a slightly more cognitive level than the skill-based level, but does not require the degree of interpretation necessary at the knowledge-based level. Subsequent analyses by Hale and Glendon (1987) and by Reason (1990) elaborated on this three-level model as a basic structure for the analysis of human error. Govindarajan (1990) commented on the need for such an analytic structure and stated that 'it would be of great use [for achieving] a classification of human errors' (p. 53). However, empirical tests of this model structure have not been reported in the literature.

In the present study, written comprehension appears to be related to rule-based tasks, where specific procedures are read, whereas deductive reasoning seems to be required of knowledge-based tasks. By contrast, both near vision and manual dexterity are perceptual and motor abilities, respectively, more related to skill-based activities. Thus, the current study provides some empirical evidence for higher human error rates for rule- and knowledge-based actions. Similarly, the abilities required of these more cognitively oriented tasks appear to generalise more readily across situations. Conversely, skill-based actions are more likely to be situation-specific, and to have lower error rates. However, it should be noted that these results were found for tasks performed under normal operating conditions. As suggested earlier, emergency conditions, where time pressure is more of a factor, might show different patterns.

15.4.3   Some practical implications

With respect to methodological contributions, the research shows the feasibility of using the Fleishman Job Analysis Survey (F-JAS) ability requirement scales as a reliable method of job task analysis to determine the personnel capabilities required for jobs in nuclear power plant situations and to assist in setting selection and training standards. These findings have substantial practical applications with respect to the *selection* of personnel for the jobs of control room operator and instrument and control technicians. Well-developed, standardised tests exist for each of the significant abilities used in the present study (Fleishman and Reilly, 1992b). The selection procedures of utility companies could be examined to determine whether these types of measures can be used and validated as part of the screening process for these jobs. Individuals responsible for *training* programmes may wish to evaluate their methods to determine if crucial written comprehension and deductive reasoning abilities are developed as a means of reducing errors in job tasks requiring these abilities.

The information provided also has implications for *job design*. Identifying tasks with requirements for high levels of certain abilities may lead to the use of job aids, redesign of job tasks, or to reallocation of these tasks to individuals who possess higher levels of these abilities. Future research should examine the impact that such actions and policies have on the reduction of error rates.

In order to conduct the present study, a Human Error Probability (HEP) scale was developed that appears to be useful to the nuclear power industry. This measure shows evidence of being reliable across different raters employed in nuclear power plant jobs, and demonstrates substantial evidence of validity, since there was a high correlation between HEPs across plants for tasks common to both plants. The HEP scale was sensitive enough to detect this correlation, despite the fact that the absolute error rate levels were higher in the PWR plant. Further evidence of the validity of this HEP scale is provided by the finding that the general results of Study 2 are very consistent with those of Study 1, in which objectively recorded HEP data were used. Thus, the HEP scale may be useful in identifying error-prone tasks in any plant. The F-JAS ability requirements scales, of course, provide information about the basis for these errors.

The results of this research indicate that quantifying the ability requirements of job tasks sheds some light on the likelihood of human error in the performance of nuclear power plant tasks. Although much remains to be done, the results of this study represent an advance in the understanding of human error in applied settings.

References

ARKIN, W. M. and HANDLER, J. (1989) *Neptune papers no. 3: Naval accidents 1945–1988*. Washington, DC: Institute for Policy Studies.

BEARE, A. N., DORRIS, R. E., BOVELL, C. R., CROWE, D. S. and KOZINSKY, E. J. (1984) *A simulator-based study of human errors in nuclear power plant control room tasks* (NUREG-3309). Albuquerque, NM: Sandia National Laboratories.

BELL, L. G. and O'REILLY, P. D. (1989) Operating experience feedback report – Progress in scram reduction (NUREG 275), Vol. 5. Washington, DC: US Nuclear Regulatory Commission.

BENTO, J. P. (1990) Collection, analysis and classification of human performance problems at the Swedish nuclear power plants. In International Atomic Energy Agency (IAEA), (eds), *Human error classification and data collection: Report of a technical committee meeting organised by the International Atomic Energy Agency* (pp. 83–94). Vienna: IAEA.

COMER, M. K., SEAVER, D. A., STILLWELL, W. G. and GADDY, C. D. (1984) *Generating human reliability estimates using expert judgment* (NUREG/CR-3688). Washington, DC: US Nuclear Regulatory Commission.

DENNIG, R. L. and O'REILLY, P. D. (1987) *Operating experience feedback report – New plants* (NUREG-1275), Vol. 1. Washington, DC: US Nuclear Regulatory Commission.

DHILLON, B. S. (1986) *Human reliability with human factors*. New York: Pergamon Press.

FLEISHMAN, E. A. (1967) Performance assessment based on an empirically derived task taxonomy. *Human Factors*, **9**, pp. 349–66.

FLEISHMAN, E. A. (1972) On the relation between abilities, learning, and human performance. *American Psychologist*, **27**, pp. 1017–32.

FLEISHMAN, E. A. (1975a) Taxonomic issues in human performance research. In SINGLETON, W. T. and SPURGEON, P. (eds), *Measurement of human resources*. New York, NY: Halsted Press.

FLEISHMAN, E. A. (1975b) Toward a taxonomy of human performance. *American Psychologist*, **30**, pp. 1127–49.

FLEISHMAN, E. A. (1982) Systems for describing human tasks. *American Psychologist*, **37**, pp. 821–34.

FLEISHMAN, E. A. (1988) Some new frontiers in personnel selection research. *Personnel Psychology*, **41**, pp. 679–701.

FLEISHMAN, E. A. (1992) *Fleishman Job Analysis Survey (F-JAS)*. Bethesda, MD: Management Research Institute.

FLEISHMAN, E. A., COSTANZA, D. P. and MARSHALL-MIES, J. C. (1996) Abilities: Evidence for the reliability and validity of the measures. In PETERSON, N. G., MUMFORD, M. D., BORMAN, W. C., JEANNERET, P. R., FLEISHMAN, E. A., and LEVIN, K. Y. (eds), *O*NET Final Technical Report*. Salt Lake City, UT: Utah Department of Employment Security.

FLEISHMAN, E. A. and MUMFORD, M. (1988) Ability requirement rating scales. In GAEL, S. (ed.), *Handbook of job analysis*. New York: Wiley.

FLEISHMAN, E. A. and MUMFORD, M. D. (1989) Individual attributes and training performance. In GOLDSTEIN, I. L. (ed.), *Training and development in organisations* (pp. 183–255). San Francisco: Jossey-Bass.

FLEISHMAN, E. A. and MUMFORD, M. D. (1991) Evaluating classifications of job behaviour: A construct validation of the Ability Requirement Scales. *Personnel Psychology*, **44**, pp. 523–75.

FLEISHMAN, E. A. and QUAINTANCE, M. K. (1984) *Taxonomies of human performance: The description of human tasks*. Bethesda, MD: Management Research Institute.

FLEISHMAN, E. A. and REILLY, M. E. (1992a) *Fleishman Job Analysis Survey (F-JAS) administrator's guide*. Bethesda, MD: Management Research Institute.

FLEISHMAN, E. A. and REILLY, M. E. (1992b) *Handbook of human abilities: Definitions, measurements, and job task requirements*. Bethesda, MD: Management Research Institute.

GOVINDARAJAN, G. (1990) An approach to human error minimisation in PHWR safety-related operations. International Atomic Energy Agency (IAEA) (eds), *Human error classification and data collection: Report of a technical committee meeting organised by the International Atomic Energy Agency* (pp. 51–60). Vienna: IAEA.

HALE, A. R. and GLENDON, A. I. (1987) *Individual behaviour in the control of danger* (Vol. 2, Industrial Safety Series). Baltimore, MD: Elsevier Science.

HAUKE, M., COSTANZA, D. P., BAUGHMAN, W., MUMFORD, M. D., STONE, L., THRELFAL, V. and FLEISHMAN, E. A. (1995) *Developing job families on the basis of ability and knowledge requirements*. Bethesda, MD: Management Research Institute.

HOGAN, J. C., OGDEN, G. D. and FLEISHMAN, E. A. (1978, June) *Assessing physical requirements for establishing medical standards in selecting benchmark jobs* (ARRO Final Report 3012/R). Washington, DC: Advanced Research Resources Organisation – 78/8.

KAMEDA, A. and KABETANI, T. (1990) Outline of the development of a nuclear power plant human factors database. In International Atomic Energy Agency (IAEA) (ed.), *Human error classification and data collection: Report of a technical committee meeting organised by the International Atomic Energy Agency* (pp. 111–26). Vienna: IAEA.

KIRWAN, B. (1990) Human reliability assessment. In WILSON, J. R. and CORLETT, E. N. (eds), *Evaluation of human work*. New York: Taylor & Francis.

LAM, P. and LEEDS, E. (1988) *Operating experience feedback report – Service water system failures and degradations* (NUREG-1275), Vol. 3. Washington, DC: US Nuclear Regulatory Commission.

LEVINE, J. M., ROMASHKO, T. and FLEISHMAN, E. A. (1973) Evaluation of an abilities classification system for integrating and generalising findings about human performance: The vigilance area. *Journal of Applied Psychology*, **58**, pp. 147–49.

LIPSCOMB, M. S. and HEDGE, J. W. (1988) Job performance measurement: Topics in the performance measurement of enlisted personnel. (AFHRL-TP-87–58: AD-A195 630).

MYERS, D. C., GEBHARDT, D. L. and FLEISHMAN, E. A. (1979, November) *Development of physical performance standards for Army jobs* (ARRO Final Report 3045/R79-10). Washington, DC: Advanced Research Resources Organisation.

O'REILLY, P. D. and PLUMLEE, G. L. (1988) *Operating experience feedback report – Technical specifications* (NUREG-1275), Vol. 4. Washington, DC: US Nuclear Regulatory Commission.

RASMUSSEN, J. (1981) Models of mental strategies in process plant diagnosis. In RASMUSSEN, J. and ROUSE, W. B. (eds), *Human detection and diagnosis of system failures* (pp. 241–58). New York: Plenum Press.

REASON, J. (1990) *Human error.* New York: Cambridge University Press.

SAMANTA, P., WONG, S., HIGGINS, J., HABER, S. and LUCKAS, W. (1988) A risk methodology to evaluate sensitivity of plant risk to human errors. In HAGEN, E. W. (ed.), *Conference record for 1988 Institute of Electrical and Electronics Engineers 4th conference on human factors and nuclear power plants* (pp. 249–58). New York: Institute of Electrical and Electronics Engineers.

SIEGEL, A. I., FEDERMAN, P. J. and WELSAND, E. S. (1980) *Perceptual psychomotor requirements basic to performance in 35 Air Force specialities* (Technical Report). Wayne, PA: Applied Psychological Services.

SKOF, M. (1990) Human characteristics affecting nuclear safety. In International Atomic Energy Agency (IAEA) (ed.), *Human error classification and data collection: Report of a technical committee meeting organised by the International Atomic Energy Agency* (pp. 95–102). Vienna: IAEA.

STILLWELL, W. G., SEAVER, D. A. and SCHWARTZ, J. P. (1982) *Expert estimation of human error probabilities in nuclear power plant operations: A review of probability assessment and scaling* (NUREG/CR-2255). Washington, DC: US Nuclear Regulatory Commission.

THORNDIKE, R. M. (1982) *Data collection and analysis.* New York: Gardner Press.

WU, T. and HWANG, S. (1989) Maintenance error reduction strategies in nuclear power plants, using root cause analysis. *Applied Ergonomics,* **20,** pp. 115–21.

ZECH, L. W., JR. (1988) Informal discussion of current issues. In HAGEN, E. W. (ed.), *Conference record for 1988 Institute of Electrical and Electronics Engineers 4th conference on human factors and nuclear power plants* (pp. 40–2). New York: Institute of Electrical and Electronics Engineers.

CHAPTER SIXTEEN

Self-assessment and learning in nuclear power plant simulation training

MIKLÓS ANTALOVITS AND LAJOS IZSÓ

Technical University of Budapest

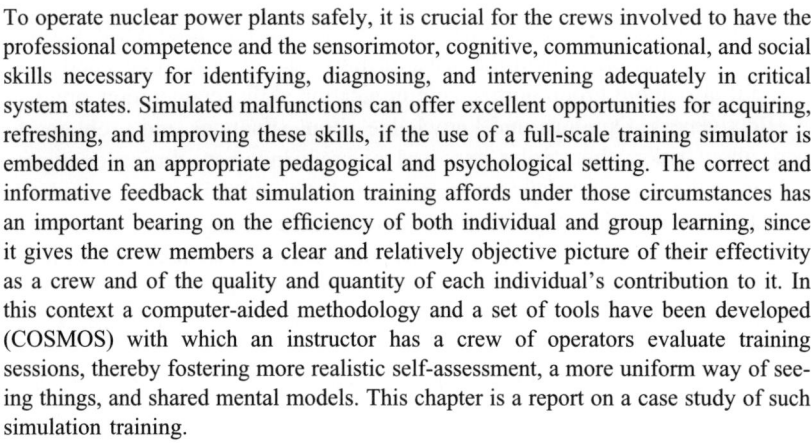

To operate nuclear power plants safely, it is crucial for the crews involved to have the professional competence and the sensorimotor, cognitive, communicational, and social skills necessary for identifying, diagnosing, and intervening adequately in critical system states. Simulated malfunctions can offer excellent opportunities for acquiring, refreshing, and improving these skills, if the use of a full-scale training simulator is embedded in an appropriate pedagogical and psychological setting. The correct and informative feedback that simulation training affords under those circumstances has an important bearing on the efficiency of both individual and group learning, since it gives the crew members a clear and relatively objective picture of their effectivity as a crew and of the quality and quantity of each individual's contribution to it. In this context a computer-aided methodology and a set of tools have been developed (COSMOS) with which an instructor has a crew of operators evaluate training sessions, thereby fostering more realistic self-assessment, a more uniform way of seeing things, and shared mental models. This chapter is a report on a case study of such simulation training.

In keeping with the philosophy of the Systematic Approach to Training (SAT) introduced by the International Atomic Energy Agency (IAEA), there has recently been an explicit call for training simulation to provide an effective learning environment capable of promoting the development of technical, communicational, and cooperation skills, and the evaluation of both individual and crew performance with

243

the greatest possible objectivity (IAEA, 1989, 1994). To meet these requirements, simulation must give trainees carefully designed, informative feedback for technical and social learning and, in addition to the usual evaluation by instructors, must utilise the possibilities for operators' self-assessment. If conducted properly, self-assessment could be a powerful tool, among other things, to increase objectivity, enhance operators' self-knowledge and understanding of their fellow operators, and improve communication skills within the group.

Experience has shown that short evaluation sessions immediately after simulation training can be unique and psychologically valuable situations that, with properly designed methods, can be used effectively to increase the preparedness of crews and, hence, the safety of operations. A number of aspects characterise the situation that exists immediately after a cognitively demanding simulation:

- The experiences and memories that the crew members have of the details of simulated malfunctions, their own behaviour, and the activities of fellow operators are quite vivid and fresh.

- In addition to facts, the emotional flavour of a largely successful or unsuccessful situation is still remembered by the crew members as tensions, which need to be acted out.

- The crew members still have quite definite opinions, whether correct or incorrect, about the expected roles and actual effectiveness of individual operators.

- Video recordings and computer protocols are still available as objective sources to be drawn upon for discussion and debate.

Such a very intense learning process can begin with technological knowledge and experience, not only about the tasks of individual crew members and those of others, but also about group norms, communication skills, cooperation, and leadership effectiveness. Opinions and knowledge about situations and problems to be solved, about risks, about roles expected of crew members during emergencies, and about optimal group behaviour are more realistic and uniform as a result of this accelerated learning than they otherwise would have been.

The basic problem, however, has been the lack of such methods with which to make use of these characteristics. For that purpose a *co*mputer-*s*upported *m*ethod for *o*perators' *s*elf-assessment (COSMOS) has been developed. It has six main steps.

1. *Carefully designing training scenarios that include identification of what are called* key situations *of the simulated emergency.* Those elements of simulated malfunctions (operating conditions) that will play a determining role in the decisions made by trainees occupying the various operator positions must be identified in advance. These elements define the key situations for which the trainees later perform group- and individual self-assessment of their performance immediately after the training session.

2. *Conducting the simulation training session; recording and observing crew behaviour.* All relevant information that characterises operators' activity, including

interventions, behaviour, communication within the crew and between the crew and the instructor, behaviour of the team leader, and group climate is recorded by means of audio/video tape, computer log, observation, and other methods.

3. *Determining and assessing perceived relative difficulties in key situations, with operators comparing pairs of situations and examining concordance within the crew.* Immediately after the training session crew members and the instructor briefly discuss the main events of the training session, redefining the key situations if necessary. They then identify and compare difficulties perceived in pairs of key situations. The computer calculates the coefficients of intra-individual consistency and group concordance data that is then projected on a large viewing screen. These coefficients are defined as

$$K = 100 - \frac{2400a}{k^3{}_{max} - k_{max}} \quad \text{if } k_{max} \text{ is odd, and}$$

$$K = 100 - \frac{2400a}{k^3{}_{max} - 4k_{max}} \quad \text{if } k_{max} \text{ is even.}$$

It did happen, however, that assessments of the key situations were inconsistent. For example, if an operator judged key situation 1 to be more difficult than key situation 2, key situation 2 more difficult than key situation 3, and then key situation 3 more difficult than key situation 1, the last assessment in this three-part chain, or triad, contradicted the initial assessment. That contradiction is what we refer to hereafter as a *decision loop*, or *inconsistent triad*. In the formulas, a is the number of decision loops, and k_{max} is the maximal number of key situations considered (actually six or seven).

After this procedure, the degree of concordance is tested on the basis of Kendall U statistics. The formula adapted to our case is

$$U = \frac{\sum\limits_{k_s, k_o}^{aggr.matrix} c^2{}_{k_s, k_o}}{\binom{n}{2}\binom{k_{max}}{2}} - \frac{n+1}{n-1}$$

where, in an aggregated matrix, k_s is the row index of perceived difficulty, k_o is the column index of perceived difficulty, n is the number of operators ($n = 5$), and c is the element of aggregated difficulty matrix in the corresponding row and column. If all coefficients of intrapersonal consistency and group concordance are high enough, the overall rank order of the perceived difficulty posed by key situations is computed on the basis of the mathematical model of COSMOS. If the coefficients are not high enough, the assessment is repeated to increase either individual consistency or group concordance.

4. *Assessing fellow trainees and their expected roles (also called* involvement) *and actual performances (also called* effectivity) *in each key situation.* Having

Figure 16.1 Arrangement of the assessment session of a simulated emergency using COSMOS (Computer Supported Method for Operators' Self-Assessment) method.

compared perceived difficulties of key situations, the trainees conduct individual situation-by-situation or operator-by-operator evaluations and self-assessments of expected roles (on a 3-point scale: *small*, *medium*, and *large*) and performance (on a 5-point scale: *unacceptable*, *poor*, *medium*, *good*, and *excellent*). Summary characteristics (and, when necessary and justified, individual characteristics) are presented to the crew in graphic form on a large projector screen. If warranted by the results of discussion, the assessment is repeated to increase objectivity and the degree of agreement within the group. The setting of the group self-assessment of performance and the individual self-assessment of performance is pictured in Figure 16.1. The hardware elements of COSMOS are seen in Figure 16.2.

5. *Playing back a two- to five-minute video recording of the simulated malfunction (the most critical key situation) as a basis for making self-assessment about their own behaviour.* Having the crew members view the recording of what they have judged to be the most difficult situation helps them recall the situation or refresh their memories. After watching the video clip, the crew members evaluate their effectivity in this critical situation along three dimensions: information-gathering, decision-making, and cooperation. They also assess the degree of their satisfaction (or dissatisfaction) with themselves on the five-point scale described above. Operators receive graphic feedback from these assessments in addition to summarised opinions formulated by their fellow operators. This aggregated feedback is pictured in Figure 16.3.

Figure 16.2 Hardware elements of COSMOS. Both the PC display and the large projector screen show the summarised results of the team evaluation in graphic form.

6. *Reviewing results and giving detailed feedback to the trainees in a discussion moderated by the instructor.* The instructor embeds the use of COSMOS into the process of evaluating performance. In other words, this method is never used in isolation. Rather, the instructor applies it as a flexible set of tools subordinated to pedagogical and didactic goals as a form of reinforcement. The instructor can also use the method to provoke debate by focusing on conflicting viewpoints and opinions, depending on the pedagogical context. In addition to the varied graphic feedback formats designed to inform operators, detailed and sophisticated numerical and tabular information about group and individual self-assessment are made available to the instructor by the mathematical model underlying the COSMOS method, the purpose being to deepen that person's understanding of, and insight into, group behaviour, dynamics, attitudes, and norms.

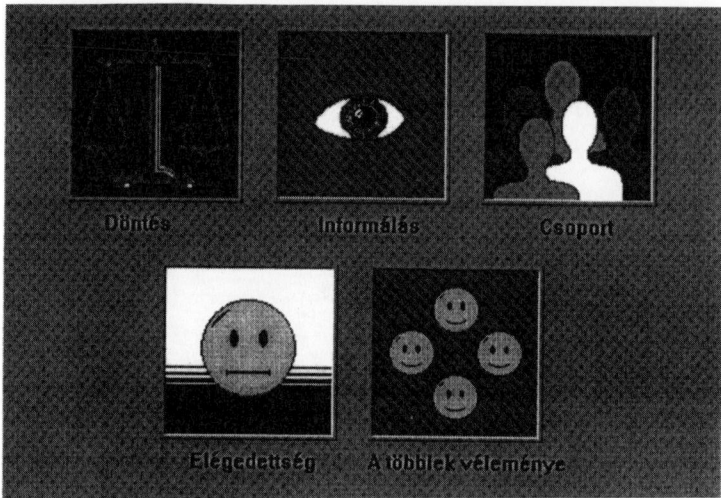

Figure 16.3   Aggregated feedback of the result of an operator's self-assessment. The colour of the icon (red = poor; green = good) corresponds to the qualitative aspects of the evaluation. The degree to which the icon's frame is filled reflects the quantitative aspects of the evaluation. Icons represent categories of self-assessment: Döntés = Soundness or effectiveness of decision-making; Informálás = Gathering or providing information within the team; Csoport = Personal impact or influence on team behaviour; Elégedettség = Self-satisfaction with recent performance; Többiek véleménye = Summarised assessment of other team members about the performance of the actual operator.

16.1   A CASE STUDY

After the COSMOS method had been tested in video-aided laboratory pilot studies involving members of the simulator training staff, it was also tested on ten trainee operator crews as part of their regular refresher simulation training sessions in April and May 1996. The aim of this series of experiments was to gain practical experience with applying the COSMOS method and to prepare for the regular use of self-assessment in simulation training sessions. (For details on all this work, see Antalovits, Izsó and Jenei, 1995; Antalovits, Izsó and Takács, 1995; Izsó and Antalovits, 1994, 1996, 1997.)

The scenario of a simulated emergency, which was developed by the staff of the simulator centre, was based upon six cognitively demanding key situations (see Table 16.1).

16.1.1   Subjects

Each of the nine five-member crews of operators taking part in the study consisted of a block electrician; a shift supervisor, who was the team leader; a reactor operator; a turbine mechanic; and a turbine operator.

Table 16.1   Key situations and the problem-solving actions required from a crew of nuclear power plant operators

Code of key situation	Problem-solving actions required of the crew
K1	Trip main circulating pump and stabilise power.
K2	Recognise rupture of steam generator and identify leakage.
K3	Keep feedwater pumps in operation and maintain water level.
K4	Realise that the leakage cannot be isolated and decrease the primary circuit pressure.
K5	Adjust the cooling speed to the appropriate level to avoid reactor shut down.
K6	Isolate steam generator to avoid spread of radioactivity.

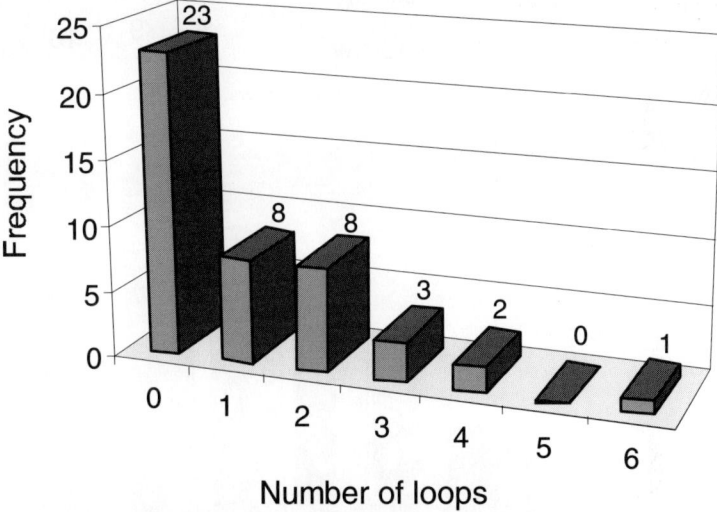

Figure 16.4   The number of inconsistent decision triads (loops) and their frequencies among 45 operators in simulated key situations.

16.1.2   Results

A technical fault prevented the results of the first session from being recorded. Across the remaining sessions, however, 39 of the 45 subjects (over 85%) produced fewer than three inconsistent decision triads, or loops (see Figure 16.4). In other words, the vast majority of the subjects were able to make consistent distinctions between the degrees of perceived difficulty posed by the key situations presented to them.

With the exception of one crew (no. 4 in Figure 16.5), there was also a significant degree of agreement on the ranking of the perceived difficulties when compared by crews as a whole.

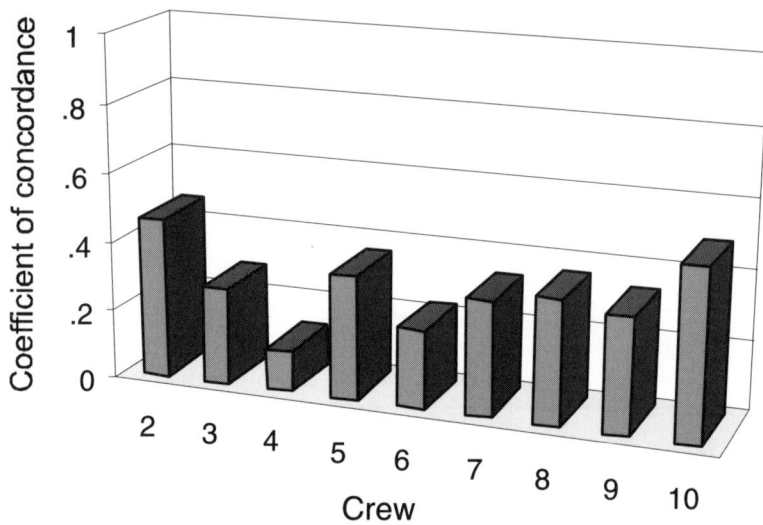

Figure 16.5   Coefficient of concordance among crews.

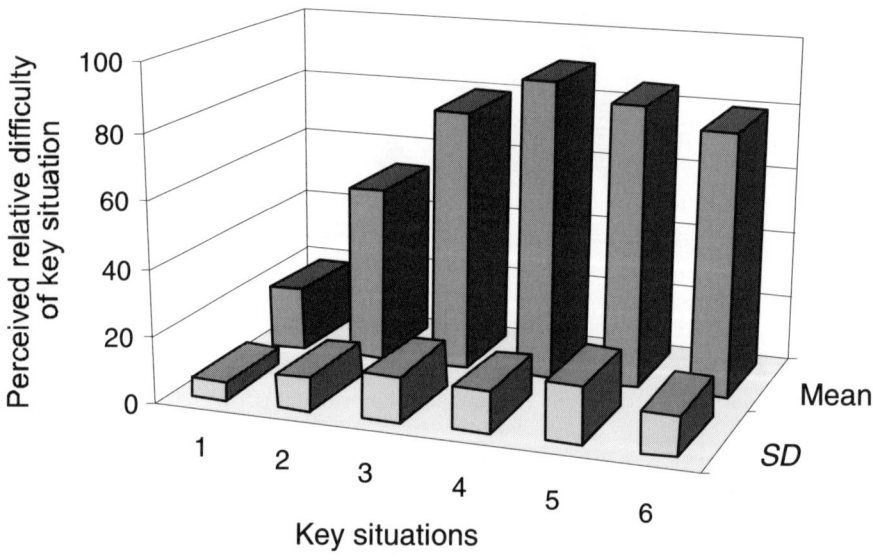

Figure 16.6   Means and *SD*s of crews' rankings of perceived relative difficulties entailed by key situations.

The rating of perceived difficulties was similar across all crews (see the relatively low *SD*s in Figure 16.6). Because the comparisons of difficulties were consistent between the members of all but one of the crews, this ranking of perceived difficulties can be taken as characteristic of most of the subjects.

Table 16.2 Comparative difficulty of key situations as rated by the five members of crew 4

Key situation	No. of crew members agreeing on 'more difficult than'						☑
	K1	K2	K3	K4	K5	K6	
K1	–	1	1	0	1	1	4
K2	4	–	2	2	3	2	13
K3	4	3	–	1	4	2	14
K4	5	3	4	–	4	4	20
K5	4	2	1	1	–	1	9
K6	4	3	3	1	4	–	15

Note: $U = 0.12$, $p > = .05$.

A close look at the assessments by individual crews, however, reveals that the comparisons of the difficulties posed by the experiment's six simulated key situations sometimes differed substantially across the given crew's five members. Such a pattern of divergence was found among the members of crew 4, who often split 2:3 in their assessments of the comparative difficulty posed by key situations – the highest possible level of disagreement in a five-member crew (see Table 16.2, particularly column K-2).

Why is this finding significant? The answer demonstrates how and why the COSMOS methodology functions as it does. In this part of the experiment, the crews had been instructed to rate the difficulties of the key situations as seen by the crew *as a whole*. However, the results reported in Table 16.2 guided follow-up study in which it eventually became clear that the reactor operator and turbine mechanic of crew 4 had responded narrowly from the vantage point of their own particular fields. The reactor operator's judgments had definitely revolved around concerns with primary circuits, and the judgments of the turbine mechanic suggested an overriding concern with secondary circuits. We designed COSMOS in a way that makes it possible to detect such excessively narrow professional thinking and reasoning, so that it can be channelled in a more desirable direction, since we are of the opinion that the basic unit of safety in terms of human factors is the crew as a whole, not the individual.

Our analysis of the ratings assigned by each of the members of crew 4 revealed that the reactor operator (see Table 16.3) and the turbine mechanic (see Table 16.4) had been perfectly consistent *individually* in their judgments of the difficulties they had perceived in the pairs of key situations that had been presented (K = 100 for both), but that each had largely disagreed with the other.

Comparing the levels of difficulty posed by key situations enabled us to identify details of technological problems that were major obstacles to crews. We were then able to design special teaching programmes to increase operators' preparedness in those areas.

Table 16.3 Perceived difficulties of key situations as rated by the reactor operator of crew 4

Key situation	No. of crew members agreeing on 'more difficult than'						ψ
	K1	K2	K3	K4	K5	K6	
K1	–	0	0	0	0	0	0
K2	1	–	0	0	1	1	3
K3	1	1	–	1	1	1	5
K4	1	1	0	–	1	1	4
K5	1	0	0	0	–	1	2
K6	1	0	0	0	0	–	1

Note: K = 100.

Table 16.4 Perceived difficulties of key situations as rated by the turbine operator of crew 4

Key situation	No. of crew members agreeing on 'more difficult than'						ψ
	K1	K2	K3	K4	K5	K6	
K1	–	1	1	0	1	1	4
K2	0	–	0	0	0	0	0
K3	0	1	–	0	1	0	0
K4	1	1	1	–	1	1	5
K5	0	1	0	0	–	0	1
K6	0	1	1	1	1	–	3

Note: K = 100.

Determining the distribution of decision loops that pertained to key situations made it possible to identify ambiguously defined key situations, that is, ones that signified different things to different crew members (see Figure 16.7).

As expected, we found that shift supervisors (team leaders) were the most consistent in their ratings of the key situations in the experiments (see Figure 16.8).

Determining the frequency of judgments distributed in a 2:3 ratio within the five-member crews, when they rated the comparative difficulties of key situations, made it possible to identify key situations in which crew members had different, but individually consistent, opinions (see Figure 16.9).

The role expected on average to be played by a given operator differed. One extreme was the position of block electrician, who was consistently expected to have a minor role. The methodological consequences of this attribution are that training scenarios should be designed in such a way that block electricians, too,

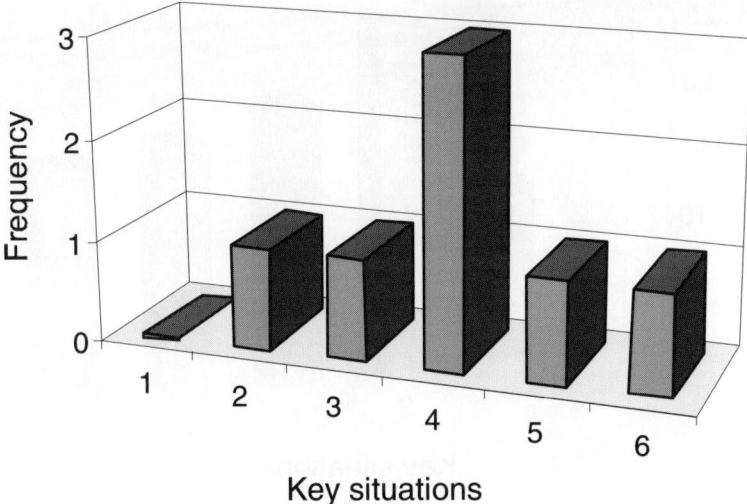

Figure 16.7 Ambiguity in the interpretation of situations by the operators indicated by the number of loops above than expected by chance.

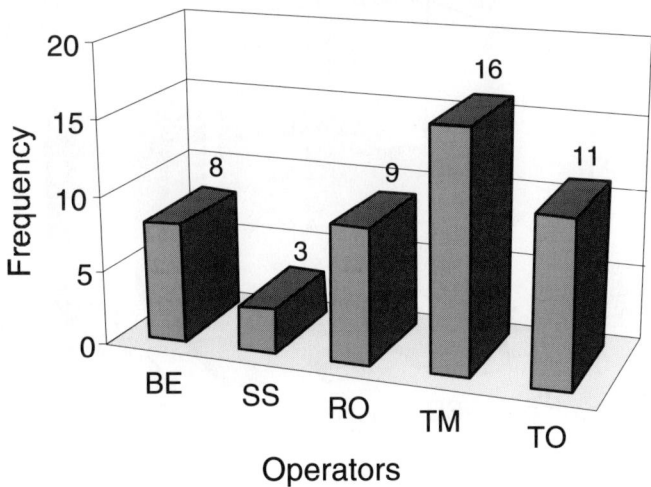

Figure 16.8 Level of consistency in assessing the perceived difficulties of key situations as indicated by the number of decision loops made by various operators. (BE: block electrician, SS: shift supervisor, RO: reactor operator, TM: turbine mechanic, TO: turbine operator)

have important tasks to perform. Turbine operators had a relatively high *SD*, a result perhaps indicating that expectations of the role to be performed by a given operator differed from one crew to the next (see Figure 16.10).

Delving further into this experiment's aggregated data presented in Figure 16.10, we analysed the relation between key situations and the roles expected of turbine

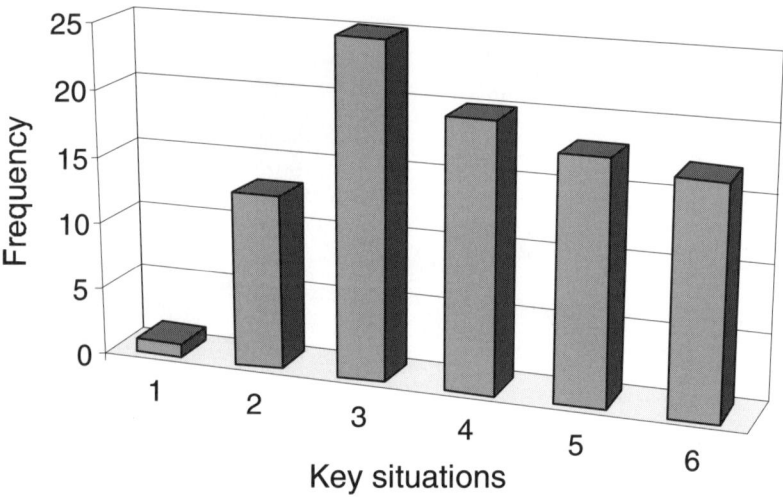

Figure 16.9 Frequency of 2:3 splits within five-member crews rating the perceived relative difficulties of paired key situations.

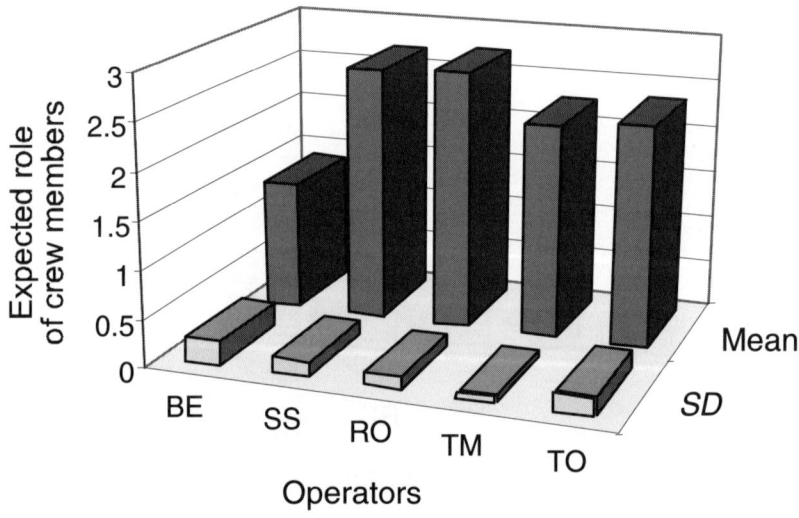

Figure 16.10 Means and *SD*s of expected roles of different operators as assessed by the members of different crews. (BE: block electrician, SS: shift supervisor, RO: reactor operator, TM: turbine mechanic, TO: turbine operator)

operators (see Figure 16.11). In key situation 3, for example, the expected role of turbine operators was the lowest rated, but had the highest *SD*s. This type of difference in perceived expected roles of crew members in nuclear power plants may reveal that identical procedures may be interpreted differently from crew to crew – a definite possibility for some procedures.

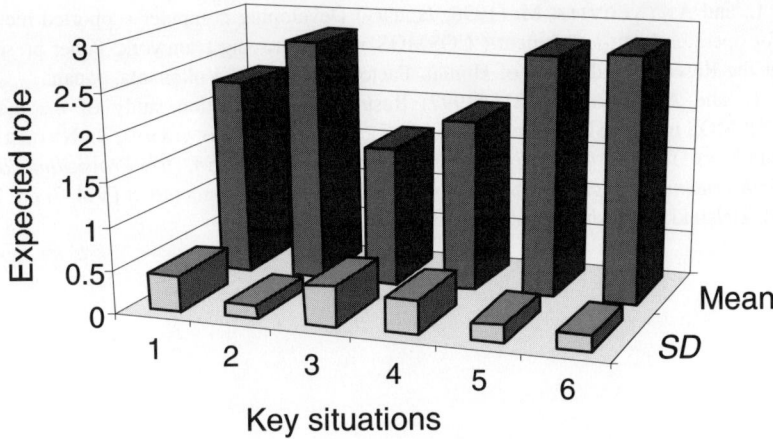

Figure 16.11 Means and *SD*s of expected roles of turbine operators as a function of the key situation.

Lastly, crew members in this experiment were generally able to utilise feedback from the computer, a capacity that sometimes led them to modify their original opinions.

In conclusion, the COSMOS method proved advantageous in its first trial application during simulator training. It helped identify sources of misunderstandings and disagreements and provided operators with quick and meaningful feedback, accelerating the learning process both technologically and socially.

References

ANTALOVITS, M., Izsó, L. and JENEI, C. (1995, April) A new method of performance assessment in NPP training simulators. Paper presented at the Seventh European Congress on Work and Organisational Psychology, Győr, Hungary.

ANTALOVITS, M., Izsó, L. and TAKÁCS, I. (1995, April) A frame for evaluation and self-evaluation of NPP personnel. Paper presented at the Seventh European Congress on Work and Organisational Psychology, Győr, Hungary.

IAEA (International Atomic Energy Agency) (1989) *Guidebook on training to establish and maintain the qualification and competence of nuclear power plant operations personnel* (IAEA-TECDOC-525). Vienna: IAEA.

IAEA (International Atomic Energy Agency) (1994) Role and responsibilities of management in NPP personnel training and competence. Presentations for the seminar of the IAEA model project in Hungary: Strengthening training for operational safety at Paks NPP (IAEA-TC-UN/9/01/1). Vienna: IAEA.

Izsó, L. and ANTALOVITS, M. (1994, October) What are the basic aims, criteria, and methods of developing application methodology for a full scope simulator? Paper presented at the First International Conference on Human Factors Research in Nuclear Power Operation, Berlin.

Izsó, L. and ANTALOVITS, M. (1996, January) Developing computer-supported methods for operators' self-assessment ('COSMOS') for improving teamwork. Paper presented at the Research and Study of Human Factors Workshop, Yokohama, Japan.

Izsó, L. and ANTALOVITS, M. (1997) Results of a validation study of the method COSMOS in NPP Simulator Sessions. In SEPPÄLÄ, P., LUOPAJÄRVI, T., NYGÅRD, C. and MATTILA, M. (eds), *From experience to innovation, IEA, '97: Proceedings of the 13th Triennial Congress of the International Ergonomics Association* (Vol. 7, pp. 231–3). Helsinki: Finnish Institute of Occupational Health.

CHAPTER SEVENTEEN

Knowledge acquisition through repeated theoretical and practical training

KUNIHIDE SASOU, TOMOHIRO SUZUKI AND SEIICHI YOSHIMURA

Central Research Institute of Electric Power Industry

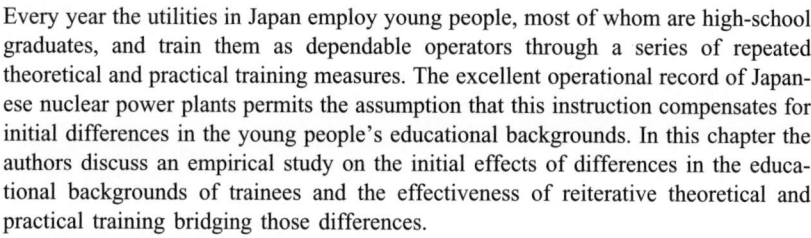

Every year the utilities in Japan employ young people, most of whom are high-school graduates, and train them as dependable operators through a series of repeated theoretical and practical training measures. The excellent operational record of Japanese nuclear power plants permits the assumption that this instruction compensates for initial differences in the young people's educational backgrounds. In this chapter the authors discuss an empirical study on the initial effects of differences in the educational backgrounds of trainees and the effectiveness of reiterative theoretical and practical training bridging those differences.

With the rapid progress of science and technology, fully automatic operational systems are being introduced into automobile assembly factories, hydroelectric power plants, thermoelectric power plants, and other major industrial facilities. Some areas of the industrial sector, however, have not begun to use fully automatic systems. One example is nuclear power plants. Ever since Japan's first nuclear power plant began commercial operation some 30 years ago, this industry has relied on teams of operators. Every year the utilities in Japan employ young people and train them to be dependable operators through a series of reiterative theoretical and practical training measures (JAIF, 1995). Some of these people are university graduates, but most are high-school graduates. Although the ratio of nuclear power plant operators who have graduated from universities is said to be lower in Japan than in western countries, the operational record of nuclear power plants in Japan is excellent (Mishiro, 1995). This fact suggests that the repeated theoretical and practical training measures that Japanese operators undergo compensate for the initial differences in their educational backgrounds. But what are the initial effects of the differences in these educational backgrounds, and how effectively does this repeated theoretical and practical training close the gap? To find out, we conducted the two experiments described and discussed below.

17.1 EXPERIMENT 1

17.1.1 Method

We conducted experiments to ascertain the initial effects of differences in the educational backgrounds of subjects (trainees). An experimental plant simulator was used, and the subjects were given the same theoretical and practical training to be operators of the simulator. Post-training pre-trial examination, and four experimental trials were also conducted.

17.1.2 Facility

For both experiments in this study, we used the experimental plant simulator at the Central Research Institute of the Electric Power Industry (CREPI) in Tokyo (Suzuki *et al.*, 1995). It consisted of a control panel, a workstation, and a recording device. The control panel measured 4 m in width by 1.8 m in height, and had 55 alarms, 40 indicators, 6 chart recorders, and 29 switches. The workstation simulated a very simplified power plant (see Figure 17.1). The communication and the behaviour of the operators were recorded on laser discs.

17.1.3 Subjects

There were two teams of three men, each with different educational backgrounds. Team A consisted of three university postgraduates majoring in nuclear engineering, aged 24 to 29. Team B consisted of three university students majoring in economics, literature, or social science, aged 19 to 25.

17.1.4 Theoretical and practical training

At the outset no one in either team knew how to operate a plant, so we gave the members two weeks of theoretical and practical training, each team separately. These measures started at 10 a.m. and ended at 4 p.m. (including a one-hour lunch break). In the theoretical part, we used a textbook we had made ourselves. It described the roles of each operator, the functions of the equipment involved,

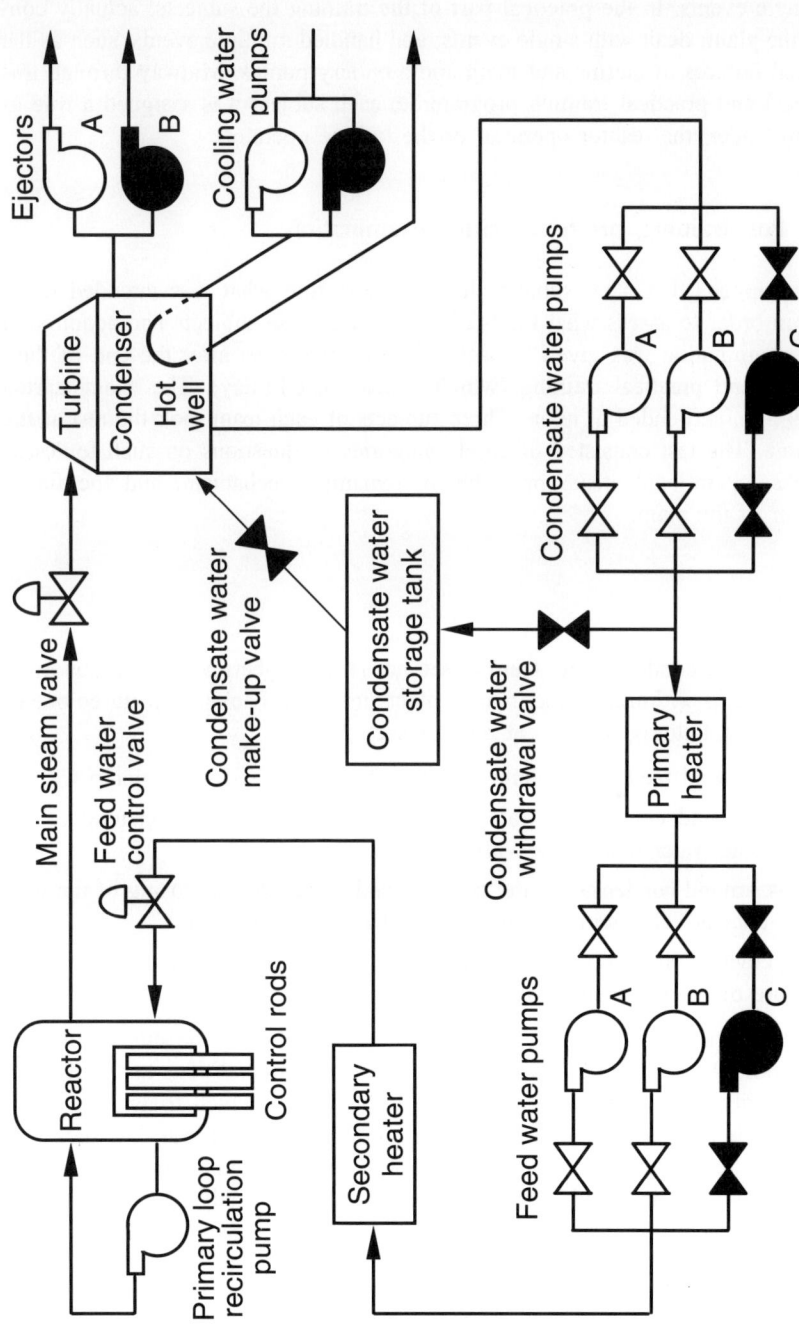

Figure 17.1 Plant model in the experimental simulator at the Central Research Institute of the Electric Power Industry.

operational procedures (e.g. changing over the lines), and procedures for dealing with single events. In the practical part of the training the subjects: actually controlled the plant; dealt with single events; and handled multiple events, such as the combined failures of alarms and main and auxiliary pumps. Midway through this theoretical and practical training programme, each subject was assigned a role as the team leader, the reactor operator, or the turbine operator.

17.1.5 Post-training, pre-trial written examination

It often happens that what subjects learn differs from what was intended to be taught. In order to assess what kinds of knowledge these subjects had acquired, a written examination was given to each of them three days after the end of their theoretical and practical training (which ended on a Friday). The exam started at 10:00 a.m. and ended at noon. Three subjects of each team took the test at the same time. The test consisted of eight categories of questions on such topics as interlocks, operational steps (procedures), dynamic mechanism, and the future behaviour of the plant.

17.1.6 Trials

Four trials were conducted per team to see how the members reacted to abnormal operating events within the experimental plant simulator. Each trial featured one of four simulated failures, a different one for each trial:

1 A leak around feed water pump B with a failure of feed water pump C.
2 A leak around condensate water pump A and an accidental opening of the condensate water withdrawal valve.
3 A leak around condensate water pump B and an accidental closing of the feed water control valve with a failure of condensate water pump C.
4 An accidental closing of feed water control valve with a failure of the warning 'Failure of reactor water level controller'.

The first two trials were conducted on the afternoon of the day the subject took the test. In addition to these two conditions, we set several conditions and combined them in various ways so that the subjects did not know which trials were to be analysed. The first trial day ended at 4:00 p.m. On the second trial day, the other two trials were conducted in the same way. The behaviour and communication of the subjects were videotaped. The tapes were later used to analyse their behaviour.

17.1.7 Results and discussion: post-training, pre-trial written examination

Table 17.1 shows the total number of incorrect responses on the examination, broken down by team and test item. The major differences between the results of

Table 17.1 Incorrect responses to written post-training questions assessing operators' ability to respond to, analyse and predict simulated behaviour of a power plant

Team	Interlocks	Sequence of action	Static mechanism	Dynamic mechanism	Causal identification	Responses for cases	Prediction of plant behaviour	Future Responses
A	0	2	0	5	0	1	2	2
B	8	4	2	b	9	4	23	2

Note: Total number of questions for each subject: 132.

the two teams occurred in the categories labelled interlocks, causal identification, and future prediction.

Interlocks

This category contained questions about conditions that activate the automatic reactor scram, feed water pump trips, condensate water pump trips, and so on. The information in Table 17.1 leads us to suspect that the members of Team B could have avoided unnecessary activations of interlocks if they had paid enough attention to relevant parameters.

Causal identification

This category contained questions designed to test the subject's ability to deduce several causes from a given condition of the plant. An example was the item 'What are the causes of the deterioration of the condenser vacuum?' The possible causes in this simulator were 'a failure of an ejector,' 'a failure of a recirculation pump,' 'the increase of seawater temperature,' or some combination of these possibilities. This result leads us to suspect that Team B had greater difficulty identifying the cause(s) of the event than had Team A.

Prediction of plant behaviour

This category contained questions designed to test the subject's ability to predict plant behaviour if no action were taken to correct the abnormal event being faced. For example, 'What would be the plant's behaviour if you did nothing against the decrease in the condensate hotwell water level?' The ideal response is that continued decrease in the condensate hotwell water level will sequentially activate the warnings 'Hotwell water level low' and 'Hotwell water level low low,' which will activate the trips of all condensate water pumps, a consequence that, in turn, will activate the warnings 'Feed water pumps suction pressure low,' 'Feed water pumps suction pressure low low,' and so forth. Table 17.1 shows that Team B did not predict this plant's behaviour well.

17.1.8　Observed behaviour

The characteristics reflected on the written examination of Team B's members were also observed in the way they dealt with abnormal operating conditions.

Unnecessary activation of interlocks

The behaviour of Team B when faced with abnormal event 3 exemplified the unnecessary activation of interlocks. One feed water pump and one condensate water pump were in operation. The feed water control valve was being operated manually. The characteristics of this plant required the reduction of reactor power. As power declined, poor manual operation of the feed water control valve caused a rapid decrease in the level of water in the reactor. In order to restore the proper water level, the members of Team B opened the feed water control valve. The opening of this valve decreased the suction pressure in the feed water pumps. The decrease triggered the warning 'Feed water pump suction pressure low low,' which activated the interlock that shut off all feed water pumps.

Difficulty identifying causes

An example of difficulty with identifying causes of abnormal operating conditions was observed during abnormal event 4. The first warning, 'Feed water pump suction flow low,' went off. An operator noticed that the reactor water level was decreasing and that the feed water control valve was closing, so the members of Team B started to open this valve manually. This response restored the plant's condition. However, they did not identify the cause of this event, a failure apparent in the leader's post-action query: 'Why was the reactor's water level decreasing so much?'

Claptrap responses

This behaviour marked the typical difference between Teams A and B. The malfunction was a leak around a feed water pump, which triggered the first warning, 'Leak in turbine building'. As shown in Figure 17.2, the member of Team A opened the condensate water supply valve immediately after identifying the point of the leak, since the team expected a decrease in the hotwell water level. The action of the team members enabled them to avoid having the interlock to shut off the condensate water pumps. By contrast, the members of Team B merely started the auxiliary pump after they had identified the point of the leak. They did not open the hotwell water supply valve until the warning 'Hotwell water level low' went off. The water level reached the double low warning level and activated the interlock that tripped the condensate water pumps. As a result, Team B faced difficulties that Team A did not.

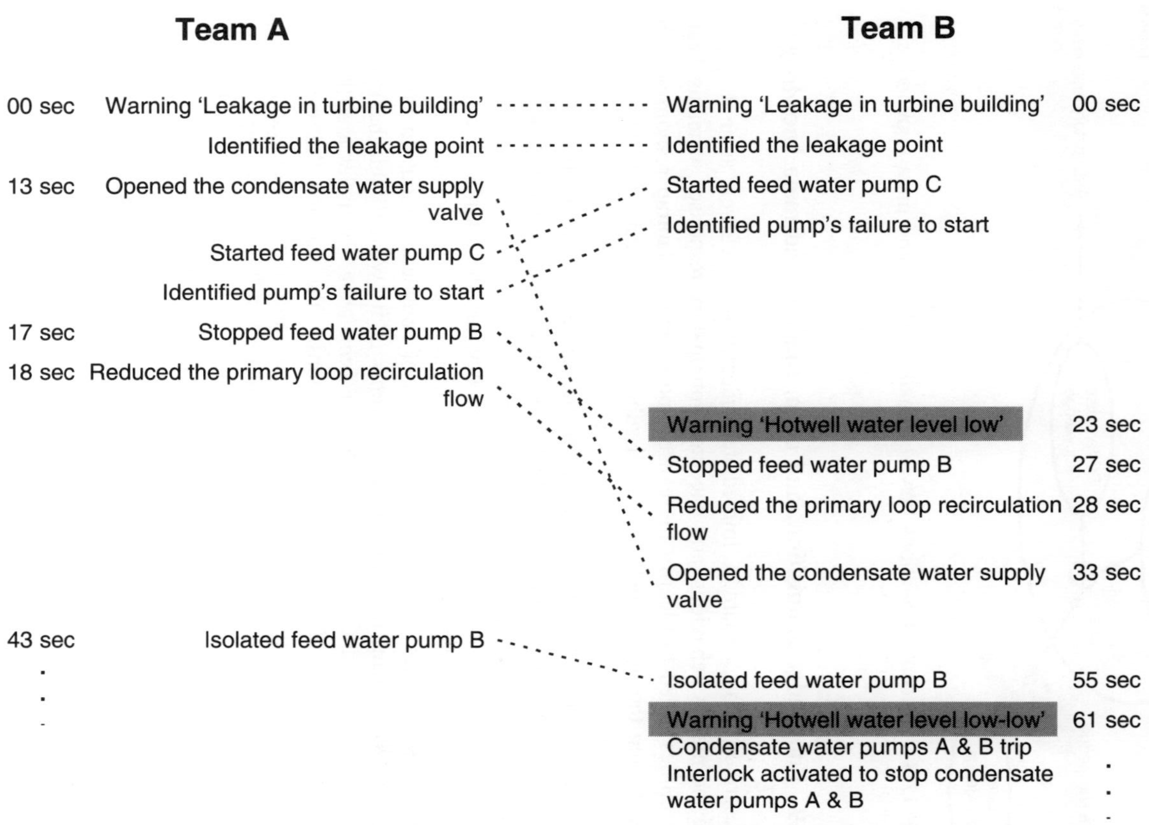

Team A

00 sec   Warning 'Leakage in turbine building'

          Identified the leakage point

13 sec   Opened the condensate water supply valve

          Started feed water pump C

          Identified pump's failure to start

17 sec   Stopped feed water pump B

18 sec   Reduced the primary loop recirculation flow

43 sec   Isolated feed water pump B

Team B

00 sec   Warning 'Leakage in turbine building'

          Identified the leakage point

          Started feed water pump C

          Identified pump's failure to start

23 sec   Warning 'Hotwell water level low'

27 sec   Stopped feed water pump B

28 sec   Reduced the primary loop recirculation flow

33 sec   Opened the condensate water supply valve

55 sec   Isolated feed water pump B

61 sec   Warning 'Hotwell water level low-low' Condensate water pumps A & B trip Interlock activated to stop condensate water pumps A & B

Figure 17.2 Differences between Teams A and B in their responses to simulated abnormal events in the experimental plant simulator.

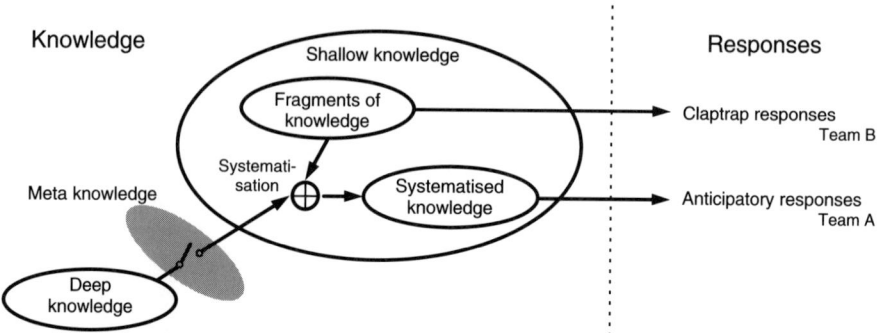

Figure 17.3   Conceptual relations between shallow, deep, and meta knowledge

17.1.9   Knowledge acquisition, and deep, shallow, and meta knowledge

As described above, Team A immediately identified the causes, accurately predicted the plant's behaviour, and executed anticipatory responses, whereas Team
B sometimes failed to identify the causes and executed claptrap responses without predicting the plant's behaviour. All subjects in both teams had received the
same theoretical and practical training, but it seems that the knowledge of the
subjects in Team A differed from that of the subjects in Team B. We attempt
now to describe these differences in terms of deep, shallow, and meta knowledge
(MITI, 1985).

Figure 17.3 shows the conceptual relations between these kinds of knowledge.
Through the theoretical and practical training given as part of this experiment,
the members of Team B acquired a great deal of knowledge about operating the
simulated power plant and dealing with events in it. We call this background
pieces of knowledge and classify it as *shallow knowledge*. Examples are 'the hotwell water level will decrease if a leak develops,' 'the warning 'hotwell water
level' will go off when the level reaches 115 cm,' and 'the hotwell water supply
valve should be opened to restore the hotwell water level'. It seems that they used
the knowledge given to them as it was. As a result, their responses lagged behind
the plant's behaviour. By contrast, the behaviour of Team A, when confronted
with a leak around a pump, seems to indicate that these subjects created their own
knowledge, which in this study is classified as *systematised knowledge*. For instance, they immediately opened the hotwell water supply valve after identifying
the leak point.

How did the members of Team A create their own systematised knowledge?
Perhaps they drew on their deep and meta knowledge from their background in
engineering. They may well have been better at understanding the dynamic behaviour of a plant than were the members of Team B. Team A's resulting systematised
knowledge enabled the members to make anticipatory responses, identify causes
quickly, and so forth.

17.2 EXPERIMENT 2

17.2.1 Method

To measure the effectiveness of the repeated theoretical and practical training measures with the leader of Team B (hereafter referred to as Subject X), two other series of experiments were performed.

17.2.2 Subjects and theoretical and practical training

Team C

After the completion of Experiment 1, another team (C) was formed with Subject X as its leader. The two other operators of this team were new to the experiment. They were also male university students, aged 22 and 25, and they were given the 30 hours theoretical and practical training with Subject X. After this phase, the written examination and the four trials were conducted in the same way as in Experiment 1. Having already participated in Experiment 1, Subject X had by now spent 80 hours on theoretical and practical training.

Team D

After the completion of the experiment with Team C, one more team (D) was formed. Subject X was made the leader of this new team as well. The reactor operator of Team D was Team C's reactor operator. The turbine operator, a 20-year-old university student, was new to the experiment. He received 20 hours of theoretical and practical training with the others. As in Experiment 1, the phase of theoretical and practical training was followed by a written examination and a series of four trials experiments. By this point Subject X had spent 100 hours on theoretical and practical training (50 hours with Team B, 30 hours with Team C, and 20 hours with Team D).

17.2.3 Results and discussion: post-training, pre-trial written examination

Figure 17.4 shows the change in Subject X's results over the three examinations he took as the leader of Teams B, C, and D, respectively. Although some categories show temporary increases in incorrect answers, his results improved overall, especially in the categories 'Interlock,' 'Causal identification,' and 'Prediction'. Figure 17.4 indicates that the teams led by Subject X can be expected to predict plant behaviour, take anticipatory responses, identify cause of events smoothly, and avoid unnecessary activations of the interlocks.

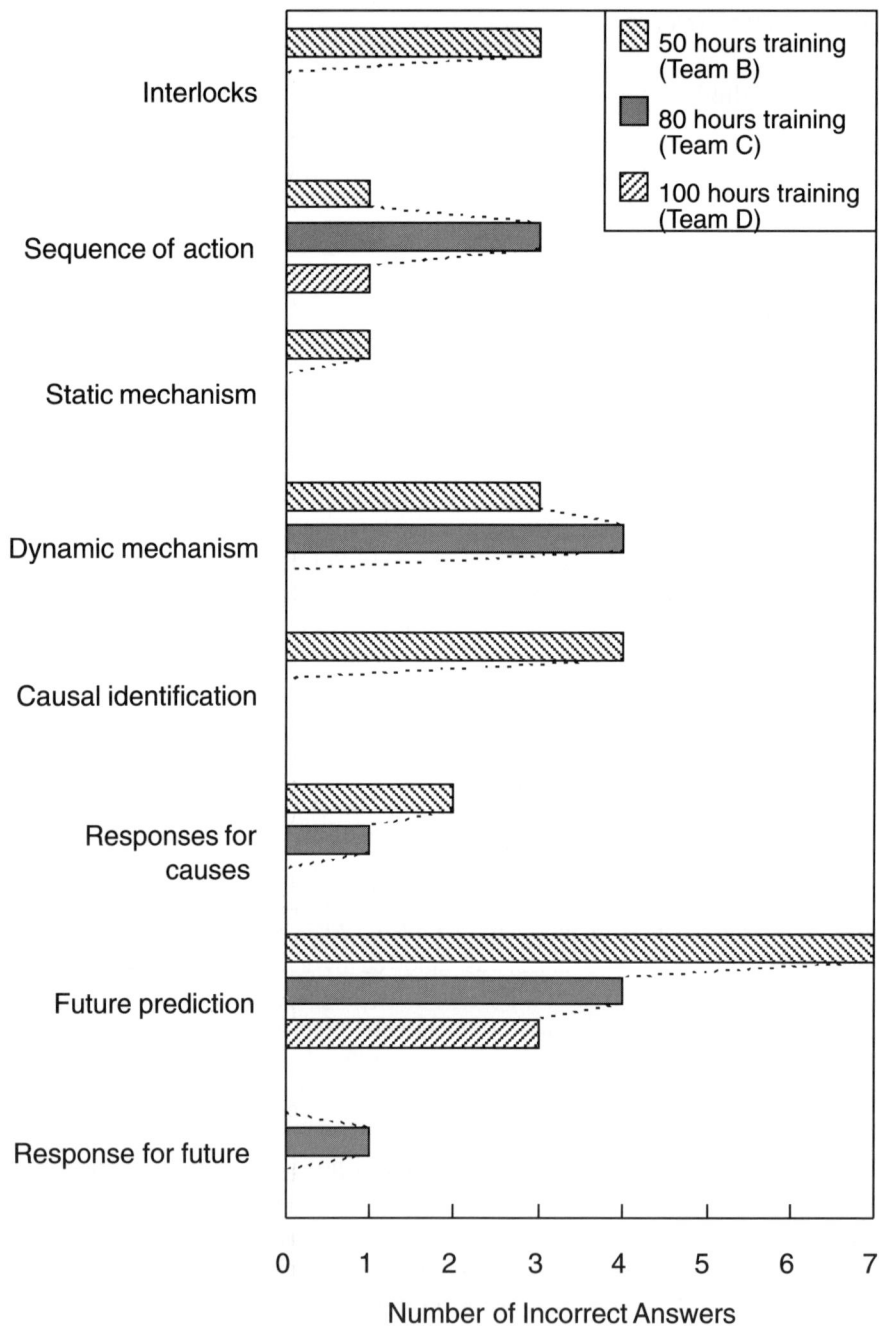

Figure 17.4 Development in the examination results of Subject X.

17.2.4 Improvement in performance observed in the experiment

Execution of anticipatory responses

When Subject X was the leader of Team B, the supply of the hotwell water was not ensured by Team B until the warning 'Hotwell water level low' went off. However, the members of Teams C and D ensured the supply immediately, as instructed by Subject X after they had started the auxiliary condensate water pump. Figure 17.5 shows an example of the changes in response described above. Because the teams were facing abnormal event 2 in this example, it could be that Subject X understood the plant's behaviour and the importance of a stable hotwell water level, because of the repeated theoretical and practical training he had received. We assume that it was that awareness that led him to order the immediate resupply of hotwell water.

Decrease in unnecessary activation of interlocks

In abnormal event 4 the members of Team B opened the feed water control valve to stop the decrease in the reactor water level. This action was appropriate in that situation. However, they paid no heed to the effects of their action: a decrease in the suction pressure of the feed water pump. They were therefore faced with the unnecessary interlock of the feed water pumps trip. By contrast, when giving the command to open the feed water control valve, and thereby restore the reactor water level, Subject X, as the leader of Teams C and D, instructed the turbine operators to pay attention at the same time to the suction pressure of the feed water pump. Subject X seemed to understand the relation between the suction pressure of the feed water pumps and the opening of the feed water control valve, by virtue of repeated theoretical and practical training. We presume it was that insight that enabled Teams C and D to avoid the unnecessary activation of interlocks.

17.2.5 Improvement in knowledge through repeated theoretical and practical training

As the leader of Team D, Subject X received the same theoretical and practical training as he had when he was the leader of Team C. Nevertheless, his results on the examination improved, and the performance of Teams C and D approached that of Team A. Subject X probably acquired accurate knowledge about the plant's configuration, operational rules, dynamic characteristics, the functions of the equipment, and so forth, via the repeated theoretical and practical training he had received. However, it is impossible to make anticipatory responses like Team A's simply by having this accurate knowledge. He needed other resources for them. Some of the possible resources for improving his expertise are deep and meta knowledge, and operational experience. His deep and meta knowledge may have grown during the instruction he received during our two experiments, and he probably

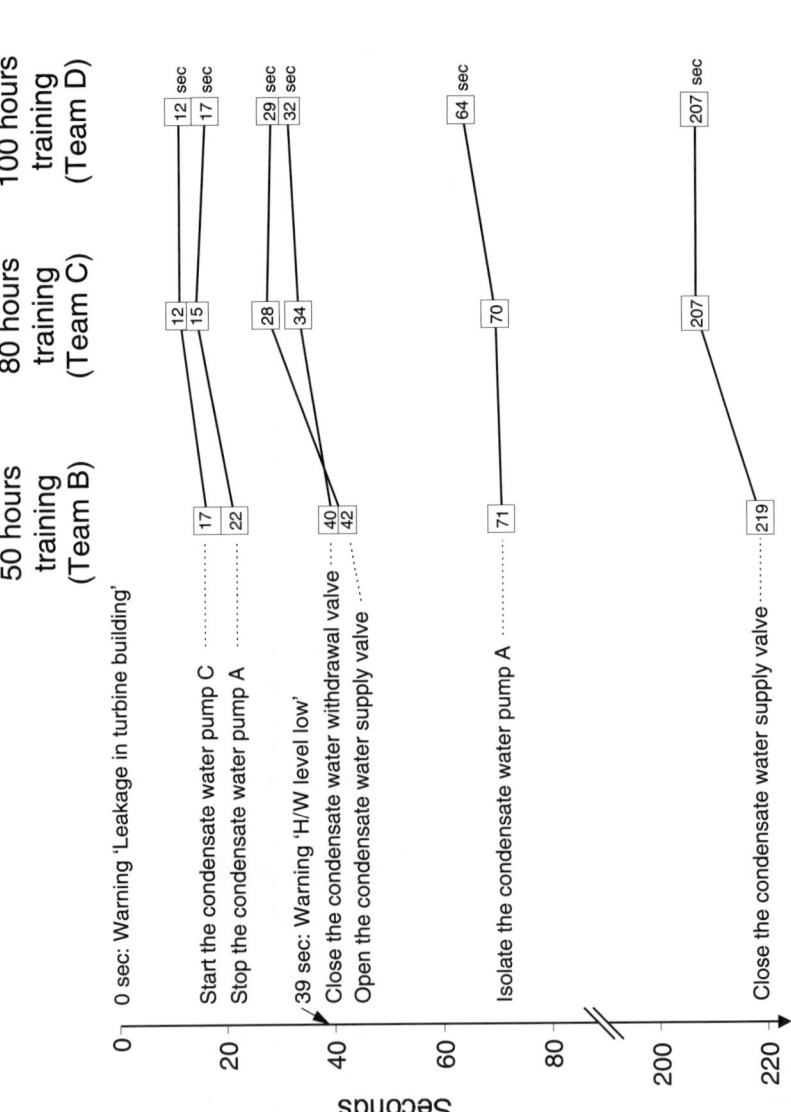

Figure 17.5 Changes in the responses of nuclear power plant operators confronted by a simulated abnormal event.

acquired the operational experience through the training given him. There might have been other resources that increased what he knew. It is impossible to clarify what they were. This experiment does, however, at least demonstrate that repeated theoretical and practical training worked well to improve Subject X's knowledge.

17.3 CONCLUSION

Some 30 years have passed since Japan's first nuclear power plant commenced commercial operation. Some operators who participated in the design, construction, and running of the plants at that time, and who had a great deal of experience with abnormal operating conditions, have retired or are going to retire soon. That loss of expertise is currently compounded by the low accident rate at Japan's nuclear power plants, a track record that affords very few accidents or abnormal events from which operators can learn. That fact makes the theoretical and practical training of nuclear power plant operators increasingly important.

In the two experiments reported in this chapter, the differences between the educational backgrounds of the operators were reflected in the differential ability to avoid the unnecessary activation of interlocks, identify the causes of abnormal operating conditions quickly, predict the plant's behaviour, and execute anticipatory responses. We have tried to describe these behavioural differences in terms of deep, shallow, and meta knowledge. It was assumed that subjects with higher educational backgrounds used not only the knowledge conveyed through repeated theoretical and practical training measures, but also the knowledge that they, the subjects, already had. We surmise that these resources enabled them to create their own systematised knowledge, take anticipatory action, easily identify the causes of events, and avoid unnecessary activations of interlocks.

In consideration of the costs of theoretical and practical training, it is advisable to compensate for differences in knowledge more quickly than has been the case in the nuclear power industry thus far. This study does not confirm, but does suggest, that those differences stem from educational backgrounds. Ensuring that the training of nuclear power plant operators takes their educational background into account might therefore enhance their effectiveness on the job.

References

JAIF (Japanese Atomic Industrial Forum) (1995) *Nuclear Yearbook, 1995.* Tokyo: JAIF.
MISHIRO, M. (1995, September) *Operational safety issues of nuclear power plants in Japan.* Paper presented at the International Atomic Energy Agency International Conference on Advances in the Operational Safety of Nuclear Power Plants, Vienna.
MITI (Ministry of International Trade and Industry) (1985) *Knowledge-based systems.* Tokyo: Ministry of Finance.
SUZUKI, T., SASOU, K., TAKANO, K., YOSHIMURA, S., HARAOKA, K., and KITAMURA, M. (1995, October) *Development of SYBORG–Manned experiments.* Paper presented at the fall meeting of the Atomic Energy Society of Japan, Tokai Village, Ibaraki, Japan.

PART FOUR

Learning from experience

Introduction

The contributions in this concluding part concern various approaches to improving the safety of nuclear installations through different strategies for analysing aspects of human factors. The chapters cover a cross-section of human factors activities and therefore cut across various systems levels in the sociotechnical system.

Yamaguchi and Tanaka outline the human-factors safety activities of the Japanese electric power industry. Kojima, Takano, and Suzuki describe a comprehensive analysis of human errors in Japanese nuclear power plants over the last 25 years. Shinohara, Kotani, and Tsukada develop an innovative method for the in-depth analysis of case reports, whereby statements drawn from written descriptions of incidents and near-incidents are clustered and compared in a process of content analysis, intended to identify the human factors variables affecting the safety of operations in large technical systems. The increasing degree to which maintenance activities affect the safety of nuclear installations is the topic of the contribution by Isobe, Shibuya, and Tabata. They describe a computer-aided support system for maintenance workers. Last but not least, Tokuine reviews the human-error prevention activities of the Kansai Electric Power Company, one of the leading power companies in Japan.

CHAPTER EIGHTEEN

An outline of human factors studies conducted by the Japanese electric power industry[1]

OSAMU YAMAGUCHI[1] AND ISAO TANAKA[2]

[1] *Japan Atomic Power Company, Tokyo*
[2] *Human Factors Research Center, CRIEPI, Tokyo*

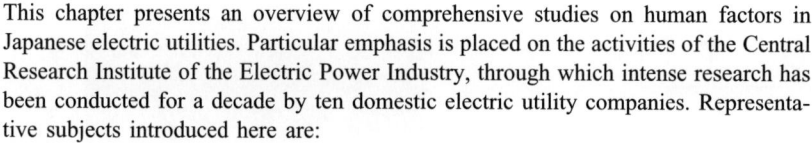

This chapter presents an overview of comprehensive studies on human factors in Japanese electric utilities. Particular emphasis is placed on the activities of the Central Research Institute of the Electric Power Industry, through which intense research has been conducted for a decade by ten domestic electric utility companies. Representative subjects introduced here are:

- the analysis and assessment of human-related events that have occurred within the last 30 years;

- the establishment of a human factors database including analysed cases, good practices, and related literature;

- the creation of a system with which to predict human behaviour;

- the creation of a behaviour simulation system of nuclear power operators coping with anomalies.

Some of the results have been judged to be effective tools for improving human performance in fields of operation and maintenance.

The effort to build social trust in nuclear power plants hinges largely on what is known as public acceptance. That is, people can well understand the importance of nuclear power generation and have no negative impressions of it, since the reliability of the operations is sufficiently high. Because public acceptance may

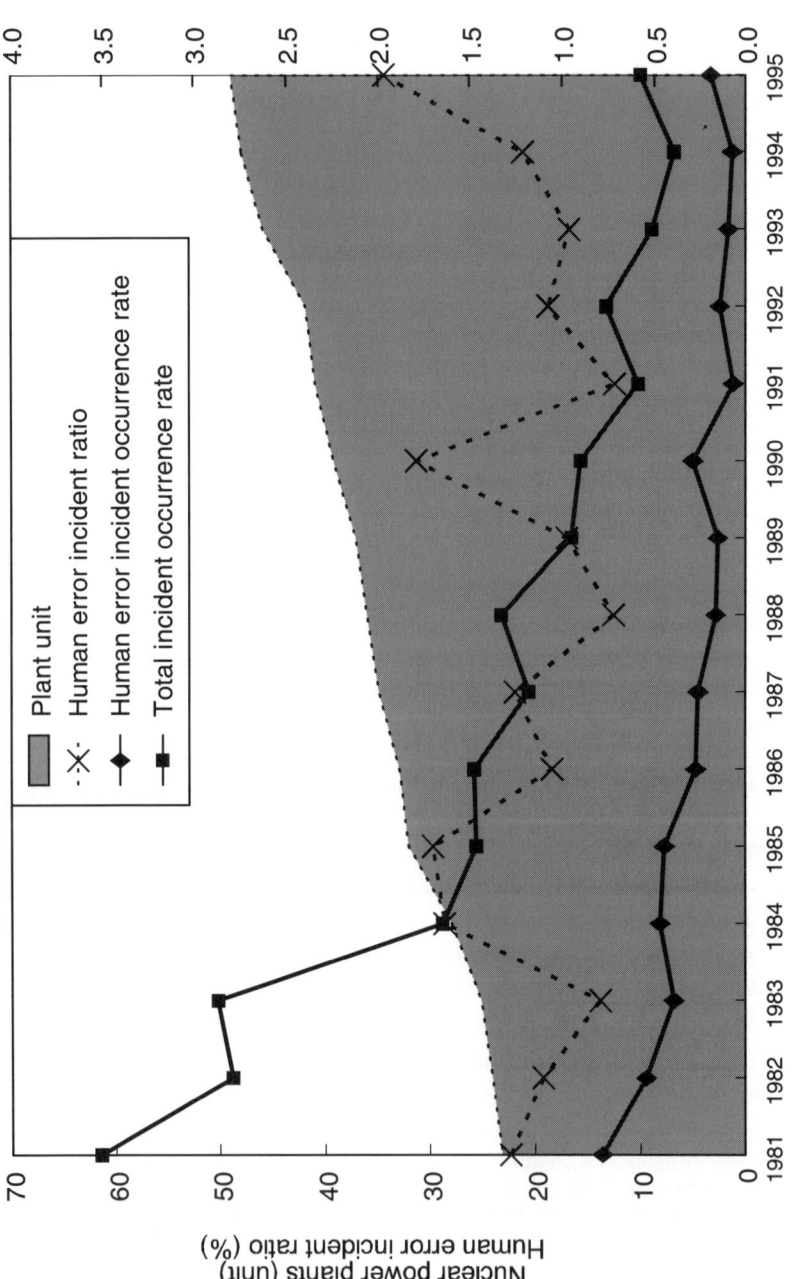

Figure 18.1 Trend of the incident frequency at nuclear power plants in Japan.

determine the direction and prospects of nuclear power's future development for commercial use, it should be given highest priority in that industry. A variety of factors impinge on social trust, however. The underlying contention in this chapter is that continuous attempts to reduce incidents in nuclear power plants is one of the most effective ways to improve public acceptance of the nuclear power.

To ensure the safety and reliability of nuclear power plants, a great deal was done in the 1970s and early 1980s to modify and standardise the designs, materials, equipment, and components of these installations. Mostly, significant improvements in hardware reduced the rate of incidents to approximately one quarter of their initial level (see Figure 18.1; Aisaka, 1988). As a result, the frequency of unplanned outages in Japanese utilities declined, and now remains at a level about one order lower than that in other nations (Aisaka, 1986). Despite the observed steady decline in the number of incidents, however, around 20 per cent of all those reported to regulatory bodies in Japan's nuclear power industry are still attributed to human error. This figure shows that continuous effort needs to be invested in anticipating and preventing human error in the operation of nuclear power plants.

Only in the last 20 years or so has analysis focused on human factors involved in the operation and maintenance of nuclear power plants. The first study in which human factors were considered in relation to the reliability of nuclear power plants was issued by the US Nuclear Regulatory Committee (NRC, 1975). Known as the Rasmussen Report, this document brought the public to a realisation of the importance of human factors. In later years it was this awareness that enabled people to recognise the role of human error in nuclear accidents, such as the one at Three Mile Island in 1979, and in incidents such as those at Ginna in 1982 and at Davis Besse and Rancho Seco in 1985 (the latter three cases involving US pressurised water reactors), Bhopal in 1984 (a chemical plant in India), and Chernobyl in 1986 (a graphite-moderated reactor). Investigations into these major accidents (Kuroda, 1986; Light Science, 1992) revealed that most of them had occurred between midnight and 4:00 a.m., when the operators' arousal level seemed to be at its lowest. With such findings identifying human factors as frequent primary contributors to incidents at relatively large plants, research on human factors has since accelerated in Japanese domestic electric utilities, especially in the field of nuclear power.

18.1 OBJECTIVES, TOPICS, AND ORGANISATION OF RESEARCH IN THE ELECTRIC UTILITIES

In Japan systematic research on human factors in nuclear power plants was begun in 1987, with the state and private industry cooperating closely to continue improving the safety and reliability of nuclear power plants. The government, for example, established the Institute of Human Factors within the Nuclear Power Engineering Corporation (NUPEC). The private sector, represented by the Federation of Electric Power Companies (FEPCO), founded the Human Factors Research Center (HFC) within the Central Research Institute of the Electric Power Industry (CRIEPI). These facilities were followed up by human factors laboratories in the Japan Atomic Energy Institute (JAERI) in 1989 and the Tokyo Electric Power Company (TEPCO)

Table 18.1   Current research areas and topics in representative organisations in Japan

Organisation	Topics[a]					Other topics
	(1)	(2)	(3)	(4)	(5)	
CRIEPI Central Research Institute of the Electric Power Industry	O		O	O	O	Sociotechnical approach
NUPEC Nuclear Power Engineering Corporation	O		O	O	O	Experimental error-rate estimation
JAERI Japanese Atomic Energy Research Institute	O		O	O		
Hitachi		O	O			
Toshiba		O	O			Computer maintenance aid
Mitsubishi		O	O	O		
TEPCO Tokyo Electric Power Company		O	O			Team efficiency human factors education
KEPCO Kansai Electric Power Company		O	O			Human factors education
Tokyo University				O		
Kyoto University				O	O	Training tool
Tohoku University			O			

[a] (1) Human reliability (including probabilistic safety assessment), (2) Improvement in the human–machine interface (including advanced interface), (3) Operator Support and Instruction System, (4) Human behaviour experiment (including incident analysis), (5) Operator simulation modelling.

in 1991. In 1992 the Kansai Electric Power Company (KEPCO) sponsored the creation of the Institute of Nuclear Safety Systems (INSS).

As in other countries, research related to human factors has been classified into roughly five areas:

- human reliability assessment (including probabilistic safety assessment);
- improvement in the human–machine interface (including the development of advanced control panels);
- operator support systems;
- the measurement, analysis, and assessment of human behaviour (including incident analysis);
- operators' cognitive models.

The ultimate objective of this work is to reduce human error in the operation and maintenance of nuclear power plants. To this end, a wide variety of topics is researched (see Table 18.1), the intention being to apply the findings directly or

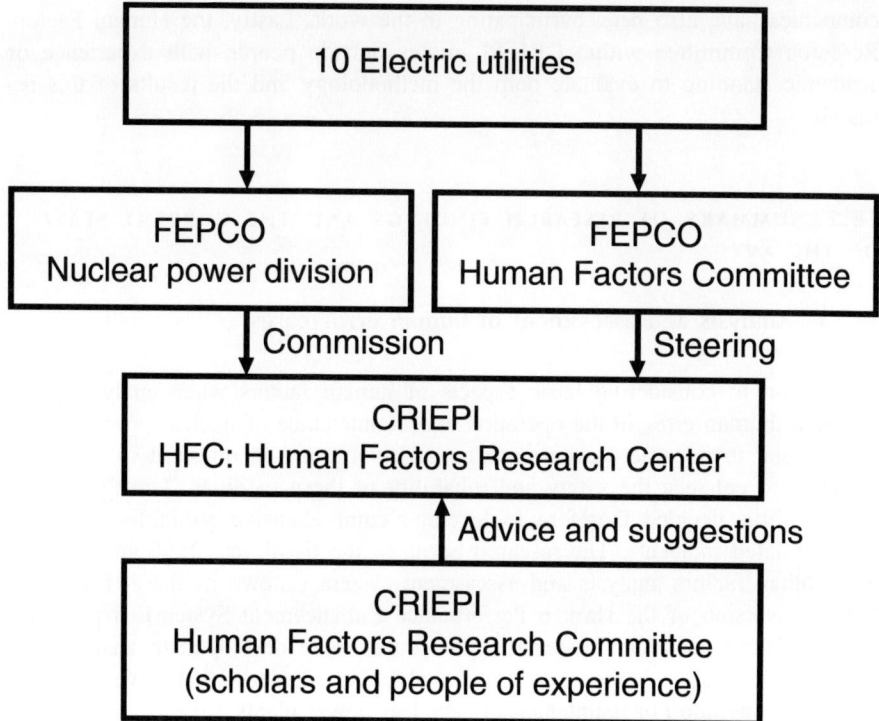

Figure 18.2 The administrative structure for conducting the human factors research projects. (FEPCO: Federation of Electric Power Companies; CRIEPI: Central Research Institute of the Electric Power Industry)

indirectly to actual fields of work in such installations. The nature of human behaviour, for example, is studied through simulator experiments, basic experiments, and human reliability analysis. The human–machine interface is the focal point of applied research that three manufacturers of the Nuclear Steam Supply System (NSSS) – Hitachi, Toshiba, and Mitsubishi – are pursuing in relation to the modification of control panels and the development of operator support systems. A feature of research in Japan on nuclear power is that universities and some non-university organisations have continued work on the development of operators' cognitive models.

Human factors research in Japan is organised by a number of key groups and is conducted in various places (see Table 18.1). FEPCO's Human Factors Committee has established joint research programmes in domestic electric utilities in Japan (see Figure 18.2). By surveying the needs of the electric utilities, gathering information, assessing the findings of the research, and promoting the practical application of those findings, the committee plays a crucial role in ensuring that the research is conducted effectively within the Human Factors Research Center in CRIEPI. Because the research activities require knowledge of, and experience with, operating and maintaining nuclear power plants, experts from Japan's electric power companies have also been participating in the work. Lastly, the Human Factors Research Committee within CRIEPI invites outside people with experience or academic standing to evaluate both the methodology and the results of this research.

18.2 SUMMARY OF RESEARCH FINDINGS AND THE CURRENT STATE OF THE ART

18.2.1 Analysis and assessment of human error cases

In addition to considering basic aspects of human factors when analysing and assessing human error in the operation and maintenance of nuclear power plants, it is crucial to document experience cumulatively, and to reflect it in measures designed to enhance the safety and reliability of these facilities. Japan's electric power utilities decided, therefore, to develop a comprehensive system for analysing human-related incidents. The research began in the fiscal year 1986 and resulted in a human factors analysis and assessment system (known as the J-HPES, the Japanese version of the Human Performance Enhancement System). It provided the foundation on which systematic procedures were developed to analyse and assess the human behaviour patterns involved in incidents that have occurred during the operation or maintenance of nuclear power plants Takano, Sawayanagi and Kabetani (1994). The J-HPES thus helps specialists formulate their proposals for specific measures to prevent the recurrence of similar incidents. To propagate the J-HPES in the utilities companies, CRIEPI's HFC has conducted an annual series of seminars for seven years, reaching 300 persons thus far. Analysis-support activities in the nuclear power plants themselves have continued as well. In 1993 the computer program for an analysis support system called the JAESS (J-HPES Analysis and Evaluation System) was completed and offered to systematise the series of analytical processes for practical use.

This technique has proved to be a useful tool not only in Japan's electric power industry but in many other Japanese industries as well, including chemicals, iron and steel, railways, and aviation. The procedure has been distributed to more than 80 companies, and similar techniques are being used in Taiwan and elsewhere.

18.2.2 The development of human factors databases

In order to support human factors activities in general, we have gathered a vast range of data that may be useful in applying human factors research and entering such information into databases. Information stored in appropriate forms (e.g. on-line, software, summaries) has been provided for use in Japanese utility companies as needed (see Fujimoto, 1996).

Case analysis database

The case analysis database consists of:

- Hiyari-Hatto case data, which currently contains details on approximately 3000 instances of near-misses experienced by operators and maintenance personnel of nuclear power plants;

- performance-shaping factors (PSF) survey data, which collects the results of surveys concerning 52 factors that may cause human error;

- J-HPES analysis data, with which incidents can be analysed and assessed by means of the previously mentioned J-HPES.

Good practice database

Collected from books, questionnaires, and field surveys, the various kinds of information in the good practice database may be useful in preventing human errors. It is used to collect:

- data on improvements in the working environments of nuclear power plants and organisations within other industries;

- information on the labelling and coding of plant instruments and equipment;

- data on foolproof or fail-safe ways that nuclear power plants and organisations in other industries have found to prevent human error;

- caution report data containing illustrated lessons learned from incidents in US nuclear power plants;

- data on, and lessons from, labour hazards in other industries.

Literature database

In this database, research papers, data, books, and other materials concerning issues of human factors mainly in nuclear power industries are collected and classified into 16 fields of research. The database consists of abstracts and the corresponding original texts.

These databases handle numerical, textual, and image data. In terms of range, the case analysis database contains about 6750 cases; the good practice database, about 480; and the literature database, approximately 1950. Effort has been made to encourage extensive use of the data. For example, the electric utilities have received some of the information as database software for personal computers. We would like to expand these databases into a comprehensive human factors database covering all the activities of the electric utilities, thermal power, hydropower, and nuclear power.

Table 18.2 Performance-shaping factors (PSFs) for survey and prediction of behavioural error

PSF	Description
	Category I: Internal human factors
1	Inflexible and inadequate response to new situations
2	Inability to evaluate operational dangers correctly
3	Little experience with the work
4	Insufficient training and education on individual tasks
5	Insufficient knowledge of the principle and structure of devices, despite much experience with them
6	Poor understanding of total flow and purpose of operations and procedures
7	Poor mental and physical condition of operator
8	Lack of active involvement in the operation
9	Advanced age of operators
10	Lack of cooperation among operators
11	Restlessness and impulsiveness of operator
12	Lack of attention and care in operators
13	Boldness and daring in operators
14	Physical fatigue of on-duty operators
15	Excessive seriousness about results and excessive fear of failure
	Category II: Human–machine interface
16	Poor visibility of indicators and pointers on devices
17	Poor audibility of oral instructions
18	Poor visibility of handles, levers, and switches
19	Unavailability of information when it is needed
20	Information overload
21	Lack or untimeliness of feedback on the results of a task
22	Unsuitable match between tools and the work place
23	Unavailability of manuals and procedural guidelines
24	Impossibility of correcting or terminating an inappropriate operation
	Category III: Work characteristics
25	Very rare need for the given task
26	Very short time available for completing the task
27	Incomplete records of previous work
28	Ambiguity of procedures, standards, and purpose of the task
29	Indirect, rather than direct, instructions and orders (involvement of a third party)
30	Impossibility of objectively checking the processes of the operation
31	Lack of alternatives for checking the operation; one method only
32	Insufficient operator experience or understanding for the task's required judgment
33	Lack of a way to check progress and procedures of the task
34	Necessity of a physically difficult action or posture in order to perform the task
35	Interference in operators' action caused by protective gear
	Category IV: Organisation
36	Atmosphere of daring workers to attempt dangerous tasks
37	Lack of breaks during monotonous work

Table 18.2 cont'd

PSF	Description
38	Ambiguity of role or responsibility of the task
39	Necessity of monotonous work for long periods
40	Interruption of work progress because of other jobs
41	Excessive number of workers compared to the work load
42	Lack of workers compared to the work load
43	Instructional difficulties with some persons because of past controversies
	Category V: External factors
44	High temperature, high humidity, or both, in the work place
45	Low temperature in the work place
46	Exposure to noise, vibration, or both
47	Lack of light in the work place
48	Narrow and unstable dimensions of the work place
49	Commencement of an operation (shift)
50	Midway point of the work time (shift)
51	Final phase of the work time (shift)
52	Late-night or early-morning work time

18.3 HUMAN BEHAVIOUR PREDICTION SYSTEM

Research on human reliability analysis has thus far been concentrated mainly on probabilistic explanations of the success or failure of task execution. Because it was virtually impossible to find a model that could identify errors that would be highly probable when one or more given PSFs are present in an operation or maintenance task, the aim of research in this field has been to develop a system that can predict such errors. With the complete cooperation of the electric power companies in Japan, surveys of 3500 experts in the operation and maintenance of nuclear power plants in all Japanese utilities were conducted, with the aim of identifying PSFs and the errors that were actually experienced by plant personnel. Five categories of PSFs (internal human factors, the human–machine interface, work characteristics, organisational factors, and external factors) comprising 52 factors were identified as input information (see Table 18.2). In addition, the surveys identified five categories of error (see Figure 18.3): sensory organs (e.g. overlooking), pattern recognition (e.g. misperception), judgment (e.g. misjudgment), decision-making and instruction (e.g. misconfirmation), and work action (e.g. inadvertent dropping of an object or taking of a wrong one). The relation between the PSFs and the errors were then quantified by statistical techniques. The model's output information consists of numerals indicating the probabilities (the rate of occurrence) of errors in the error categories. By checking the 52 PSFs at a given work site, one is now able to predict the rate of error occurrence for a given task and worker. The notifications it gives to workers about possible danger-points seem to be valuable for reducing errors, so we have put it to practical use (Yoshino, 1996; Yoshino and Inoue, 1993).

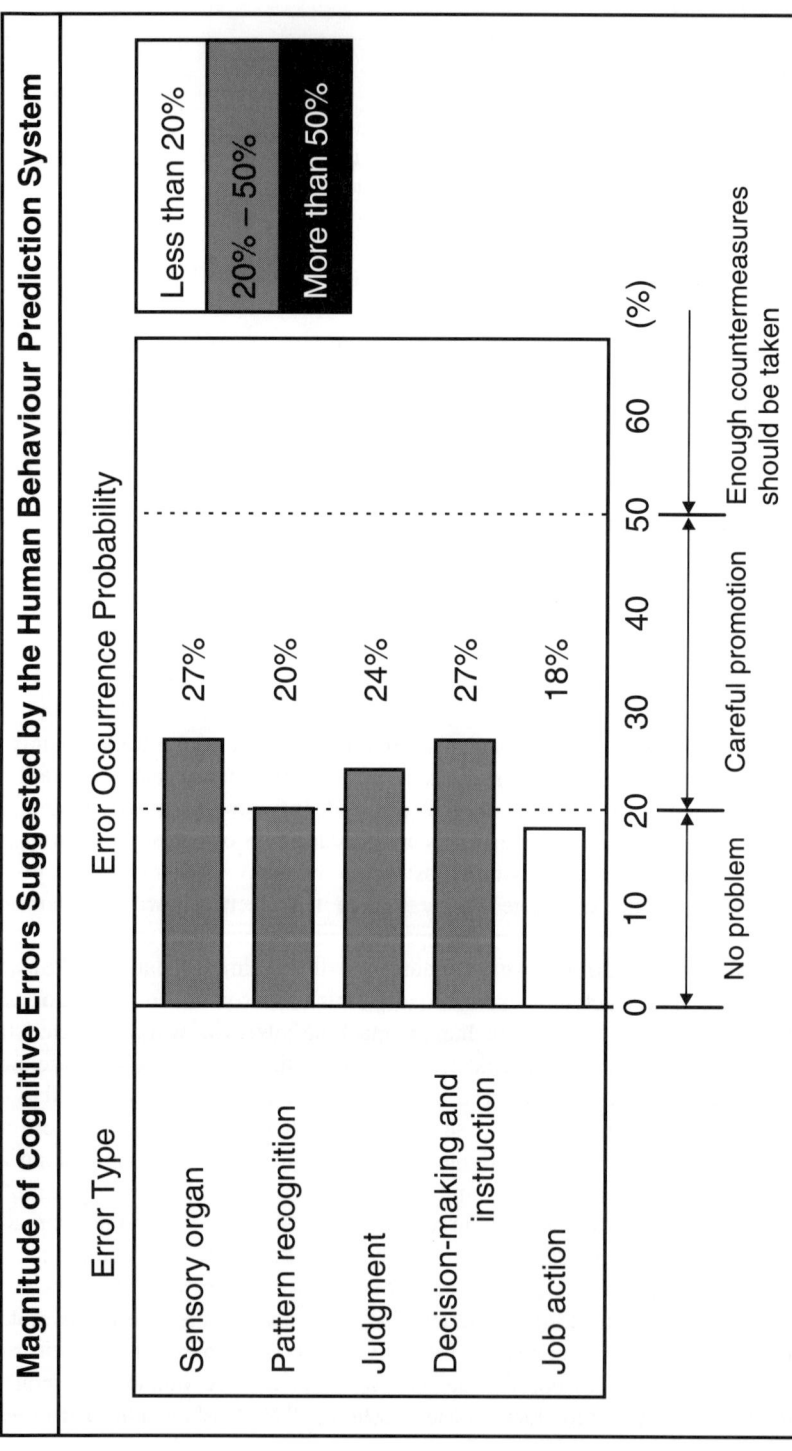

Figure 18.3 Intensity and probability of errors predicted by the Human Error Prediction System.

The applicability of this human-behaviour prediction system was tested further by having approximately 150 experts in operating and maintaining nuclear power plants use it to make predictions for about 250 cases of actual work. Around 70 per cent of the experts judged that the predictions made by the system were appropriate, a result that has led to its extensive use as a tool for the preliminary checking of work conditions, and as a source of concrete information about industrial safety at nuclear and thermal power plants and similar installations. To increase the utility of this system, we are conducting research to make its predictions more of a concrete example. For example, a predicted human error is now being accompanied by a detailed drawing of a past similar kind of instance and the effective precautionary measures.

18.4 SIMULATION OF OPERATORS' BEHAVIOUR

Technical aspects of operational environments for large systems such as nuclear power plants have been changing rapidly as computer technologies have developed. To identify human-related issues that arise when operators are coping with abnormal situations, it is still important not only to analyse the behaviour of the individual operator, but to clarify the decision-making process of the operators. To reveal various accident paths that could be generated by a combination of mechanical and human failure, and to prevent recurrence of accidents along those paths, researchers at CRIEPI's HFC have been developing an operator-team behaviour model in which each operator in the team thinks individually according to his or her own mental model, and all the operators make a team decision through discussion (Sasou, Takano, Yoshimura, Iwai and Sekimoto, 1994; Takano, Sasou and Yoshimura, 1995; Takano, Sasou, Yoshimura, Iwai and Sekimoto, 1994; Yoshimura, Takano and Sasou, 1995; Takano, Iwai and Hasegawa, 1994).

As shown in Figure 18.4, the operator-team behaviour model is composed of operator behaviour models and a human–human interface (HHI) model. The operator behaviour model simulates the cognitive process of an individual operator, with each operator processing information and behaving according to task allocation. The HHI model simulates the operator team's decision-making process through conversation. The HHI model also coordinates the opinions of the operators, and assigns jobs to the operators according to the prescribed task allocation principles. In the study by Takano, Sasou, and Yoshimura (1995), there were three operators (one leader and two followers having different tasks). The operator-team behaviour model was connected to a plant simulator via a human–machine interface. Hence, the model could monitor dynamic plant conditions and access the plant directly.

The thinking process and the associated knowledge of each operator were important issues in the simulation of operator-team behaviour. To clarify them, we conducted extensive simulator experiments in which sequential utterances and control actions of expert operators were documented in communication-flow diagrams (Takano, Sasou et al., 1994). This information was then used as the basis for the development of the model. Detailed analysis of the results in close cooperation with

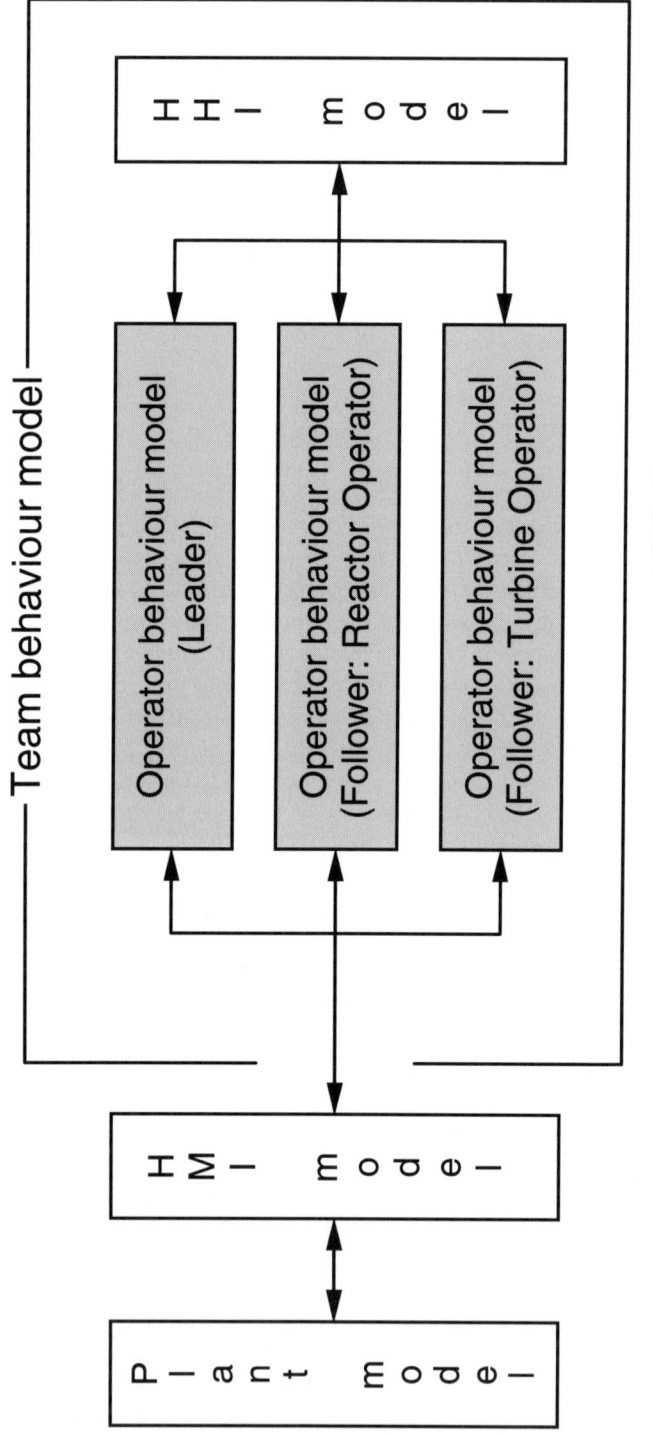

HMI: Human–Machine interface
HHI: Human–Human interface

Figure 18.4 Structure of the developed Operator Team Behaviour Simulation Model.

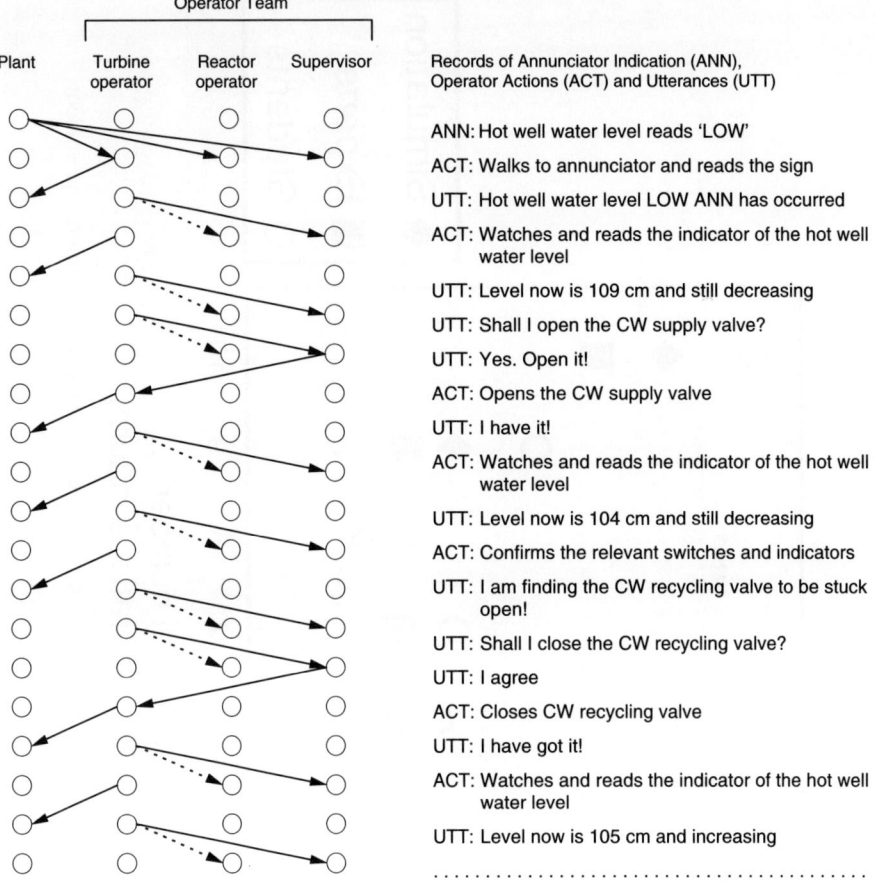

Figure 18.5   Typical example of a simulated result of failure to open the condensate water (CW) recycling valve.

ex-operators clearly demonstrated that the thinking and behaviour of the operators was consistent with a mental model created from the plant situation and corresponding accumulated knowledge and experience.

On the basis of these findings, we developed a simulation code for the operator-team behaviour model (the languages are C++ and C). The simulation code was developed at an engineering work station connected to a simplified boiling-water reactor simulator. An example of the simulation results appears in Figure 18.5, which shows a time sequence simulation of the dialogue and actions of the operator team, and of the annunciator's indications of the plant's operating conditions after a malfunction was signalled ('failure of CW recycling valve to open'). Simulation results obtained for possible malfunctions in this simplified simulator (24 malfunctions at most) showed reasonable agreement (70% of the operators' utterances heard in the experiments and 95% of their actions could be reproduced by simulation) with those obtained by the subjective experiments with the expert operators

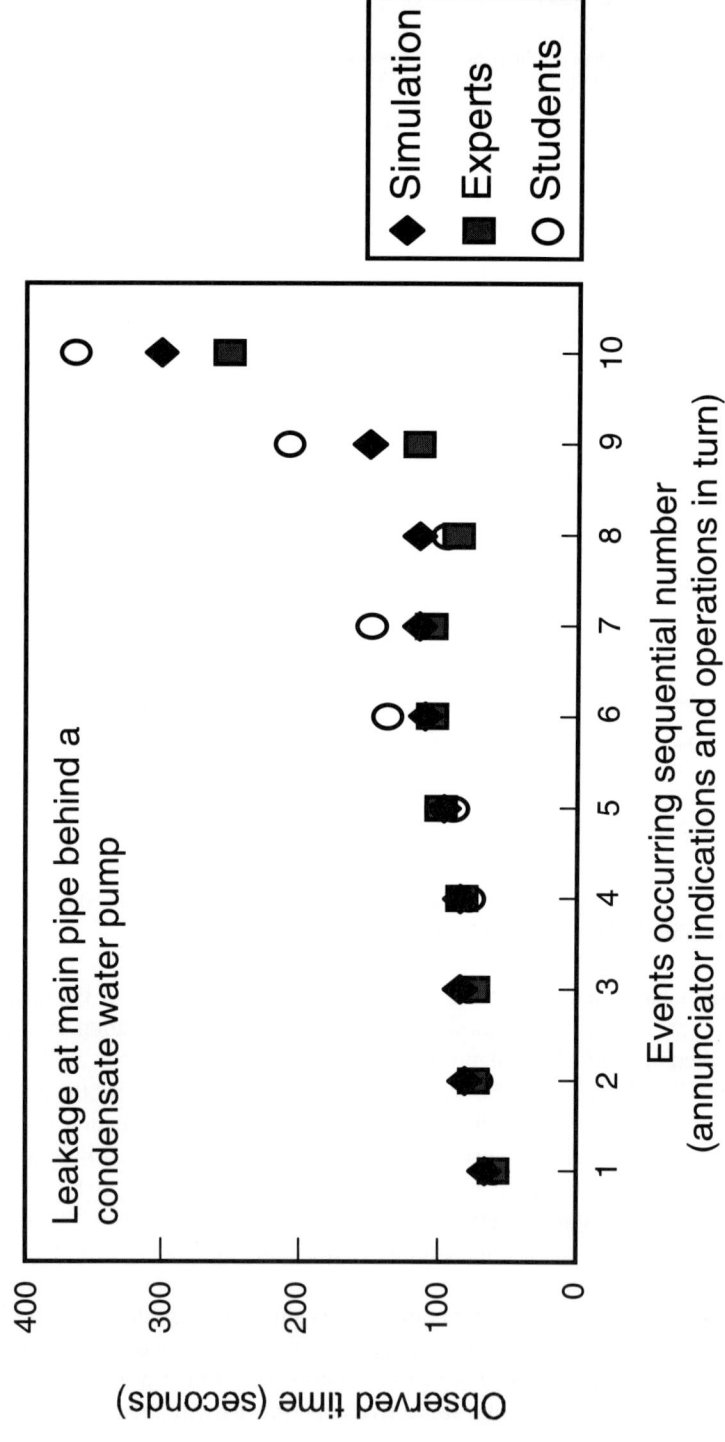

Figure 18.6 Comparison between the result of a simulation and the results of experiments conducted by expert and student participants.

and students. In other words, comparison between the simulation results and the experimental results of the communication and operation sequences suggested that modelling could mostly reproduce the actual behaviour of the human operators (see Figure 18.6).

We are planning to extend this operator-team behaviour model into a model that can simulate operator-team behaviour for the same level at an actual nuclear power plant. A large knowledge-base necessary for a full scope simulator has been prepared in cooperation with ex-operators, so that we can attempt to apply the operator-team behaviour model to more specific problems, such as the assessment of (a) operator team performance in relation to the modification and sophistication of control panels and (b) accident paths generated by the overlapping of plant and human failure.

18.5 CONCLUDING REMARKS

Human factors activities by Japanese electric utilities is not limited to CRIEPI's HFC. The Tokyo Electric Power Company and the Kansai Electric Power Company have been promoting research and development from more practical points of view. Through close contact with the previously named institutes in NUPEC, JAERI, and TEPCO, and with the INSS, CRIEPI will continue to pursue research on topics common and basic to all electric utilities. Moreover, we do not intend to limit our attention to the field of nuclear power, but to turn to various other technical fields as well, such as engineering and the distribution and transmission of electric power. Because human factors research should relate to practice, continuous discussion with persons in actual work settings are essential, especially when new tools and findings are being introduced there. Given the productive relationship that has been established between the theoretical realm and the practical world, human-factor research can now progress into a new stage in which the products and findings of research can contribute to improving actual daily work.

Note

1   This chapter summarises the research conducted in the Joint Research Projects of the ten electric power companies in Japan.

References

AISAKA, K. (1986, April) Efforts and strategies for reducing reactor scram frequencies in Japan. OECD symposium on reactor scram frequency, Tokyo.

AISAKA, K. (1988, February) Current status of and future prospects for the man–machine interface in Japan. Paper presented at the conference of the International Atomic Energy Agency on the Man–Machine Interface in the Nuclear Industry, Tokyo.

FUJIMOTO, J. (1996, May) Construction and use of a human factors database system. Paper presented at the International Electric Research Exchange [IERE] Workshop on Human factors in nuclear power plants, Yokohama.

KURODA, I. (1986) *Considering current disasters and accidents* (Kanagawa-koatsugasu-kyokai Report No. 124). Yokohama: Kanagawa.

LIGHT SCIENCE (1992) *Lighted visor explanation materials.* Boxborough, MA: Light Science.

NRC (US Nuclear Regulatory Commission) (1975) *Reactor safety study – An assessment of accident risks in US commercial nuclear power plants* (NRC, WASH-1400; NUREG-75/014). Washington, DC: Government Printing Office.

SASOU, K., TAKANO, K., YOSHIMURA, S., IWAI, S. and SEKIMOTO, Y. (1994) *Simulation of operator team behaviour in nuclear power plants – Detailed design of an HHI model that simulates decision-making and task allocation in the team* (CRIEPI Report No. S93002). Tokyo: Central Research Institute of the Electric Power Industry (CRIEPI).

TAKANO, K., IWAI, K. and HASEGAWA, F. (1994) Development of a human-related event analysis support system using a personal computer. *Journal of the Atomic Energy Society of Japan,* **36**, pp. 1059–67.

TAKANO, K., SASOU, K. and YOSHIMURA, S. (1995, June) Simulation system for behaviour of an operating group (SYBORG) – Development of an individual operator behaviour model. Paper presented at the 14th European Annual Conference on Human Decision-making and Manual control, Delft, the Netherlands.

TAKANO, K., SASOU, K., YOSHIMURA, S., IWAI, K. and SEKIMOTO, Y. (1994) *Simulation of operator team behaviour in nuclear power plants – Development of an operator behaviour model (individual model)* (CRIEPI Report No. S93001). Tokyo: Central Research Institute of the Electric Power Industry (CRIEPI).

TAKANO, K., SAWAYANAGI, K. and KABETANI, T. (1994) System for analysing and evaluating human-related nuclear power plant incidents. *Journal of Nuclear Science Technology,* **31**, pp. 894–913.

YOSHIMURA, S., TAKANO, K. and SASOU, K. (1995) A proposal for an operator model and operator's thinking mechanism. *Transactions of the Society of Instrument and Control Engineers,* **31**, 1754–61.

YOSHINO, I. (1996, May) Practical development of a human behaviour prediction system (KY-ASSTST). Paper presented at the International Electric Research Exchange [IERE] workshop on human factors in nuclear power plants, Yokohama.

YOSHINO, K. and INOUE, K. (1993, May) Practical development of a human error and behaviour prediction system from the viewpoint of psychological and statistical methods. Paper presented at the European Safety and Reliability (ESREL) Conference, Munich.

CHAPTER NINETEEN

Human errors in Japanese nuclear power plants: a review of 25 years[1]

MITSUHIRO KOJIMA, KEN'ICHI TAKANO AND TOMOHIRO SUZUKI

Central Research Institute of the Electric Power Industry

The Japanese version of the Human Performance Enhancement System and the computer program used to analyse incidents in Japanese nuclear power plants are described and the terms used in this instrument are defined. Preliminary results of an analysis of 188 incidents caused by human error in these installations over the last 25 years are presented and compared in the light of design alterations that were made in two types of nuclear power plants in the late 1970s. Suggestions for improvements in the presentation and accessibility of the information gathered by means of this analytical tool are noted.

In Japan all major trouble in nuclear power plants must be reported to the Ministry of International Trade and Industry (MITI). These reports are reviewed and classified according to their causes. There have been 188 incidents caused by human error in various nuclear power plants in 25 years. To analyse these incidents, the Human Factors Research Center (HFC) of the Central Research Institute of the Electric Power Industry (CRIEPI) proposed a human-error analysis method named J-HPES (the Japanese version of the Human Performance Enhancement System) (Takano, Kabetani, and Iwai, 1994). J-HPES is a kind of root-cause analysis method based on procedure manuals. This method was programmed into a computer application named the J-HPES Assistance and Evaluation Support System (JAESS).

The HFC has already analysed all 188 reports on incidents due to human error. With the JAESS program the results of the analysis are compiled in machine readable formats. These data are now converted into more general formats, such as

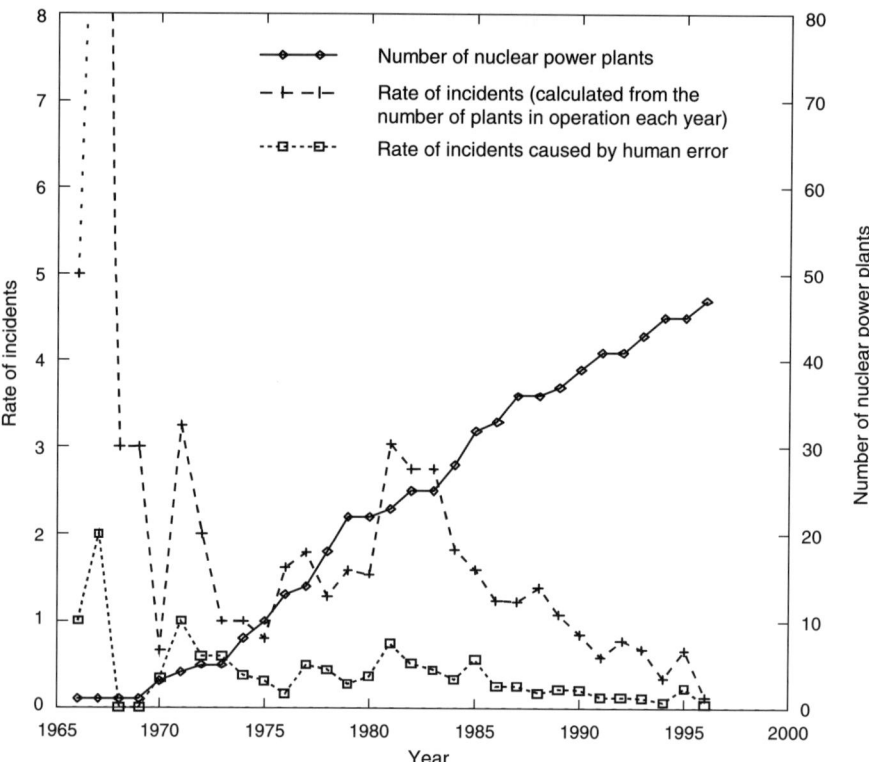

Figure 19.1   Japanese nuclear power plants, error rates and human error rates. (In 1981 report standards were changed to include less serious incidents, so the rate jumped in that year.)

the Comma Separated Values (CSV), and can be read with any kind of application for statistics and spreadsheets. The HFC is now analysing these results from various perspectives. In this chapter we show the framework of the J-HPES/JAESS and some preliminary results of our analysis.

19.1   INCIDENTS IN JAPANESE NUCLEAR POWER PLANTS

The Tokai Power Plant was the first commercial nuclear power plant in Japan and is owned by the Japanese Atomic Power Company. The plant started commercial power generation in 1966. Starting that year, all serious incidents in the country's nuclear power plants were to be reported to the Ministry of International Trade and Industry (MITI). By the end of the 1995 fiscal year, there were 47 nuclear power plants in Japan and a collection of 863 incident reports. Examination of these reports for causes has shown that 188 incidents were due to human error, and that the rate of such errors has been gradually decreasing, dropping in the 1990s below the level of one per year for each nuclear power plant (see Figure 19.1).

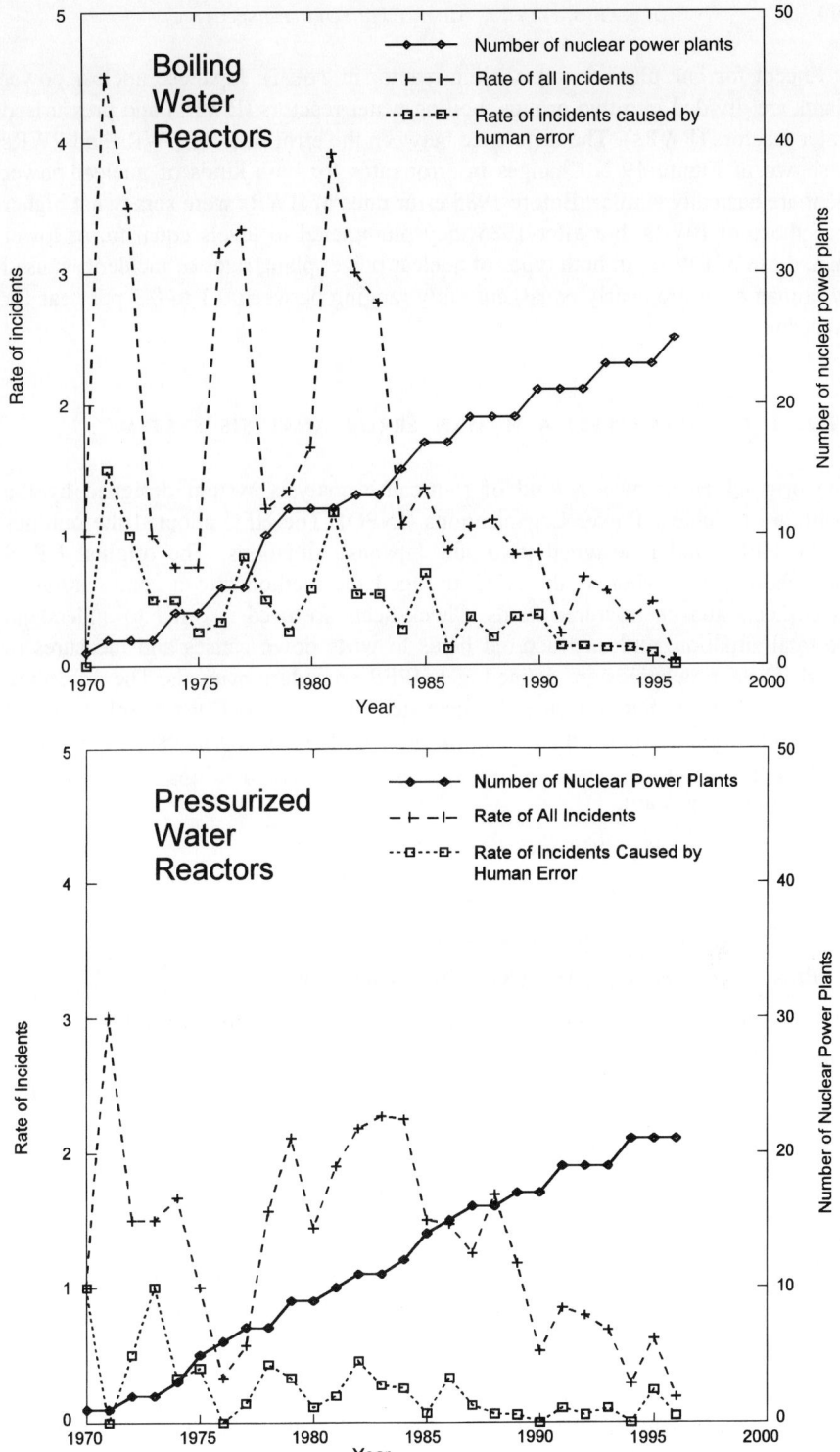

Figure 19.2 Differences between error rates of boiling water reactors (BWRs) and pressurised water reactors (PWRs).

Except for one plant (a gas cooling reactor in Tokai), Japanese nuclear power plants are divided into two groups, boiling water reactors (BWRs) and pressurised water reactors (PWRs). The difference between the error rates of BWRs and PWRs is shown in Figure 19.2. Changes in error rates for both kinds of nuclear power plant are basically similar. Before 1985 error rates of BWRs were somewhat higher than those of PWRs, but after 1985 they plummeted to levels equal to, or lower than, those of PWRs. In both types of nuclear power plant, rates of incidents caused by human error are nearly equal, currently ranging between 0.1 to 0.3 per year for each unit.

19.2   THE J-HPES/JAESS: A HUMAN ERROR ANALYSIS SYSTEM

The original HPES was a kind of root-cause analysis system designed by the Institute of Nuclear Power Organizations (INPO). The HFC adopted the outlines of the HPES and redesigned it to suit Japanese situations. The original HPES used checksheets, whereas the HFC arranged the method into a combination of checksheets and unstructured fields. Checksheets are used in order to understand the total situation, and unstructured fields to write down causes and measures in detail. These procedures are defined in J-HPES procedure manuals. The procedure of the J-HPES has four stages and fifteen steps. Coordinators (the people who use the J-HPES to analyse incidents) can follow the manuals and investigate all items thoroughly.

The four stages are:

1 Understanding the incidents.

2 Analysing the situation of the incidents.

3 Analysing the causes of the incidents.

4 Proposing measures to be taken against each cause.

In the J-HPES, actions that brought about the incidents directly are called *inappropriate actions*. Causes that brought about these inappropriate actions directly are called *direct causes*. Similarly, causes that brought about direct causes are called *indirect causes*, and causes that brought about indirect causes are called *latent causes*. Latent causes are similar in concept to root causes.

19.2.1   An example of J-HPES analysis

Let us now consider an analysis flow based on the J-HPES, with the incident being the tripping of a pump during operations.

1 Why was the pump tripped?
   Because personnel shorted electrical equipment with screwdrivers when checking wires (*inappropriate* action)

2 Why did the screwdrivers short the electrical equipment?
Because (a) the staff member checked the wires with two screwdrivers held in one hand, and (b) the screwdrivers were not insulated (*direct* causes of the inappropriate action)

3 Why did these direct causes happen?
The *indirect* causes of direct cause (a):
a(i) The working space was limited.
a(ii) The staff member was not wearing a working belt to hold the screwdrivers.
The *indirect* cause of direct cause (b):
b(i) The manuals did not prescribe the use of insulated tools.

4 Why did these indirect causes happen?
The *latent* causes for each indirect cause:
a(i) The workplace was not designed to be worked in.
a(ii) The staff member was behind schedule and in a hurry.
b(i) The order of the site foreman was ambiguous.

5 The coordinator proposes measures to deal with inappropriate actions and with each cause.
- The measure to counter the inappropriate action: do not check live wires.
- The measure to counter the direct cause: do not use uninsulated tools.
- The measure to counter the indirect and latent causes: establish design principles ensuring that the workplace has enough workspace
- Inspect the required tools before using them.
- Train personnel to check the wiring.

The flow of analysis is conceived of as shown in Figure 19.3.

19.2.2 Terms used in J-HPES

In adapting the outlines and format of the HPES to the Japanese context, the HFC also changed some of the terms and categories used. For example, inappropriate action, the consequences of an incident, and the means of detecting an incident or inappropriate action were modified as follows:

Inappropriate actions

In the J-HPES, the concept of inappropriate action was broken down into three classes (designated by Arabic numerals in the outline below) and 13 categories (designated by bullets).

1 *Omission*:
- Personnel failed to perform one or more required acts.
- Some other aspect of work or duty has not been performed.

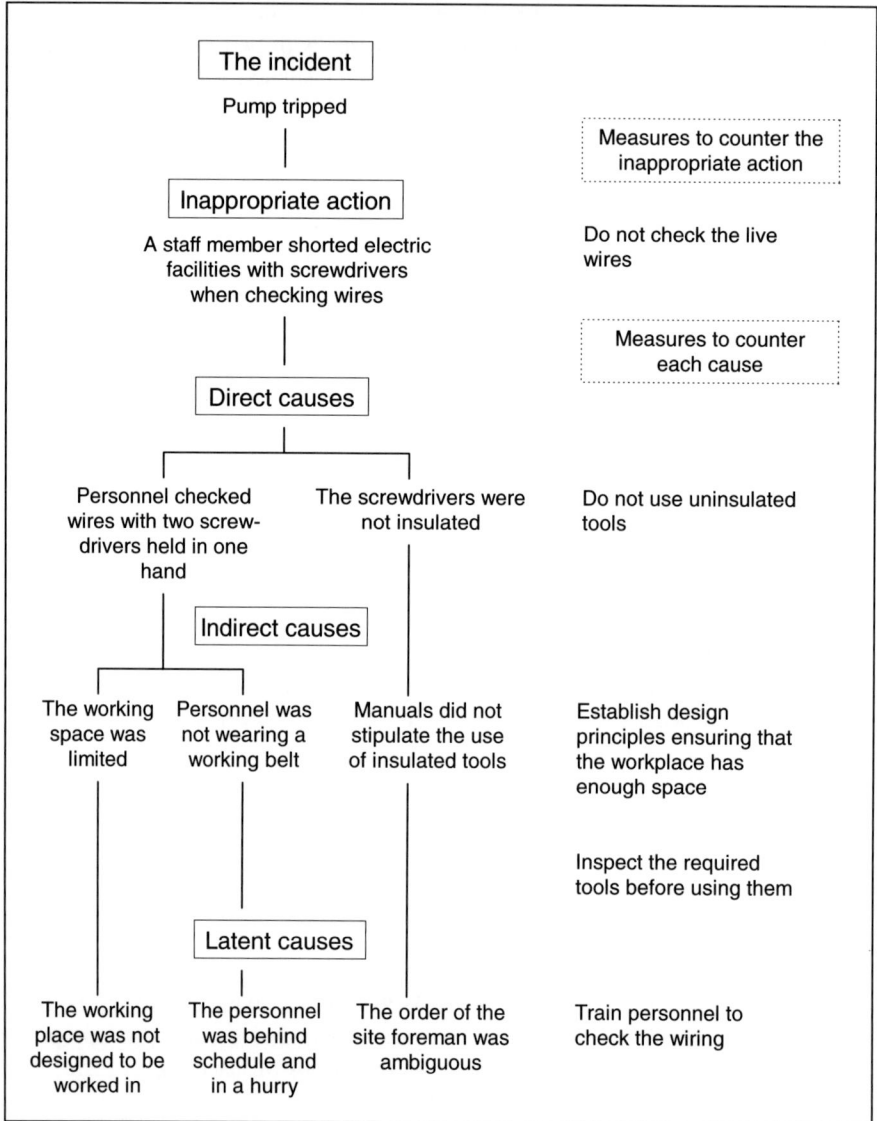

Figure 19.3   Example of analysis flow based on J-HPES.

2   *Extraneous acts*:
   - Personnel fell from scaffolds, made contact with one or more pieces of equipment, tumbled at the workplace, or left behind something (such as waste) that resulted in an incident.
   - Personnel has carried out a prohibited act.
   - Personnel has carried out operations not required by procedures or training.
   - Some other unnecessary act has occurred or has been performed.

3 *Inadequate acts*:
- Personnel has performed the necessary operations, but not at or within the proper time (untimely act).
- Personnel mistook equipment (transposition).
- Personnel has performed correct acts but in the wrong order (faulty sequence).
- Personnel selected incorrect operational values to set (qualitative mistake).
- Personnel selected correct operational values but overset or underset them (quantitative mistake known as excess or insufficient operational values).
- Personnel performed inadequate operations not classifiable into the above categories.

Consequences of the incident

Consequences of the reported incident are classified into one or more of the following 12 categories:

- Reactor scram or plant electric output change (the reactor scrammed or electricity output fluctuated because of the incident).
- Shutdown of the systems or components (the plant or reactor did not stop, but some systems or components in the plant did because of the incident).
- Failure of the systems or components (systems or components failed to function).
- Malfunction of the systems or components (systems or components did not stop, but their output or performance could not be forecast).
- Deterioration or loss of function of the systems or components (systems or components continued working as designed, but their output or performance was somewhat reduced).
- Damage to the systems or components (the systems or components were damaged by the incident).
- Failure to remain within operating limits (the systems or components continued working, but their status or power output somewhat exceeded prescribed levels).
- Increased outage time or increased periodic inspection period (the plant's outage time was somewhat extended because of the incident).
- Impacts on the environment (surrounding environments or working environments were contaminated because of the incident).
- Impacts on personnel (personnel were injured, contaminated, or otherwise affected because of the incident).
- Other impacts.
- No adverse consequences.

The means for detecting an incident or inappropriate action

There are 11 categories for classifying what signalled or enabled someone to detect an incident or inappropriate action:

- Recorders, indicators (such as gauges and meters), cathode-ray tubes (CRTs), or video monitors.
- Alarms.
- Labels or other markings attached to the system or component.
- Tags attached to the system or component.
- The external appearance of the system or component (noticed mostly by daily patrols).
- Internal inspections (verification, routine rounds, and surveillance).
- Improper equipment responses (abnormal sound, smell, vibration, and the like).
- Improper working mode of the equipment (running, stopping, pausing, and so on).
- Working conditions (work progress, consequences of the actions, and so on).
- Work situations (change in the temperature, brightness, or other aspect of the workplace).
- Staff meetings, staff reports, discussion, records.

These terms and methods of the J-HPES were defined in the J-HPES procedure manuals. The methods of the J-HPES were also programmed into the JAESS computer application running on MS-Windows. The HFC distributed J-HPES procedure manuals and the JAESS program to all Japanese electric power companies, and has for the past five years conducted one-day training courses, which more than 200 persons have meanwhile attended.

19.3 SOME FEATURES OF HUMAN ERRORS IN JAPANESE NUCLEAR POWER PLANTS

Based on 188 incidents involving human error in Japanese nuclear power plants, Figure 19.4 shows the rate for each category of how the inappropriate action was found. Up to 41 per cent of the inappropriate actions were signalled by a sound system or alarm, whereas the external appearance of the systems or components, along with recorders, gauges, meters, or CRTs, accounted for the second largest percentage of detections. The relatively low rate of detections by recorders, gauges, meters, and CRTs seems somewhat strange, and is one characteristic finding of our analysis. The incidents on which our analysis is based had been reported to MITI, so they were all serious. Recorders, gauges, or meters should have been able to alert personnel at an early stage, never allowing the situations to have escalated into something serious enough to warrant submission of an official report to MITI.

Figure 19.5 shows what the first impact of the incident was. Nearly half of the initial impacts were classified as malfunctions of the systems or components, or as shutdowns of the systems or components. This result is a second characteristic finding of our analysis. Of the serious incidents that occurred over the 25 years covered by our records, 130 resulted in reactor scram or a change in

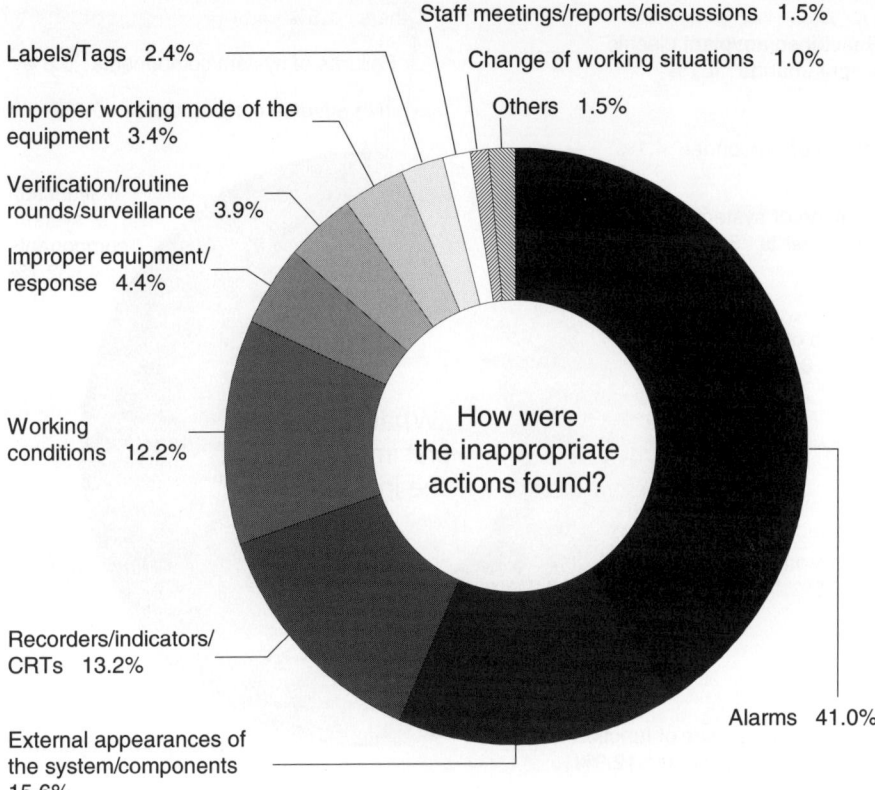

Figure 19.4 How have inappropriate actions been detected in Japanese nuclear power plants?

the plant's output of electrical power. But this figure shows that, even in such serious cases, most incidents were found from very few malfunctions or shutdowns of the systems or components. Had the operators been able to take proper measures at that time, some serious incidents could have been prevented in their early stages.

The HFC is now examining the results of our analysis in detail. One focal point is the differences between the types of nuclear power plants. In Japan, BWR plants operating before 1983 were built according to the original designs of General Electric's BWR plant. But in the late 1970s, plant designs were modified somewhat to take account of Japanese operational experience that had accumulated in the meantime. The new specifications were accepted by MITI as 'new standards of plant design,' and they are the basis for the construction of BWR plants that began operating in Japan after 1984.

Figure 19.6 shows the differences between these two groups of BWRs in terms of the inappropriate actions that brought about the incidents we have studied. The number of such inappropriate actions as the selection of incorrect operational

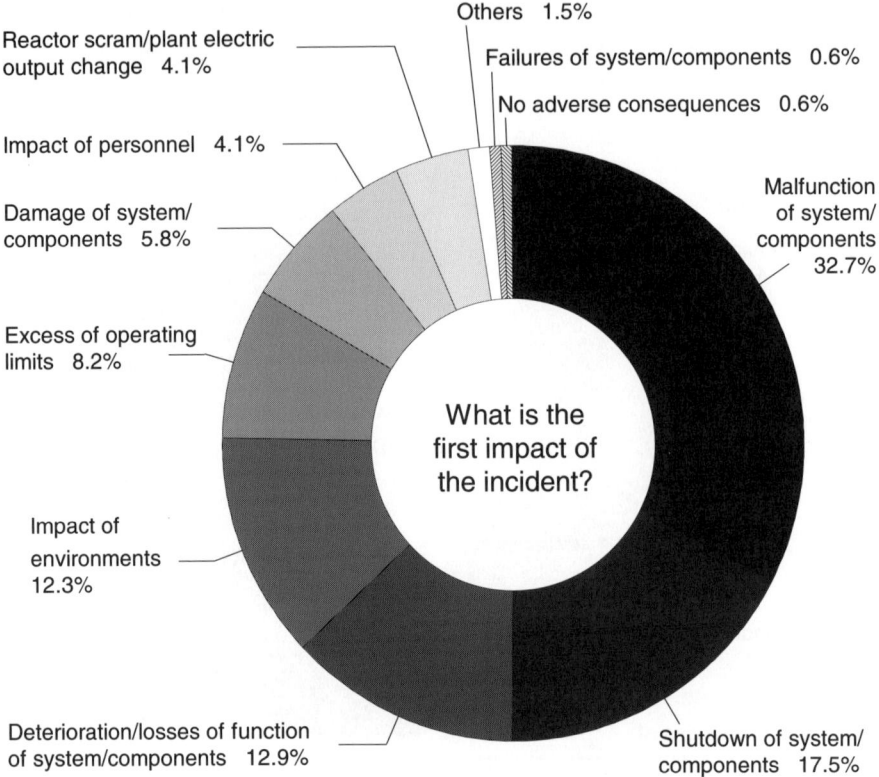

Figure 19.5 What has been the first impact of reported incidents in nuclear power plants?

values to set, and the oversetting or undersetting of correctly selected operational values, has been vastly reduced in plants built according to the revised design standards (represented by the lines in Figures 19.5 and 19.6), making it especially clear that the new design has improved the operator–machine interfaces for setting operational values. Falling, tumbling, and making contact with equipment has also been reduced in plants of the revised design, an indication that the work spaces in such installations have been improved. On the other hand, the number of untimely and other kinds of inadequate acts were not greatly reduced in either type of nuclear power plant. This category of causes was mostly due to human factors, so it seems unrealistic to reduce them by redesigning hardware.

Figure 19.7 shows the differences between these two types of BWR plants in terms of where the inappropriate actions have happened. It also shows what parts of the plants were improved by the change in design. In particular, the number of inappropriate actions that had been occurring in the containment or drywell, and in the reactor building or annulus, have been greatly decreased. Lastly, two main purposes of the revised design were to increase the efficiency of inspection and to enlarge the workplace in the containment. The results show just how

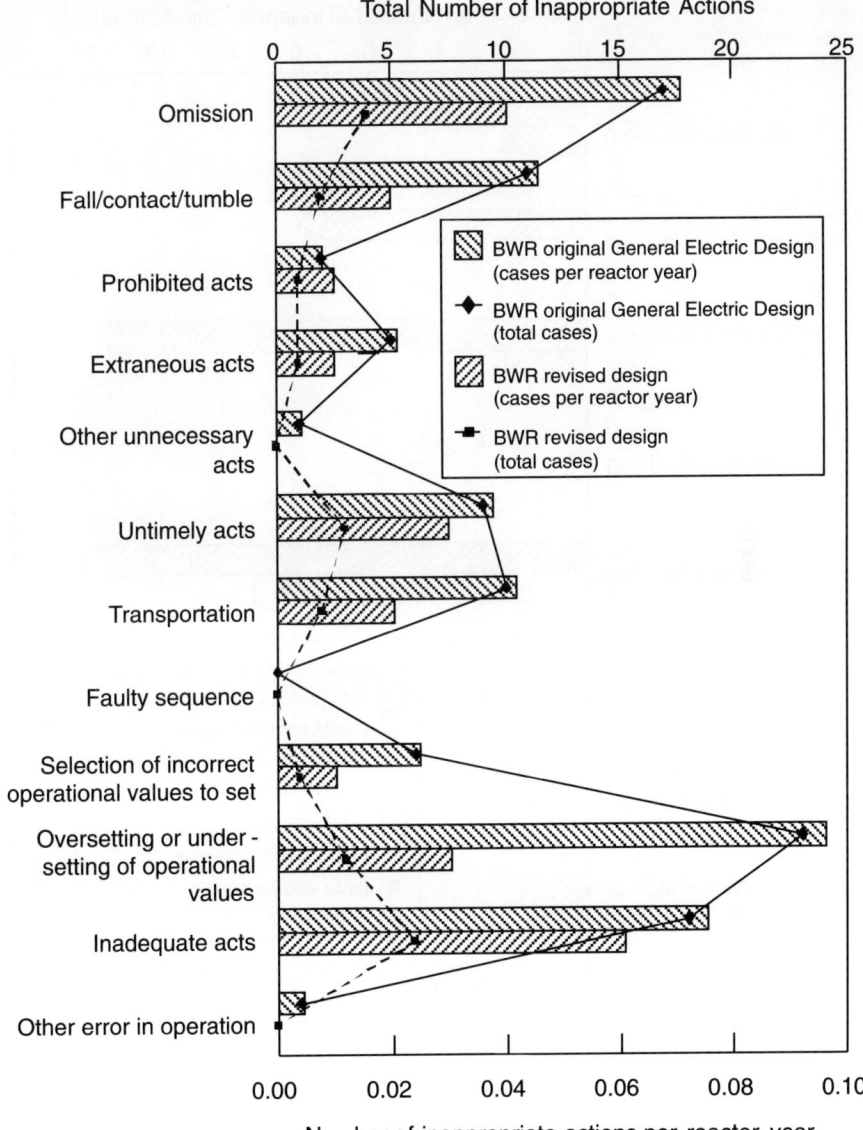

Figure 19.6 Differences between two types of Japanese boiling water reactors (BWRs). ('What was the inappropriate action?')

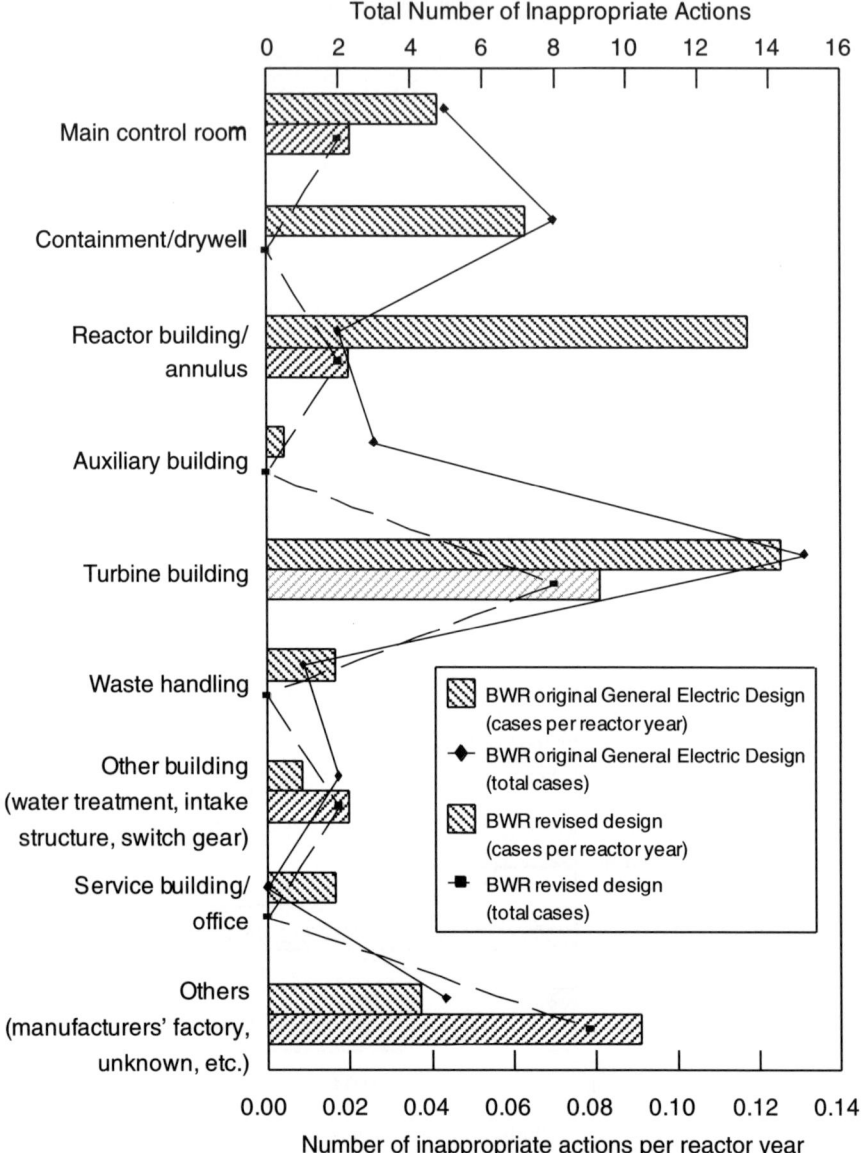

Figure 19.7 Differences between two types of Japanese boiling water reactors (BWRs). ('Where did the inappropriate action occur?')

effective the design changes have been in reducing human error in Japanese nuclear power plants.

19.4 FUTURE RESEARCH

The HFC finished analysing 188 incident reports with J-HPES/JAESS in the spring of 1997 and is now investigating the results from a variety of perspectives. One characteristic of J-HPES is that it had unstructured columns to write down the causes and countermeasures for each incident. These fields contain a wealth of useful information for improving plant maintenance. We plan to organise this information in databases and make it available over computer networks. If the results can be accessed directly by the nuclear power plants, they could be useful for prechecking tasks for the human errors they might easily cause.

Note

1   This report summarises the work performed in the Joint Research Program of the ten electric power companies in Japan. They assume no responsibility for any damage resulting from use of the information disclosed herein.

Reference

TAKANO, K., KABETANI, T. and IWAI, S. (1994) System for analysing and evaluating human-related nuclear power plant incidents – Development of a remedy-oriented analysis and evaluation procedure. *Journal of Nuclear Science and Technology*, **31**, pp. 894–913.

CHAPTER TWENTY

Human factors as revealed by the use of natural language in near-incidents at nuclear power plants

HIROFUMI SHINOHARA[1], FUMIO KOTANI AND
TETSUYA TSUKADA[2]

[1] *Faculty of Education, Kumamoto University*
[2] *Institute of Nuclear Safety System*

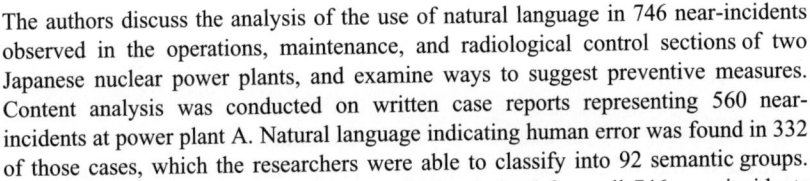

The authors discuss the analysis of the use of natural language in 746 near-incidents observed in the operations, maintenance, and radiological control sections of two Japanese nuclear power plants, and examine ways to suggest preventive measures. Content analysis was conducted on written case reports representing 560 near-incidents at power plant A. Natural language indicating human error was found in 332 of those cases, which the researchers were able to classify into 92 semantic groups.

A separate list of 90 human-factor variables derived from all 746 near-incidents was prepared and was found to reduce to 44 categories. Analysis of these 44 categories by means of Quantification Method III showed the identified human factors to fall along six axes.

20.1 INTRODUCTION

Analysis of near-incidents likely to be caused by human error has led to numerous important suggestions for ways to prevent human error (Labour Safety Institute, 1994). Nevertheless, near-incidents reported in recent years have shown a trend involving four problems in particular:

1   Differences between work places tend to bias the frequency with which cases are reported.

2   Expressions used in reports describing near-incidents tend to be standardised.

3   Management that has received a near-incident report does not always make effective use of the information.

4   Because it is not easy for staff members to express their real feelings in reports, the content of near-incident case reports tends to be composed of superficial descriptions.

Although suggestion systems and near-incident reporting systems have resulted in many improvements in equipment at nuclear power plants (Quality Control Centre of Nuclear Power of the Kansai Electric Power Company, 1996), it cannot be said that management is consistent in implementing effective measures for countering human error. With this background in mind, we took 746 near-incidents reported by the personnel of two Japanese nuclear power plants and analysed them for problems caused by human error. We had three purposes:

■   to clarify the causes of errors other than those caused by machinery and equipment, and to propose effective policies for accommodating each factor;

■   to devise feasible methods to improve the process flow, ranging from the collection of cases to the extraction of relevant factors and the development of countermeasures;

■   to devise methods for effectively and efficiently accumulating cases of near-incidents.

The research reported in this chapter focuses primarily on the analysis of natural language used in written case reports of near-incidents, especially regarding our third purpose.

20.2   STUDY 1: ANALYSIS OF NATURAL LANGUAGE USED IN NEAR-INCIDENTS

For our purposes, content analysis involved the study of key expressions (words and phrases) and other characteristics of natural language actually used in the Japanese case reports on the near-incidents selected for investigation. To facilitate current and future storage and retrieval of information on near-incidents at nuclear power plants (see problem 2 as listed above), the texts were analysed specifically for the types of key words that are suitable, and the types of classification categories that should be used.

20.2.1   Procedure for analysis using natural language

Analysis of the text data on near incidents began with the six basic questions about each near-incident: who? what? when? where? why? how? In other words, it was necessary to understand the original description of a near-incident without establishing specific standards for classifying answers to the following queries:

1   When did it happen?

2   Where did it happen?

3 Who caused it?

4 What happened?

5 Which equipment was being operated?

6 How did it happen?

To understand the characteristics of a case, one must therefore read the case report and extract key words used in the natural language relating to the six basic interrogatives. Then the key words relating to each of the interrogatives must be classified under a single synonym and arranged according to meaning.

One of the problems encountered when classifying this natural language is that the circumstances surrounding the appearance of the key words are not understood in advance, making it impossible to establish appropriate classification categories beforehand.

For this reason it is important to sort key words after extraction from their original natural language. They will then be easy to group into semantic groups. The next step is to count the number of times each key word appears in the written text. Furthermore, words, phrases, sentences, and so forth are considered for preparing words. Unlike the written forms of languages used in Europe and North America, written Japanese has no space between words, so it will most likely be necessary to produce computer programs that leave spaces between words in phrases and sentences in Japanese.

Once key words have been determined in this manner, the following steps can be followed to create an in-house internet (intranet) that can be used to create a database on near-incidents:

1 Use Internet's hypertext markup language (HTML) to facilitate the production of key words.

2 Use HTML language to construct an internet database for near incidents.

This database must make it easy for individual work groups (leaders and their subordinates) engaged in routine tasks to access data from toolbox meetings and other sources, reflect on the procedure used in routine work, and reconfirm work precautions for that day. It is thought that analysis of the natural language relating to near-incidents will become an effective, computerised support tool for the prevention of human error.

20.2.2 Results of content analysis of natural language: classification example of key words expressing human error

Of the 560 analysed near-incidents observed in the operations, maintenance, and radiological control sections of power plant A, 332 (59.3%) were reported in language using words that expressed human error. There were 157 such incidents in the operations section, 125 in the maintenance section, and 50 in the radiological section. The words and phrases used to describe these 332 near-incidents were classified into 92 semantic groups as shown in Table 20.1.

Table 20.1 Semantic groups of expressions indicating human error in 332 near-incidents at Power Plant A

No. of group	No. of expressions	No. of appearances	Representative example
1	2	2	In a hurry
2	6	9	Was careless
3	1	1	Error in interpretation of abstract words (e.g., *this*, *over there*, *that*, and *which one*)
4	1	1	Possibility of overflowing
5	2	2	Proceeded without noticing
6	1	1	Forget to turn on, turn on the wrong one, omission, mistake
7	1	1	Do not understand clearly, difficult to understand
8	1	1	Happened to notice
9	1	1	Forget what one is doing, end up forgetting
10	2	1	Make a mistake when not fully understanding
11	1	1	Possibility of tripping
12	1	1	Possibility of mistaking the drain
13	3	5	Do something because of human error
14	1	1	Occurs easily because of repeating over and over
15	1	1	Easily causes a mistake
16	8	11	Incorrect unit
17	2	2	Possibility of performing work easily
18	1	1	Absence of jigs for attaching safety belts, possibility of falling
19	1	1	Dark
20	1	1	Pushed the wrong button, was about to stop
21	1	1	Possibility of something
22	3	3	Was about to open
23	1	1	Possibility of coming off
24	1	1	Presumed risk of causing a disturbance
25	7	7	Easy to operate without checking
26	1	1	Easily isolated
27	10	13	Misunderstanding, preconception
28	1	1	Possibility of operating out of habituation or misunderstanding
29	56	134	Do something by mistake
30	1	1	Dangerous
31	4	5	Did not notice
32	1	1	Was not stated
33	5	1	On backwards, forgot, operated based on preconception
34	1	1	Risk of something
35	6	6	Difficult to distinguish
36	4	5	Illegible, difficult to confirm
37	27	50	Mistakenly
38	2	2	Possibility of misunderstanding and releasing the power supply

Table 20.1 cont'd

No. of group	No. of expressions	No. of appearances	Representative example
39	1	1	Mistaken contact
40	4	5	Easy to make operational error
41	3	3	Perceived incorrectly
42	3	5	Went incorrectly, made a mistake
43	1	1	Non-uniform description
44	1	1	Forgot to restart
45	7	18	Easily confused into thinking something
46	1	1	Easy to make an error
47	19	21	Misconception, possibility of starting error
48	2	2	Possibility of making a mistake if directions are poor
49	1	1	Risk of stopping
50	1	1	Inadequate prior discussion
51	1	1	Operated incorrectly because of similarity, confusing names
52	1	1	Risk of doing something
53	3	3	Mixed up
54	2	2	Error attaching, risk of making a mistake
55	1	1	Inadequate consideration
56	1	1	Small and difficult to see
57	1	1	Became impatient, very similar, possibility of interchanging
58	1	1	Incorrect location
59	3	3	Same colour, easily mistaken
60	3	3	Opened because thought switching was completed
61	1	1	Possibility of contact, erroneous operation
62	1	1	Possibility of forgetting to remove the ground
63	7	7	Possibility of making a mistake in operation
64	1	1	Were different
65	3	3	During busy periods
66	1	1	Possibility of pushing the wrong button out of habit
67	1	1	Incorrect target equipment
68	1	1	No means of informing
69	3	23	If caution is not taken
70	1	1	Risk of falling
71	2	2	End up using the wrong power supply
72	2	2	No uniform
73	1	1	Same name, easy to make a mistake
74	1	1	Located in the same place
75	1	1	Same colour and shape, easy to make a mistake
76	1	1	Difficult to read, high possibility of releasing
77	1	1	Difficult to recognise
78	2	2	Difficult to discriminate
79	1	1	Ended up using the wrong number

Table 20.1 cont'd

No. of group	No. of expressions	No. of appearances	Representative example
80	1	1	No name plate
81	1	1	Thought it was necessary
82	6	7	No indication, easy to make a mistake
83	1	1	Inadequate, forgot to attach
84	8	19	Difficult to understand
85	2	2	Heard something wrong
86	1	1	Ended up stopping
87	1	1	Reversed the direction
88	1	2	Ended up forgetting
89	1	1	Busy, hurried, made a mistake between the left and the right
90	2	4	Subconsciously
91	3	3	Name was not indicated, misconception
92	1	1	Did not notice, ended up causing a misunderstanding

20.3 STUDY 2: ERROR MODES AND ERROR OCCURRENCE MECHANISM OF NEAR-INCIDENTS AT POWER PLANT A

The number of near-incident cases involved in the analysis of the modes and mechanisms of the errors was 455 (81:3%), out of the total number of 560. Descriptions of the work involved in these incidents and the percentage of incidents that each type of work was found to be involved in are shown in Table 20.2. As it indicates, there are differences between task types, that is, the task involving a high number of operations seems to be characterised by a higher percentage of near-incident cases. Tables 20.3, 20.4, and 20.5 show the error modes (classes of errors) involved in these 455 near-incidents. The distribution of their occurrence mechanisms across the three sections of Power Plant A is shown in Table 20.6.

The modes and mechanisms of the errors involved in these near-incidents at Power Plant A are summarised in the following outline, along with their proportional significance (in percentages) for the incidents.

20.3.1 Error modes

Errors caused by an omission of action (omission errors)

- Omission of a checking action (86.2%; the most common omission error).
- Omission of part of a task (5.7%).
- Omission of a piece of information or of instructions (1.1%).

Table 20.2 Involvement of tasks in 455 near-incidents in three sections of Power Plant A

Task description	Frequency	%
Operating section (n = 219)		
Isolation and isolation restoration operations	110	50.2
Plant output and operation supervisory operations	49	22.4
Auxiliary equipment start-up and shutdown operations	19	8.7
Regular testing operations	11	5.0
Trial run operations	3	1.4
Inspections during abnormalities	7	3.2
Auxiliary equipment regular switchover operations	7	3.2
Surveillance inspections	7	3.2
Plant start-up operations	6	2.7
Plant shutdown operations	0	0
Other	0	0
Maintenance section (n = 177)		
Disassembly inspection	106	59.9*
Testing and inspection	21	11.9
Part replacement and repair	12	6.8
Assembly	9	5.1
Trial runs	9	5.1
Abnormality inspection and survey	7	4.0
Instrument calibration	5	2.8
Removal construction	3	1.7
Supervision and result confirmation	2	1.1
Equipment replacement	2	1.1
Routine inspection	1	0.6
Cleaning and clean up	0	0
Radiological control section (n = 59)		
Health physics	31	52.5
Chemistry	28	47.5
Outside negotiation duties (engineering department)	0	0
Other duties (engineering)	0	0
Nuclear fuel and safety control	0	0
Office administration	0	0

* Rounding causes the sum of these maintenance section percentages to exceed 100.

Execution errors (commission errors)

- Incorrect selection of equipment or parts (40%).
- Wrong position or location (40.2%).
- Procedural errors (5.5%).
- Insufficient quality (5.5%).

Table 20.3   Errors due to omission of action (omission errors) in 455 near-incidents in three sections of Power Plant A

Error description	Operations n = 219		Maintenance n = 177		Radiological Control n = 59		Total N = 455	
	No.	%	No.	%	No.	%	No.	%
Omission of an entire task	0	0	0	0	0	0	0	0
Omission of part of a task	13	5.5	13	7.3	0	0	26	5.7
Omission of checking action	194	88.6	148	83.6	50	84.7	392	86.2
Omission of correspondence or instructions	3	1.4	1	0.6	1	1.7	5	1.1
Other	0	0.0	0	0.0	0	0.0	0	0.0
Not applicable	9	4.1	15	8.5	8	13.6	32	7.0
Total (excluding not applicable)	210	95.9	162	91.5	51	86.4	423	93.0

Table 20.4   Execution (commission) errors in 455 near-incidents in three sections of Power Plant A

Error description	Operations n = 219		Maintenance n = 177		Radiological control n = 59		Total N = 455	
	No.	%	No.	%	No.	%	No.	%
Error in work (too much or too little)	5	2.3	0	0	0	0	5	1.1
Error in settings (too high or too low)	0	0	0	0	0	0	0	0
Error in manual control (too much or too little)	0	0	0	0	0	0	0	0
Operation timing error (too early or too late)	0	0	0	0	0	0	0	0
Incorrect selection of equipment or parts	101	46.1	56	31.6	27	45.8	184	40.4
Wrong position or location	80	36.5	87	49.1	16	27.1	183	40.2
Procedural errors	14	6.4	9	5.1	2	3.4	25	5.5
Errors in correspondence or instructions	6	2.7	6	3.4	6	10.2	18	4.0
Insufficient quality	10	4.6	12	6.8	3	5.1	25	5.5
Other	0	0.0	0	0.0	0	0.0	0	0.0
Not applicable	3	1.4	7	4.0	5	8.5	15	3.3
Total (excluding not applicable)	216	98.6	170	96.0	54	91.6	440	96.7

Table 20.5 Execution (commission) errors due to unrelated actions in 455 near-incidents in three sections of Power Plant A

Error description	Operations $n = 219$		Maintenance $n = 177$		Radiological control $n = 59$		Total $N = 455$	
	No.	%	No.	%	No.	%	No.	%
Execution of unscheduled work	9	4.1	7	4.0	3	5.1	19	4.2
Fall or incorrect contact	0	0	4	2.3	1	1.7	5	1.1
Other	0	0	0	0	0	0	0	0
Total	9	4.1	11	6.3	4	6.8	24	5.3

Table 20.6 Distribution of error occurrence mechanisms (multiple choice) in 455 near-incidents across three sections of Power Plant A

Error description	Operations $n = 219$		Maintenance $n = 177$		Radiological control $n = 59$		Total $N = 455$	
	Cases	%	Cases	%	Cases	%	Cases	%
Information checking								
Overlooking of trigger information	1	0.5	1	0.6	0	0	2	0.4
Loss of intent	0	0	2	1.1	0	0	2	0.4
Subtotal	1	0.5	3	1.7	0	0	4	0.9
Skill-related errors								
Detection level								
Lack of detection of situation because of habituation	58	26.5	48	27.1	12	20.3	118	25.9
Misinterpretation of trigger information	0	0	3	1.7	0	0	3	0.7
Action level								
Vagueness of intent of action	7	3.2	0	0	1	1.7	8	1.8
Substitution with familiar action	10	4.6	4	2.3	7	11.9	21	4.6
Differences between internal maps and the outside world	61	27.9	29	16.4	7	11.9	97	21.3
Careless operation	0	0	0	0	0	0	0	0
Subtotal	136	62.2	84	47.5	27	45.8	247	54.3

Table 20.6 cont'd

Error description	Operations $n = 219$		Maintenance $n = 177$		Radiological control $n = 59$		Total $N = 455$	
	Cases	%	Cases	%	Cases	%	Cases	%
Rule-related errors *Detection and identification level*								
Simplification	36	16.4	25	14.1	14	23.7	75	16.5
Misinterpretation of input information	12	5.5	23	13.0	4	6.8	39	8.6
Judgment and decision-making level								
Simplification of rules	0	0	2	1.1	2	3.4	4	0.9
Incorrect selection	14	6.4	9	5.1	6	10.2	29	6.4
Incorrect recollection of information	2	0.9	1	0.6	0	0	3	0.7
Forgetting isolated items	5	2.3	3	1.7	0	0	8	1.8
Subtotal	69	31.5	63	35.6	26	44.1	158	34.7
Knowledge-related errors *Detection and identification level*								
Overlooking habitual patterns	1	0.5	0	0	0	0	1	0.2
Incorrect selection of input information	4	1.8	6	3.4	2	3.4	12	2.6
Substitution of memory for information that should have been input	30	13.7	22	12.4	5	8.5	57	12.5
Judgment and decision-making level								
Error in reasoning related to judgment	4	1.8	6	3.4	0	0	10	2.2
Error in reasoning related to decision-making	3	1.4	0	0	0	0	3	0.7
Subtotal	42	19.2	34	19.2	7	11.9	83	18.2
Total	248		184		60		492	

20.3.2 Error mechanisms

- Skill-related errors (54.3%).

- Rule-related errors (34.7%).

- Knowledge-related errors (18.2%).

In Table 20.6 all the percentages of skill-, rule-, and knowledge-related errors in the operations, maintenance, and radiological control sections of Power Plant A were consistent with the overall trend. Note the particularly high rate of skill-related near-incidents, 62.2%, in the operations section. In addition, the rates of skill- and rule-related near-incidents are roughly the same in the radiological control section: 44.1 and 45.8%, respectively. The results imply that the longer one's experience, the higher the rate of the skill-related near-accidents.

Examination and comparison of certain facts in Table 20.6 have practical implications. Take, for instance, the skill-related near-incidents caused by failure to detect a situation because of familiarity (25.9%) and differences between internal maps and the outside world (21.3%), the rule-related near-incidents caused by simplification (at 16.5%, the highest percentage for this error mode), and the knowledge-related near-incidents caused by the substitution of memory for information that should have been input (12.5%, also the highest percentage). The main characteristic of all four error modes is that they are factors at the level of information detection. Clearly, measures to prevent human error must be focused on avoiding a decrease in the level of caution when it comes to the detection and identification of information.

20.4 STUDY 3: ANALYSIS OF HUMAN FACTORS USING QUANTIFICATION METHOD III

Of the 746 near-incidents at the two Japanese nuclear power plants we studied, we looked into those that had occurred in the operations, maintenance, and radiological control sections (see Hayashi, 1956; Hayashi, Higuchi and Komazawa, 1970; Misumi, 1985). We analysed them for nine sets of human factors: person–machine interface factors, manual-related factors, management factors, environmental factors, human interaction factors, competence factors, personality factors, work attitude factors, and physical factors. In all, these sets of human factors comprised 90 variables:

1   Person–machine interface factors (12 variables): form, strength, layout, signals and display, size, ease of operation, compatibility, safety, inspections, repairs, allocation, and clothing.

2   Manual-related factors (6 variables): existence of manuals, convenience, difficulty, approachability, validity, and revision.

3   Management factors (17 variables): quality-related policies, targets of work, work organisation, work shifts, commissioning of work, work supervision, work-planning, deployment of work personnel, contact network, error-prevention policies, information, work instructions and methods, tests and examinations, training planning methods, parts management, special construction and measuring, and technology management.

4   Environmental factors of the work site (12 variables): consisting of weather, temperature, humidity, noise, vibrations, odour, dust, lighting, size, height, signs, and atmosphere.

5   Human interaction factors (5 variables): the relationships between superiors and subordinates, between senior and junior personnel, and between co-workers and rivals, degree of social affinity, and familial relations.

6   Individual competence factors (6 variables): lack of basic knowledge, insufficient training, insufficient experience, insufficient knowledge, insufficient English language ability, and inadequate revision.

7   Personality factors (10 variables): laziness, carelessness, hastiness, impatience, incoherence, overconfidence, forgetfulness, timidity, overfocus, and self-centredness.

8   Attitude factors (12 variables): burden of time-related pressures, overmotivation, stress, emotional instability, spite, preoccupation, evaluation, familiarity, monotony, intention, loss of confidence, and panic.

9   Physical factors (10 variables): fatigue, lack of sleep, latent fault, illness, drug use, alcoholism, undernourishment, poor condition, biorhythm, and disability.

Quantification Method III assigns values (called category scores) to the items in each set in order to express the degree of similarity between the variables (categories) (see Table 20.7). The higher the correlation coefficients between the category score and the sample score, the more the computed category scores and sample scores become a one-dimensional scale. Six axes were calculated for the category scores. In Figures 20.1, 20.2, and 20.3 category scores are plotted along two axes in order to facilitate interpretation of what the axes mean. The first two axes, which represent the compatibility of work and inspections, relate to the suitability of technology management and quality control (axis 1), and the validity of transfers and manuals (axis 2). On the third axis, which represents equipment inspections, we have plotted the compatibility of manuals and other sources of information. The fourth axis was named 'strict quality control – inadequate work system'. The fifth axis is our plotting of technology management procedures in terms of the ease of tests and examinations. On the sixth axis, which deals with the expedience of parts-management procedures, we have plotted the scores according to the manner in which materials are allocated and inspections are made.

As shown in Table 20.7, the ten items observed most frequently in connection with the 746 near incidents that occurred at the two Japanese plants we studied were:

Table 20.7 Category scores of 44 variables according to Quantification Method III

Code	Description of variables[a]	Freq.	X1	Y2	X3	Y4	X5	X6	d[b]
21	Existence of manuals	6	3.034	-1.543	-1.418	0.983	-1.683	0.225	17.466
3A	Information	34	1.973	-0.970	-0.953	0.756	-0.248	0.015	6.375
39	Contact network	51	1.876	-0.864	-0.929	0.562	-0.071	0.201	5.490
77	Forgetfulness	1	1.713	-0.488	0.010	-1.391	-0.766	1.540	8.066
64	Insufficient knowledge	10	1.696	-0.545	0.629	-0.312	-1.036	0.542	5.033
33	Work organisation	10	1.570	-0.705	-0.739	-0.352	-0.407	0.616	4.177
44	Noise	3	1.46	-0.930	-1.233	0.736	-0.153	0.202	5.129
34	Work shifts	2	1.434	-0.325	-0.184	-1.287	-0.627	1.174	5.624
37	Work-planning	41	1.399	-0.438	-0.029	-1.088	-0.533	1.201	5.060
49	Size	3	-1.380	-0.011	-0.488	-0.070	-0.594	-0.395	2.656
3G	Technology management	1	-1.325	-0.009	-0.458	-0.131	-0.565	-0.387	2.452
16	Ease of operation	13	1.311	-0.481	1.057	-1.039	-0.913	1.016	6.013
4A	Signs	33	-1.274	-0.003	-0.398	0.051	-0.452	-0.274	2.063
35	Commissioning of work	40	1.088	-0.222	0.003	-0.376	-0.284	0.323	1.560
13	Layout	456	-0.854	0.012	-0.191	-0.038	-0.207	-0.128	0.827
48	Lighting	4	-0.771	-0.056	-0.551	-0.418	0.688	-0.194	1.587
14	Signals and display	219	-0.461	-0.027	-0.268	-0.202	0.388	-0.066	0.481
81	Burden of time-related pressure	18	0.409	-0.215	-0.264	-0.269	-0.059	0.073	0.364
17	Compatibility	1	5.058	24.159	-6.616	0.734	-0.521	0.392	653.976
89	Monotony	3	2.997	13.035	-3.060	0.257	-0.225	0.139	118.393
88	Familiarity	16	0.719	0.779	0.153	-0.588	-0.046	0.243	1.554
19	Inspections	2	1.977	1.395	15.523	-1.414	7.352	-4.205	320.551
12	Strength	2	1.247	0.474	3.517	-0.246	0.453	-0.042	14.416
75	Incoherence	1	0.753	0.694	2.403	-0.511	0.648	-0.576	7.836
3B	Work instructions and methods	21	1.059	-0.022	2.278	-0.257	0.207	0.186	6.432
72	Carelessness	18	0.579	0.420	2.016	-0.583	0.591	-0.371	5.403
61	Lack of basic knowledge	3	0.458	0.829	1.702	-0.132	0.978	-1.359	6.615

Table 20.7 cont'd

Code	Description of variables[a]	Freq.	X1	Y2	X3	Y4	X5	Y6	d[b]
62	Insufficient training	10	0.844	0.147	1.488	-1.023	-0.105	0.600	4.366
42	Temperature	1	1.041	-0.757	-1.070	0.510	0.122	-0.068	3.081
36	Work supervision	102	0.486	0.447	0.716	-0.008	0.110	-0.216	1.007
38	Deployment of work personnel	9	0.571	-0.388	-0.641	-0.342	0.120	0.237	1.075
71	Laziness	19	-0.148	0.259	0.527	-0.223	0.064	-0.323	0.523
25	Validity	2	4.168	-2.863	-5.966	9.594	0.388	-0.807	154.009
3E	Parts management	7	-0.790	0.992	4.139	9.504	-2.697	3.884	131.328
31	Quality-related policies	17	-1.162	0.242	1.114	2.882	-1.215	1.032	13.497
63	Insufficient experience	2	0.329	-0.111	0.451	-1.106	-0.206	0.820	2.262
18	Safety	4	0.068	0.006	0.110	-0.712	-0.274	0.458	0.809
86	Preoccupation	65	0.407	0.044	0.330	-0.422	-0.048	0.054	0.460
11	Form	10	-0.050	-0.432	-1.949	2.238	9.575	2.237	105.681
3C	Tests and examinations	6	0.652	0.114	0.048	0.760	2.317	0.282	6.440
40	Height	2	-0.709	-0.072	-0.638	-0.532	1.015	-0.194	2.266
1A	Allocation	3	2.624	-0.497	0.318	4.612	-0.061	-17.360	329.877
26	Revision	2	2.198	-0.784	-0.004	-2.957	-1.317	3.750	29.987
4B	Atmosphere	2	0.437	-0.396	-0.231	-1.544	-0.941	1.682	6.500
eigenvalue (maximising correlation)			0.819	0.770	0.713	0.687	0.684	0.669	

[a] Variables are placed based on the level of their absolute magnitude of category-score in the axis.
[b] The square root of this value expresses the distance from the intersection of the six axes.

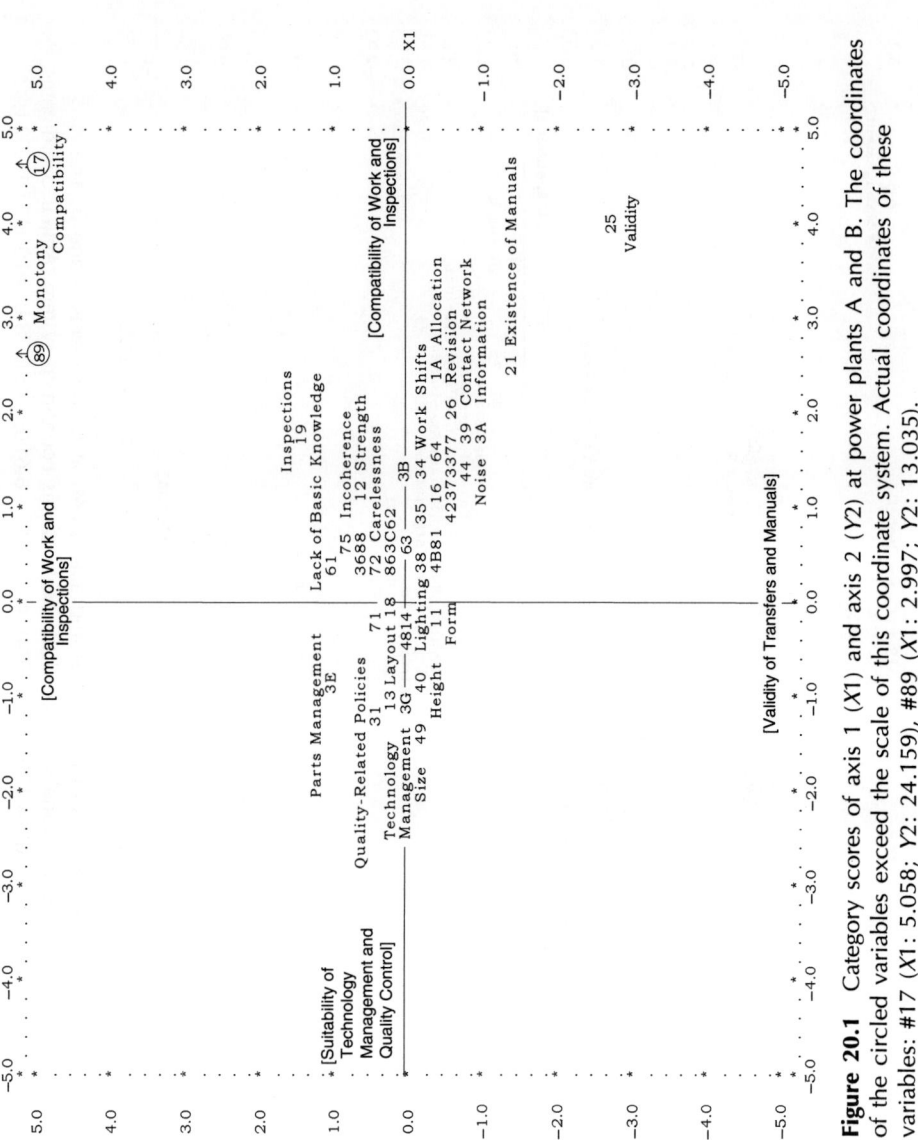

Figure 20.1 Category scores of axis 1 (X1) and axis 2 (Y2) at power plants A and B. The coordinates of the circled variables exceed the scale of this coordinate system. Actual coordinates of these variables: #17 (X1: 5.058; Y2: 24.159), #89 (X1: 2.997; Y2: 13.035).

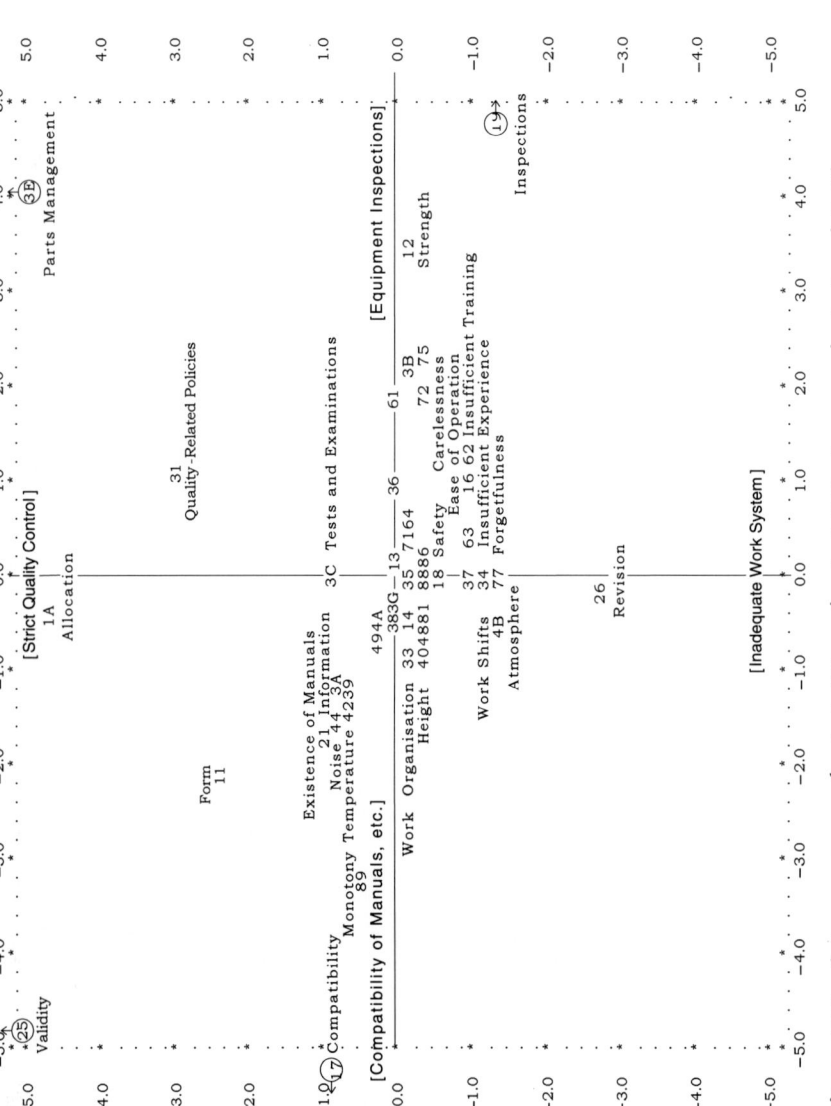

Figure 20.2 Category scores of axis 3 (X3) and axis 4 (Y4) at power plants A and B. The coordinates of the circled variables exceed the scale of this coordinate system. Actual coordinates of these variables: #3E (X3: 4.139; Y4: 9.504); #17 (X3: −6.616; Y4: 0.734), #19 (X3: 15.523; Y4: −1.414), #25(X3: −5.966; Y4: 9.594).

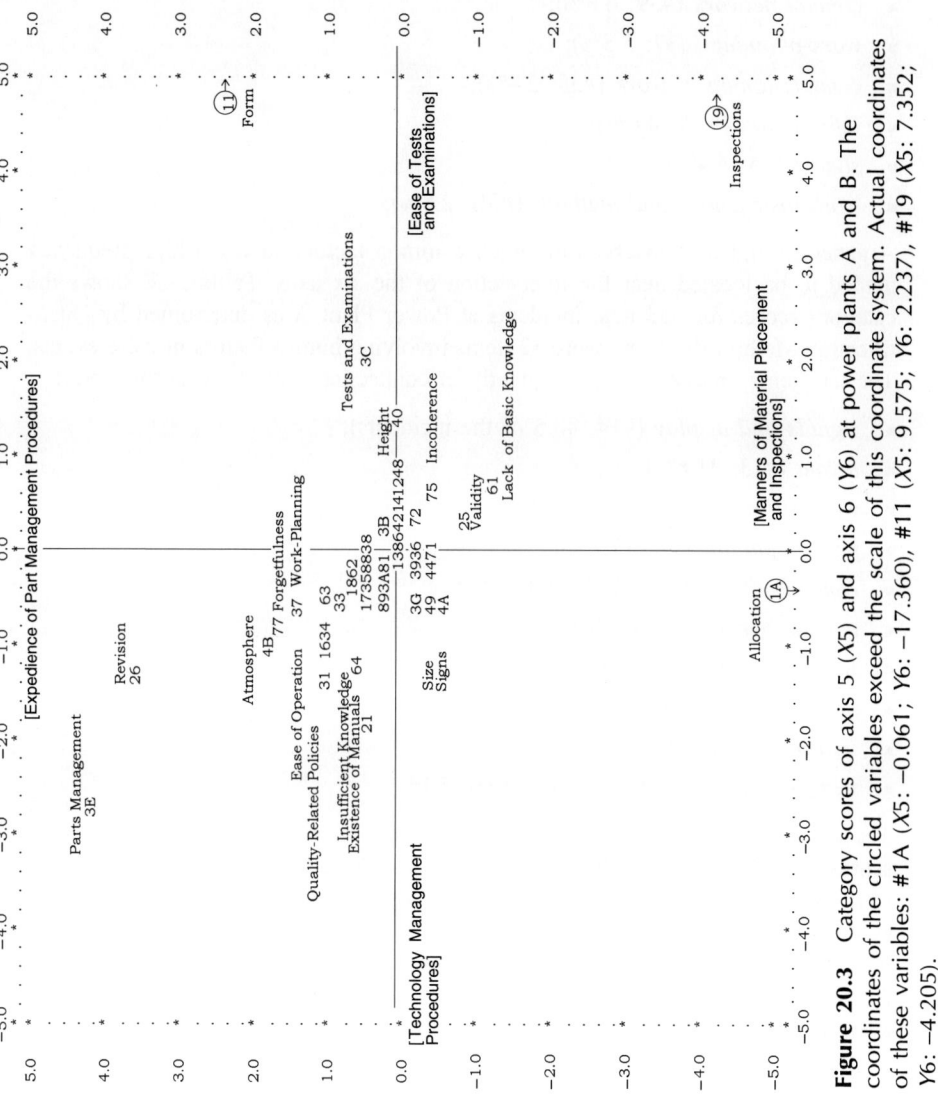

Figure 20.3 Category scores of axis 5 (X5) and axis 6 (Y6) at power plants A and B. The coordinates of the circled variables exceed the scale of this coordinate system. Actual coordinates of these variables: #1A (X5: −0.061; Y6: −17.360), #11 (X5: 9.575; Y6: 2.237), #19 (X5: 7.352; Y6: −4.205).

- *layout* (#13: 61.1% of the cases);
- *signals and display* (#14: 29.4%);
- *work supervision* (#36: 13.7%);
- *preoccupation* (#86: 8.7%);
- *contact network* (#39: 6.8%);
- *work-planning* (#37: 5.5%);
- *commissioning of work* (#35: 5.4%);
- *information* (#3A: 4.6%);
- *signs* (#4A: 4.4%);
- *work instructions and methods* (#3B: 2.8%).

A general trend was observed in which common factors having a high frequency tended to be located near the intersection of the six axes. Table 20.8 shows the category scores for 333 near-incidents at Power Plant A as determined by Quantification Method III. There were 32 items involving human factors in these events. The ten items observed most frequently in connection with these events were:

- *signals and display* (#14: 40.5 of the incidents);
- *layout* (#13: 31.8%);
- *work supervision* (#36: 27.6%);
- *preoccupation* (#86: 17.1%);
- *commissioning of work* (#35: 12.0%);
- *contact network* (#39: 10.5%);
- *work-planning* (#37: 8.4%);
- *laziness* (#71: 5.7%);
- *information* (#3A: 5.7%);
- *work instructions and methods* (#3B: 5.1%).

These variables, too, were plotted along six axes of human factors. The category scores in Tables 20.8 and 20.9 can be plotted as done for Table 20.7, but for lack of space they will not be presented. The first was labelled work difficulty – quality control and basic knowledge; the second, axis inspection system – intricate work; the third, parts-management system – inspection system; the fourth, strict quality control; the fifth, inadequate material allocation – inadequate manuals and inspection system; and the sixth, inadequate work system – inadequate material allocation.

   Table 20.9 shows the category scores for 413 near-incidents at Power Plant B as determined by Quantification Method III. The eight variables observed most frequently in connection with these events were:

- *layout* (#13: 84.7% of the incidents);
- *signals and display* (#14: 20.3%);
- *signs* (#4A: 7.7%);

Table 20.8 Category scores of 32 variables at Power Plant A according to Quantification Method III

Code	Description of variables[a]	Freq.	X1	Y2	X3	Y4	X5	Y6	d[b]
11	Form	6	9.973	-2.925	0.696	0.424	-0.006	0.219	108.729
3C	Tests and examinations	1	0.978	4.163	-0.651	-0.863	-2.980	-2.534	34.757
3A	Information	19.	0.607	2.400	-0.367	-0.290	-1.018	-0.806	8.033
39	Contact network	35	0.689	1.673	-0.263	-0.504	0.277	-0.002	3.674
48	Lighting	1	-0.385	-1.509	-0.476	-0.800	-0.970	-1.478	6.417
14	Signals and display	135	-0.250	-0.871	-0.222	-0.284	-0.314	-0.461	1.262
18	Safety	2	-0.370	-0.811	0.026	0.180	-0.004	-0.364	0.960
13	Layout	106	-0.321	-0.753	-0.190	-0.211	-0.231	-0.314	0.903
71	Laziness	19	-0.362	-0.702	-0.004	-0.010	-0.231	-0.601	1.039
35	Commissioning of work	40	0.035	0.648	-0.156	-0.336	0.012	0.538	0.848
3E	Parts management	4	-0.621	0.719	12.486	-1.354	-1.208	-0.377	160.238
31	Quality-related policies	1	-0.230	0.612	0.554	11.332	1.085	2.247	135.375
12	Strength	2	-0.288	0.199	0.600	5.871	1.098	1.406	38.131
3B	Work instructions and methods	17	-0.167	0.340	0.179	4.198	0.279	0.740	18.424
16	Ease of operation	7	0.071	0.954	-0.184	1.913	-1.484	-1.477	8.992
62	Insufficient training	9	-0.252	-0.011	0.079	1.860	0.310	0.740	4.173
1A	Allocation	3.	0.531	2.155	0.344	-0.261	11.769	-8.099	209.216
21	Existence of manuals	6	0.772	3.496	-0.536	-0.035	-3.542	-3.453	37.575
61	Lack of basic knowledge	3	-0.418	-0.730	0.467	0.243	1.867	0.976	5.423

Table 20.8 cont'd

Code	Description of variables[a]	Freq.	X1	Y2	X3	Y4	X5	Y6	d[b]
75	Incoherence	1	−0.369	−0.343	0.406	0.301	0.870	0.589	1.613
89	Monotony	2	0.305	−0.175	0.370	0.084	0.819	0.764	1.522
72	Carelessness	15	−0.324	−0.347	0.185	0.164	0.501	0.397	0.695
36	Work supervision	92	−0.236	−0.134	0.325	0.135	0.363	0.158	0.354
38	Deployment of work personnel	3	0.335	1.753	−0.172	−1.360	1.190	3.105	16.122
33	Work organisation	8	0.427	1.988	−0.330	−1.349	0.748	2.598	13.372
77	Forgetfulness	1	0.104	1.174	−0.322	−0.960	0.519	2.492	8.894
34	Work shifts	2	0.011	0.847	−0.278	−0.809	0.476	2.133	6.226
64	Insufficient knowledge	10	0.145	1.474	−0.152	1.590	−2.058	−2.112	13.441
37	Work-planning	28	0.094	0.953	−0.264	−0.685	0.522	1.918	5.407
88	Familiarity	16	−0.189	−0.079	−0.068	−0.239	0.316	0.686	0.674
81	Burden of time-related pressures	12	−0.063	0.138	−0.153	−0.550	0.317	0.624	0.839
86	Preoccupation	57	−0.166	−0.041	−0.084	−0.128	0.200	0.409	0.260
	eigenvalue (maximising correlation)		0.836	0.790	0.756	0.686	0.674	0.661	

[a] Variables are placed based on the level of their absolute magnitude of category score in the axis.
[b] The square root of this value expresses the distance from the intersection of the six axes.

Table 20.9 Category scores of 33 variables at Power Plant B according to Quantification Method III

Code	Description of variables[a]	Freq.	X1	Y2	X3	Y4	X5	Y6	d[b]
3B	Work instructions and methods	4	3.248	0.105	-0.397	-2.407	-1.881	-0.775	20.651
37	Work-planning	13	2.985	0.158	-0.351	-0.011	2.898	1.796	20.683
3E	Parts management	3	-0.794	-0.421	-0.359	-0.440	0.145	0.699	1.640
31	Quality-related policies	15	-0.660	-0.347	-0.360	-0.811	0.094	0.325	0.897
49	Size	3	-0.602	-0.354	-0.362	-0.243	0.046	0.028	0.681
3G	Technology management	1	-0.596	-0.332	-0.362	-0.248	0.064	0.028	0.663
4A	Signs	32	-0.543	-0.323	-0.361	-0.174	-0.002	0.030	0.561
13	Layout	350	-0.444	-0.197	-0.360	-0.133	0.034	0.015	0.385
36	Work supervision	10	-0.689	6.232	-0.339	-0.096	0.064	-0.176	39.472
3C	Tests and examinations	5	-0.634	6.210	-0.338	-0.017	-0.009	-0.171	39.110
11	Form	4	-0.302	1.143	-0.351	0.185	-0.258	0.002	1.622
17	Compatibility	1	-0.247	-0.074	16.316	-0.332	-0.020	-0.039	266.384
89	Monotony	1	-0.247	-0.074	16.316	-0.332	-0.020	-0.039	266.384
48	Lighting	3	-0.251	-0.211	-0.355	0.199	-0.298	0.045	0.364
14	Signals and display	84	-0.136	-0.125	-0.355	0.190	-0.215	0.041	0.244
86	Preoccupation	8	0.106	-0.202	-0.353	0.244	0.216	0.263	0.352
19	Inspections	2	5.622	0.429	-0.507	-9.651	-9.895	3.114	214.008
25	Validity	2	3.356	0.241	-0.262	6.664	-4.347	-3.195	84.903
72	Carelessness	3	3.470	0.212	-0.416	-3.934	-2.465	2.035	37.953

Table 20.9 cont'd

Code	Description of variables[a]	Freq.	X1	Y2	X3	Y4	X5	Y6	d[b]
39	Contact network	16	2.036	0.017	-0.321	2.420	-0.841	-0.333	10.923
44	Noise	3	1.553	-0.042	-0.324	2.285	-1.203	-0.549	9.488
3A	Information	15	1.964	0.002	-0.326	2.153	-1.340	-0.631	10.793
42	Temperature	1	1.150	-0.067	-0.333	1.741	-1.239	-0.532	6.287
33	Work organisation	2	0.466	-1.161	-0.341	1.101	-0.496	-0.156	1.842
38	Development of work personnel	2	0.145	-0.193	-0.347	0.764	-0.686	-0.192	1.270
81	Burden of time-related pressures	6	0.185	-0.207	-0.347	0.755	-0.515	-0.200	1.073
26	Revision	2	4.084	0.365	-0.342	0.095	5.888	4.454	71.445
62	Insufficient training	1	2.919	0.220	-0.343	0.129	4.799	5.399	47.694
63	Insufficient experience	2	1.672	0.061	-0.348	0.102	3.039	2.682	19.094
4B	Atmosphere	2	1.744	0.016	-0.352	-0.076	2.976	2.241	17.050
18	Safety	2	1.744	0.016	-0.352	-0.076	2.976	2.241	17.050
40	Height	2	-0.175	-0.190	-0.354	0.328	-0.425	0.044	0.482
16	Ease of operation	6	2.567	-0.171	-0.417	-3.168	3.446	-8.268	97.047
eigenvalue (maximising correlation)			0.867	0.810	0.798	0.778	0.750	0.709	

[a] Variables are placed based on the level of their absolute magnitude of category score in the axis.
[b] The square root of this value expresses the distance from the intersection of the six axes.

- *contact network* (#39: 3.9%);
- *quality-related policies* (#31: 3.6%);
- *information* (#3A: 3.6%);
- *work-planning* (#37: 3.1%);
- *work supervision* (#36: 2.4%).

The first axis of human factors that served for plotting the data in Table 20.9 was labelled work inspection – technology management; the second, supervisory work; the third, compatibility with situation – technology management; the fourth, validity of communication and information – careless inspection; the fifth, training suitability – suitability of instructions; and the sixth, the suitability of work procedures – validity of operation.

20.5 METHODS OF PRESENTING EFFECTIVE MEASURES FOR PREVENTING HUMAN ERRORS

Using key words and the properties associated with the 746 near-incidents analysed by means of Quantification Method III, we are now structuring those cases and determining their various types, so that we can propose methods by which to arrive at measures that will eschew human factors in errors. The next step is to discuss the typed cases with power plant personnel and experts in the field of human factors. The factors thought to be present in these cases and the phenomena related to them will then be determined by a fault-tree diagram of the 746 near incidents, which has yet to be developed. A fault-tree diagram illustrating the phenomenon of 'mistaking the unit,' for example, might include the following human factors:

- Lack of specific overall planning and policies for improving equipment.
- Lack of communication among relevant personnel about aspects of equipment that are difficult to understand.
- Lack of training and guidance regarding aspects of equipment that are difficult to understand or operate.
- Absence of safety incentives.
- Misconception of the relationship between the volume of work and the deployment of personnel.
- Low level of safety awareness among personnel.
- Shortage of highly qualified workers.
- Lack of effort to improve skills of the personnel.
- Lack of a specific management on ways to transfer and organise information.
- Difficulty with use of procedures and checklists.

Measures likely to be effective against the identified factors will be proposed mainly through group interviews with local personnel. At this stage, the feasibility of suggested countermeasures will not be given particular consideration. Instead, emphasis will be placed on thoroughly organising the steps needed if the listed human factors are to be eliminated. The human factors and countermeasures will be examined in the light of experience with similar measures at power plants and formal case studies of efforts in other kind of organisations and other industries. We then plan to summarise specific measures to be implemented at power plant A.

20.6 DISCUSSION

Analysis of natural language describing 560 near-incidents in three sections at Power Plant A revealed that words and phrases expressing human error were used in 332 of the cases. These words and phrases were broken down into 92 semantic groups, with care being exercised to avoid subjective classification. The relationship that these words and phrases have to actual patterns of error behaviour has yet to be clarified. It will also be necessary for key words and phrases within each of the 92 semantic groups to be examined for similarity so that they can be clustered.

Analysis of the error modes (Tables 20.3, 20.4 and 20.5) and the mechanisms of error occurrence (Table 20.6) revealed that omission of a checking action accounted for a large percentage of errors (86.2% of the 455 near-incidents on which Tables 20.3, 20.4, 20.5 and 20.6 are based) across the three sections of Power Plant A. Reports about near-incidents were prompted by this finding, but the reports eventually revealed that this statistic resulted from such errors as 'carelessly turned off the switch of a transformer protective circuit during operation'. In other words, when the administrative office requested each member of the staff to report one near-incident, the respondents first thought in terms of careless errors, a mindset in which many persons reported similar cases. The same explanation accounts for the 40.4% of the commission errors attributed to 'incorrect selection of equipment or parts' and the 40.2% attributed to 'wrong position or location'. These careless errors seem to explain why skill-related mistakes accounted for more than half the 455 reported errors in the three sections of Plant A, with 'failure to detect a situation because of familiarity' (25.9%) and 'differences between internal maps and the outside world' (21.3%) being particularly high. Thus, making sure that work is performed thoroughly at the detection level, and that checks are reliably made at the action level, would most likely be effective in preventing errors by skilled personnel. In order to prevent mistakes, then, it is necessary to perform and check work with all the care of a person new to a given task.

Among the 90 human-factor variables used in our analysis of the 746 near-incidents at two Japanese nuclear plants in this study, 3 of the 10 variables found to be most frequently involved in these events were:

- *layout* (#13: 61.1%);

- *signals and display* (#14: 29.4%);

- *signs* (#4A: 4.4%).

This finding shows the importance of (a) the arrangement of objects and (b) the ease with which signs and displays at the work place are understood. In other words, erroneous behaviour occurs when physical objects are 'difficult to detect'. The contribution that unreliable intergroup communication made to the near-incidents is indicated by five other human-factor variables on the top-ten list:

- *work supervision* (# 36: 13.7%);

- *contact network* (#39: 6.8%);

- *commissioning of work* (#35: 5.4%);

- *information* (#3A: 4.6%);

- *work instructions and methods* (#3B: 2.8%).

In terms of the nine sets of human factors used for the analysis with Quantification Method III, all these variables are part of management factors. Errors stemming from the latter five communication-related, human-factor variables can most likely be prevented if one improves the person–machine interface at the work place, so that work can be performed more easily, and also the reliability of on-the-job management, supervision, and work instructions. Because the axes derived from analysis with Quantification Method III were grouped by work difficulty (axis 1), intricate work (axis 2), parts-management system (axis 3), inadequate work system (axis 4), and inadequate material allocation – inadequate manuals (axis 5), solving problems relating to the person – machine interface, the management and supervision system (work organisation, contact network, information) and manuals (existence, validity, revision) is important in preventing errors. Naturally, work-planning and the existence of manuals are also important in preventing errors in the planning stage before work commences.

In the studies presented in this chapter, the 90 human-factors variables and error modes and error mechanisms we listed were analysed independently, so that the relationship between them could not be clearly determined. To shed light on this aspect of safety, further analysis using Quantification Method III is needed.

References

HAYASHI, C. (1956) Theory and examples of quantification (II). *Proceedings of the Institute of Statistics and Mathematics*, **4**, pp. 19–30.

HAYASHI, C., HIGUCHI, I. and KOMAZAWA, T. (1970) *Information processing and statistical mathematics*. Tokyo: Sangyo-Tosho. [In Japanese]

LABOUR SAFETY INSTITUTE (1994) *Report of survey with Analysis and evaluation of 'Hatto-hiyari cases*. Tokyo: Author.

MISUMI, J. (1985) Hayashi's Quantification III. In MISUMI, J. (ed.), *The behavioural science of leadership* (pp. 335–9). Ann Arbor, MI: University of Michigan Press.

QUALITY CONTROL CENTRE OF NUCLEAR POWER OF THE KANSAI ELECTRIC POWER COMPANY (1996) *Analysis of occurrence of human factors by fault-tree diagram*. Mihamo-Cho: Author.

Human factors in nuclear power plant maintenance – an empirical study[1]

KATSUJI ISOBE, SHINYA SHIBUYA, AND NOBUYUKI TABATA

Human Factors Department, Nuclear Power R&D Center,
Tokyo Electric Power Co., Inc.

In human factors research, more attention has been devoted to the operation of nuclear power plants (NPP) than to their maintenance. However, more NPP incidents are caused by inadequate maintenance rather than by faulty operation. Therefore, the human factors study of NPP maintenance was begun.

In this chapter the problem areas that researchers have identified in the maintenance of NPPs are presented. Such R&D topics as support systems for maintenance work, the transfer of maintenance skills from experienced to less experienced maintenance personnel, and information management systems are described. The chapter concludes with a review of discussion about future R&D and management factors.

In the literature on human factors research on NPPs, the operation of these installations has received more attention than their maintenance. However, inadequate maintenance accounts for a greater share of the incidents in NPPs than does faulty operation (see Figure 21.1). It is thus likely that measures to improve maintenance could reduce trouble in NPPs. Yet several problems seem to have thwarted the emergence of research results on which to base such action. First, the components and pieces of equipment to be maintained are dissimilar, each having its own specifications and difficulties. Secondly, trouble caused by inadequate maintenance tends to occur long after the maintenance has been carried out,

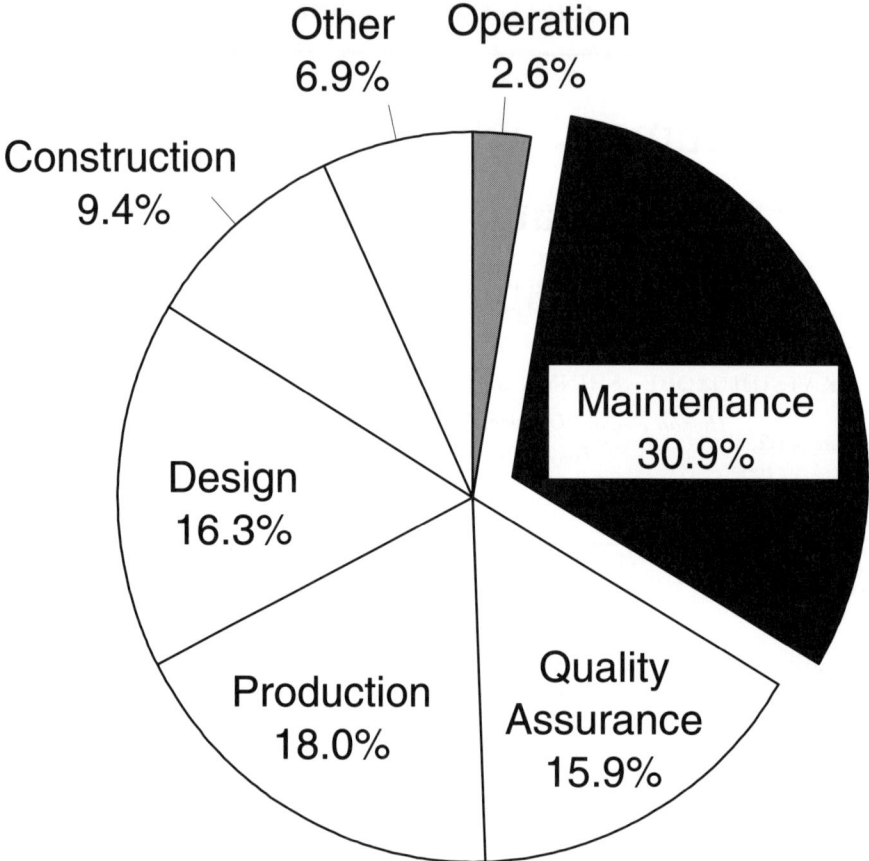

Figure 21.1   Cause of incidents in nuclear power plants in Japan, 1969–94. (Adapted from *Nuclear power plant operation annual report* [p. 261] by the Thermal and Nuclear Power Engineering Society, Japan [ed.], 1995, Tokyo: Thermal and Nuclear Power Engineering Society)

complicating thorough investigation into the root cause(s) of incidents, especially in terms of human factors. Thirdly, many companies, including electric power utilities, plant vendors, subcontractors, and component vendors, are involved in outage maintenance.

In short, identifying root causes of trouble in NPPs and finding a suitable approach to investigating them has been considered a daunting task. It seemed best to have an initial study identify problem areas involving the factors named above. Remedial action and more detailed studies would then follow as needed. The objective of the research reported in this chapter was, therefore, to formulate empirically grounded measures for reducing the number of NPP incidents caused by inadequate maintenance.

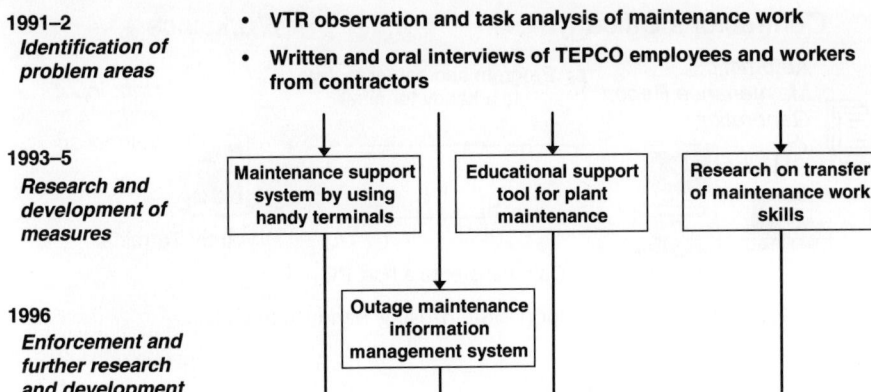

| 1991–2
Identification of problem areas | • **VTR observation and task analysis of maintenance work**
• **Written and oral interviews of TEPCO employees and workers from contractors** |

Figure 21.2 Project overview of the maintenance studies in the Tokyo Electric Power Company (TEPCO) Human Factors Department.

21.1 PROJECT OVERVIEW

The project is illustrated in Figure 21.2. To identify problems, video tape recordings were made of tasks being performed by employees of the Tokyo Electric Power Company (TEPCO) during field observations in NPPs. Maintenance experts and human factors researchers then studied the recordings by means of error mode and effect analysis (EMEA). Parallel to this observation and analysis, oral and written interviews were conducted with TEPCO employees and workers from contractors, in order to gather information about workplace conditions as well as organisational and social factors affecting NPP maintenance.

After the central problems had been identified, developments of two different kinds of support systems and research on the transfer of maintenance work skills began in 1993. In one of the systems, handy terminals were used to aid maintenance personnel (Shibuya, Isobe, Tanaka, Yoshimura and Kanai, 1995). The other system was an educational support tool for plant maintenance (Nagata *et al.*, 1995). Both systems were constructed and tried out in outage maintenance work in 1995. Operational strategies for these systems are being considered. In addition, development began on an information management system for outage maintenance.

The video tape recordings were made mainly of the maintenance of the clean-up water pump, the control rod drive (CRD), and the main steam isolation valve (MSIV) in the NPP outage workplace. Further guidance for improvement was sought in short interviews designed to elicit the opinions of the personnel. Task analysis of the video tape recordings suggested that maintenance work was affected by several factors, such as pre-work instructions to less-experienced maintenance personnel, communication among different work groups, and training programmes. The written and oral interviews also revealed that measures to improve communication, education, and skill transfer would improve maintenance work.

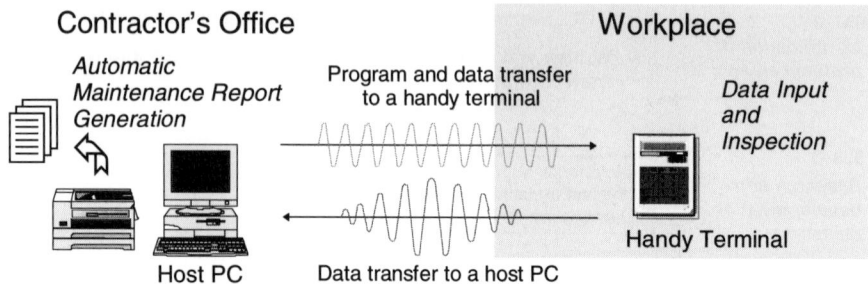

Figure 21.3   Maintenance support system using handy terminals.

21.2   RESEARCH AND DEVELOPMENT OF MEASURES

21.2.1   Maintenance support system using handy terminals

A support system for CRD maintenance was developed (Shibuya *et al.*, 1995). Objectives of the system are (a) to reduce workload during outage maintenance by supporting data entry and automatic generation of maintenance reports, and (b) to reduce human error in data recorded by hand. This objective was achieved by means of an electronic checklist with an inspection data entry interface (for numeric, text, and graphics) on handy terminals and by a host personal computer (PC) with an automatic report-generation function. The system is also equipped with a bar-coded tool registration feature. As Figure 21.3 shows, the system consists of a host PC, which stores maintenance records, and a handy terminal that is carried into the actual workplace. An employee (inspector class) is supposed to download the data on that day from the host PC to the handy terminal and carry it into the workplace. At the workplace inspection data (numeric, text, graphics) are entered into it and checked simultaneously with regulated values and histories.

Researchers tested the system at an actual maintenance site and confirmed that the workload of the maintenance foreman was reduced, especially the preparation of maintenance reports. Subjective evaluation by personnel has suggested that inspection data entry and the checking function had worked effectively, because employees did not have to look up regulation data for each measure by themselves. This result was expected to help prevent human error during inspections.

21.2.2   Educational support tool for plant maintenance work

One of the findings from the identification of problem areas suggested that a foreman of a maintenance group had a relatively high workload. He had to order or instruct beginners frequently during their work, and sometimes it was necessary to perform tasks himself instead of the beginners. Effective instruction to beginners at a prework session, called a toolbox meeting, seemed to reduce the foreman's workload. To build on these observations, an educational tool for improving plant maintenance work was developed (Nagata *et al.*, 1995).

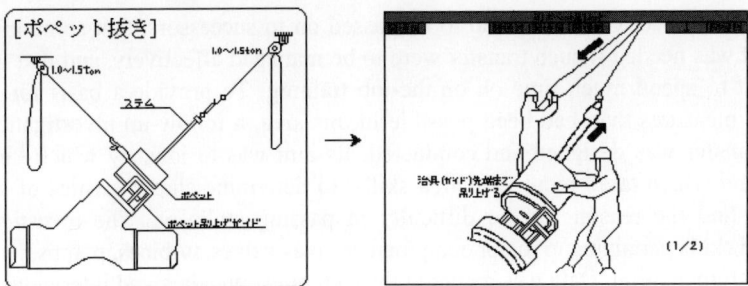

Figure 21.4 Visualisation of procedural instructions: Conventional document on the left, illustrated picture on the right.

The aim of this system was to give employees an overview of the day's work. They were to know, for example, about each task's place in the overall maintenance procedure, the work items that were involved, specific work procedures required by each item, unusual notices, and hold points (the points that need inspection by TEPCO employees during maintenance procedures). Usually, such communication takes place at a toolbox meeting, where personnel exchange information on the day's work. Instructions were generally given in a variety of ways, including system specification documents and procedural manuals. However, it was not easy enough, especially for beginners, to grasp the required information in a short meeting. The developed system therefore featured an illustrated instruction interface, which used pictures to help personnel grasp the assigned work or visually memorise the material (see Figure 21.4). Photographs and videos had advantages when it came to conveying real-life images of actual workplaces, whereas illustrations were expected to emphasise important points of maintenance work. Having the pictures include illustrations of personnel executing a task seemed to enhance the ability of maintenance to contextualise and memorise their roles and movements.

This educational tool was tried out and evaluated at an actual maintenance site. Subsequent subjective evaluation suggested that tasks and procedures illustrated by the system were more easily understood than when they were presented as conventional instructions. The visualised information served as a retrieval key or mnemonic device for recalling important related information. The system allowed the instructor to explain clearly without having to omit essential information. It was recognised that notes attached to each illustrated procedure would help the employees themselves contribute to error prediction.

21.2.3 Research on transfer of maintenance skills from veterans to younger personnel

The questionnaires and interviews in previous research suggested a need to improve the transfer of maintenance skills from veterans to younger personnel. It was mentioned that there were difficulties with hiring and retaining skilled workers, that maintenance work skills had to be passed on to successors, that some kind of support was needed if such transfer were to be managed effectively, and that it was difficult to spend much time on on-the-job training. To provide a basis for more precise measures than had been possible in this area, a follow-up investigation on skill transfer was designed and conducted. Its aim was to identify which kind of work and which tasks required which skills, to determine characteristics of skills, and to find the reason for the difficulty in passing skills on. The investigation covered skills pertaining to major equipment such as valves, turbines, pumps, motors, and instrumentation. Data was gathered through questionnaires and interviews with experts and younger employees who had experience as maintenance personnel.

The results indicated that skill characteristics and reasons for problems with transferring skills varied from one kind of equipment to the next (see Figure 21.5). Certain findings were common to all equipment, however.

1   Inherently difficult skills were reported as being hard to learn.

2   Skills that required repetition of work experience were difficult to pass on, even if they were not difficult to carry out.

3   Skills for dealing with equipment encountered in only one kind of context, such as an NPP, were evaluated as difficult to pass on.

4   Skills requiring special knowledge that is not recorded in written form were also difficult to pass on.

It was also found that NPP maintenance personnel used the word *skill* to mean both knowledge and skill, between which they generally distinguished. Accordingly, an effective method of knowledge acquisition will also promote the transfer of skills.

As Figure 21.5 shows, there seemed to be various factors surrounding skill transfer. It appears that factors of difficulty fall into clusters, or metafactors. One such metafactor is the difficulties with hiring and retaining skilled workers; a second, the inherent difficulty of a skill or its performance; and a third, the tendency of young workers to move out of limited domains of activity. The metafactor of 'difficulties in hiring and retaining skilled workers' consists of three subsidiary problems – the declining number of new employees, the difficulty in hiring maintenance workers and the loss of maintenance workers to other kinds of jobs. The metafactor of 'learning difficulties' stems from such subsidiary problems as the inherent difficulty of a skill or its performance, impasses to learning skills and knowledge not presented in written form, intergenerational change of attitudes and expectations about acquiring skills, and the impossibility of learning the requisite skills through the manufacturing process, because new orders for equipment have declined. Learning difficulties are exacerbated, too, by the high speed of technology change, the ensuing misgivings about acquiring skills that may quickly become useless, and the tendency to avoid hiring or retaining professionals who can operate only in a limited domain. All these factors complicate the transfer of maintenance skills in NPPs.

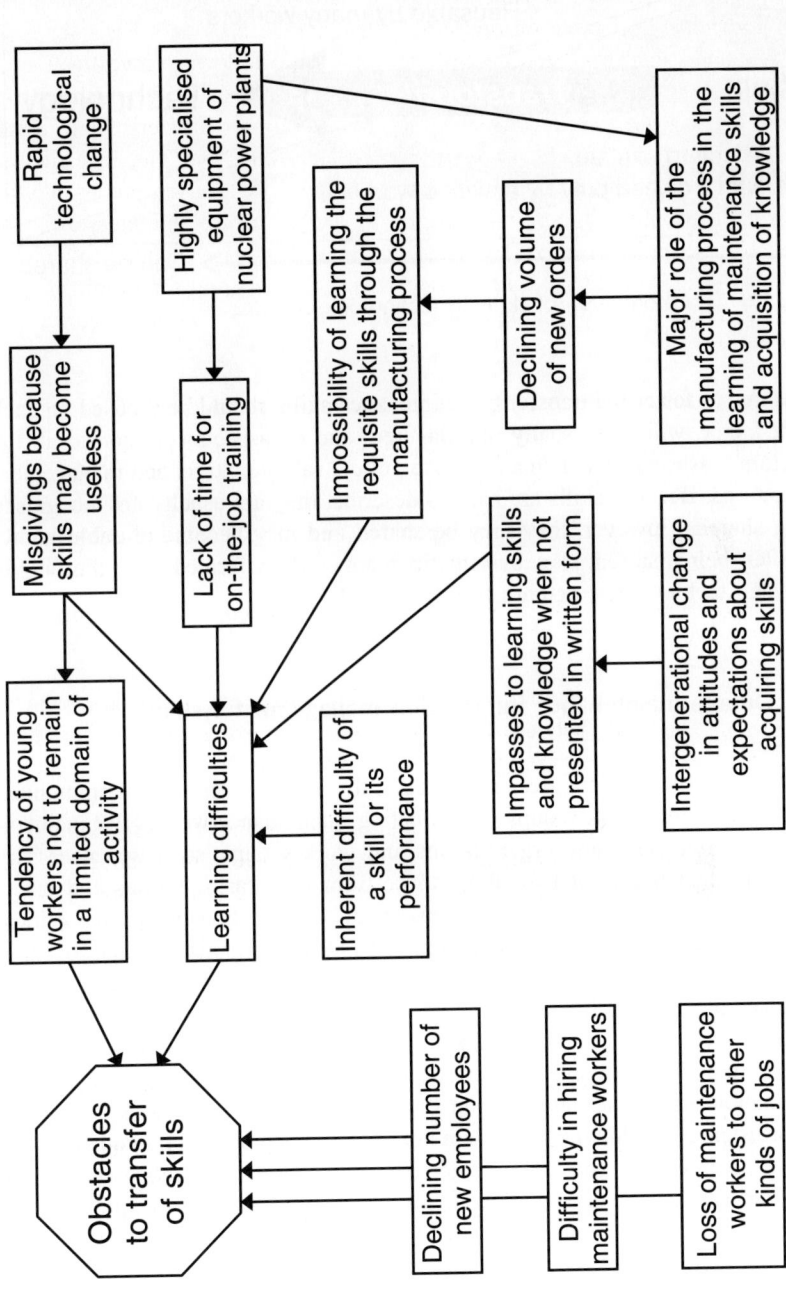

Figure 21.5 Structure surrounding the transfer of skills of NPP maintenance work.

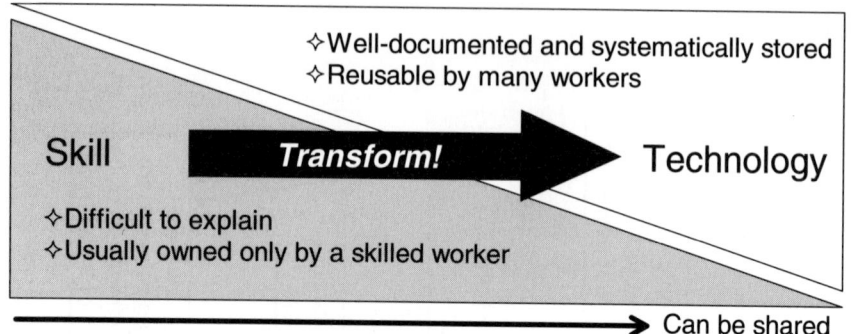

Figure 21.6   Transformation of skills to technologies.

Measures to foster the transfer of maintenance skills should be focused on each different factor, which is usually multilayered, and on aspects common to all the metafactors – aspects such as maintenance policy, safety culture, and management considerations. Because skills are hard to describe, they are usually not recognised as being shared. However, they may be shared and may become reusable knowledge after being stored in a systematic manner, transformed into thoroughly documented technologies, or both (see Figure 21.6).

21.2.4   Outage maintenance information management system

As indicated by the problem areas identified at the outset of the study, communication among different work groups was affecting maintenance work. NPP maintenance usually involves many organisations and work groups, so it was necessary to negotiate what is required for them to perform their tasks. Lapses and errors in communication, such as missing or inaccurate information, may have critical consequences, because an NPP is complex in its human organisation and mechanical structures, all of which must function correctly if safe operation and reliable maintenance are to be achieved.

Presumably, the difficulties described above stem from the fact that the status of the system and of maintenance work changes every day, requiring vast amounts of information and numerous different organisations. The resulting relationships and inconsistencies within the system are not easy to grasp or detect. In other words, the factors pointed out above may interfere with negotiation among the related work groups and cause information to go astray. To reduce these probabilities and effectively aid maintenance personnel as they perform their duties, a system for managing information pertaining to such work is under development (see Figure 21.7). Its conceptual design calls for all requisite information to be collected in all the organisations engaged in outage maintenance.

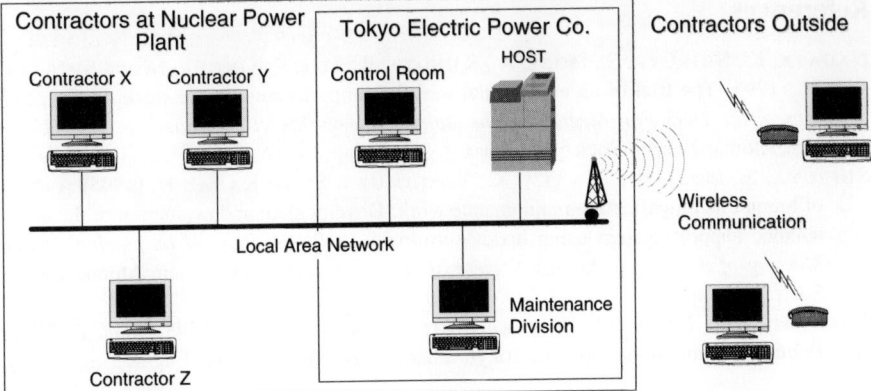

Figure 21.7 Overview of the outage maintenance information management system.

21.3 FUTURE RESEARCH AND DEVELOPMENT

Because of the variety of equipment in NPPs, measures concentrated on discrete pieces of equipment will probably not be able to cover all the factors that complicate maintenance work in such installations. The focus of future research and development should therefore soon be shifted to the management domain. In that context, the outage maintenance information management system will be the key to improving NPP outage maintenance. Measures focused on such management factors as maintenance policy and safety culture will be crucial as well.

For measures pertaining to individual pieces of equipment, there is a need for additional research and development based on the concept of human factors or, more broadly, on an anthropocentric perspective. Many different utilities, plant vendors, equipment vendors, maintenance agents and other organisations should contribute their respective approaches to each factor and each piece of equipment. The variety of equipment and organisations related to NPP maintenance will also require cooperative approaches in order to achieve a total solution to the issues addressed in this chapter.

Note

1   We express our appreciation to the people who took part in our research, especially those who cooperated in the questionnaires and interviews. We also thank those who used and examined our trial systems even though they may have taken more time than conventional work practices. The videotaped observations, the task analysis of error mode and effect, and the development of the support systems described in this chapter were carried out as cooperative studies with the Tokyo Electric Power Company, the Toshiba Corporation, and Hitachi Ltd.

References

NAGATA, K., SHIBUYA, S., ISOBE, K., KUBOTA, R., ITO, Y., OKUDA, N. and SHIBATA, M. (1995) The trial of an educational tool for support maintenance work. In *Proceedings of the 1995 Fall Meeting of the Atomic Energy Society of Japan* (p. 83). Tokyo: The Atomic Energy Society of Japan.

SHIBUYA, S., ISOBE, K., TANAKA, K., YOSHIMURA, S. and KANAI, H. (1995) A study of human factors affecting maintenance work: Development and evaluation of the maintenance support system using handy terminals. In *Proceedings of the Spring Annual Meeting of the Atomic Energy Society of Japan* (p. 283). Tokyo: The Atomic Energy Society of Japan.

THERMAL AND NUCLEAR POWER ENGINEERING SOCIETY (1995) *Nuclear Power Plant Operation Annual Report*. Tokyo: Thermal and Nuclear Power Engineering Society.

CHAPTER TWENTY-TWO

A review of human error prevention activities at Kansai Electric's nuclear power stations

TERUO TOKUINE

The Kansai Electric Power Co., Inc., Osaka, Japan

Although the Kansai Electric Power Company has taken various measures to improve hardware and software in order to prevent human errors at nuclear power plants, trouble caused by human error sometimes occurs. Even a minor incident at a nuclear power plant is usually reported as an 'occurrence of an accident' in the mass media. Furthermore, the general public's view of human error has recently become more severe than in the past. An accident caused by human error therefore decreases the public's confidence in the reliability of nuclear power plants. In this context, it is essential to prevent human errors.

Recently, a power decrease caused by an error in discriminating between units occurred at Kansai's nuclear power plant. Within a few months after the incident, a reactor shutdown was caused by an error while an operator was discriminating between units at another plant. These occurrences have spurred Kansai to review current measures for guarding against human errors and to investigate additional ones. This chapter summarises the current status of Kansai's preventive measures.

22.1 TACKLING THE HUMAN-FACTORS ASPECT OF SAFETY IN NUCLEAR POWER PLANTS

In 1986, when the discussion of how human factors affect safety in nuclear power plants hit the headlines after the accidents on Three Mile Island and at Chernobyl, three incidents caused by human error occurred within three months in Kansai's

341

Table 22.1 Issues identified and measures implemented after three incidents in Kansai's nuclear power stations

Issues	Measures
1 Consciousness-raising	Vitalisation of small-group activities, such as toolbox meetings and danger anticipation training Enhancement of human factors education Distribution of a guideline manual for human factors Standardisation of the activities to prevent narrow escape from danger (Utilisation of reported cases: reflection of the danger anticipation training sheet) Implementation of a safety education: key points of toolbox meetings
2 Weaknesses detected through case analysis	Changes in operational responses (e.g. adoption of radio paging) Improvement of work-related knowledge (education on how to use the service water)
3 Analysis	Development of accident factor analysis techniques, compilation of education tools, and promotion of education
4 Leadership	Implementation of leadership education for maintenance personnel (a reflection of the result of the motivation improvement survey)
5 Manuals	Compilation of models of manuals for maintenance and operation
6 Work Environment	Investigation into the actual status of and demands for the working environments using specialists outside Kansai
7 Human behaviour and human reliability	Study of human reliability through use of operation simulators (joint study by electric power companies that are operating pressurised water reactor plants) Evaluation of and research on human factors at nuclear power plants (joint study by all electric power companies) Development of human-action prediction system Development of evaluation methods for human reliability Development of man–machine simulator

nuclear power stations. Immediately, company managers started systematic efforts to reduce human errors. First, past problems were analysed in detail for the issues to be dealt with. Next, these items were classified into two categories: issues closely related to the site, and issues related to human nature. The district office and nuclear power stations were in charge of site issues, and the head office was in charge of questions pertaining to the substantive issues of human nature. We examined these issues and implemented the measures. Table 22.1 indicates major extracted items and the measures implemented.

22.2 OCCURRENCE OF INCIDENTS CAUSED BY HUMAN ERROR IN KANSAI'S NUCLEAR POWER PLANTS

Since 1970, when the Kansai Electric Power Company started the commercial operation of its first nuclear power plant, the number of incidents caused by human errors decreased gradually because of the personnel's accumulated experience and efforts to reduce human errors. They declined to a low annual level of approximately 0.1 incidents per unit several years ago. This figure, however, has been increasing over the last couple of years. Furthermore, because an incident due to unit confusion occurred at the Ohi and Takahama power stations in 1995 and 1996, the management of Kansai had a sense of crisis, so the new activities described in the following pages were initiated.

22.2.1 Outline of recent incidents at the Ohi and Takahama power stations

Generator output decrease caused by degraded condenser vacuum at Ohi Unit No. 1

Ohi units No. 1 and No. 2 have almost the same design, and the operators of auxiliary equipment take care of both units.

On 27 October 1995, Unit No. 1 was operating at a rated power, and Unit No. 2 was shut down for periodic inspection. The condenser vacuum in Unit No. 1 suddenly decreased from 712 mmHg to 670mmHg, and, in addition, the generator output fell from 1175 MW to 1100 MW. Investigation showed that an operator had opened the condenser drain valve of Unit No. 1 instead of the one in Unit No. 2, because he had confused the two units.

Automatic Reactor shutdown caused by actuation of transformer protective relay at Takahama Unit No. 2

Takahama Units No. 1 and No. 2 have also almost the same design.

On 15 March 1996, Unit No. 1 was shut down for annual inspection, and Unit No. 2 was operating at the rated power. Maintenance personnel then mistakenly short-circuited the protective relay circuit in Unit No. 2 instead of the circuit in Unit No. 1, causing the transformer protective relay in Unit No. 2 to be actuated. As a result, Unit No. 2 was automatically shut down.

22.2.2 Investigation and analysis of the incident at Ohi Unit No. 1

Immediately after the incident at Ohi Unit No. 1, we managers conducted detailed interviews of those involved. Using the fault-tree analysis technique, we then analysed the factors that caused the errors.

Incident description

1   Operator B gave instructions to Operator A to prepare a regular test operation of the auxiliary feed water pump of Unit No. 1 and to open the condenser drain valve of Unit No. 2.

2   On the way to the place where the condenser drain valve of Unit No. 2 is located, Operator A decided to prepare in advance for the test operation of the auxiliary feed water pump of Unit No. 1.

3   After finishing preparation for the test operation of the Unit No. 1 auxiliary feed water pump, Operator A intended to go to the place where the condenser drain valve of Unit No. 2 is located. Instead, he erroneously went to the place where the corresponding valve of Unit No. 1 is located.

4   Operator A checked the valve number described on the nameplate of the valve. At the time, however, he did not check the unit number, which precedes the valve number. The valve number in Unit No. 1 is identical to that in Unit No. 2.

5   Operator A intended to open the drain valve of Unit No. 2, but he opened that of Unit No. 1 by mistake.

6   The vacuum level of the condenser in Unit No. 1 decreased, and the generator output decreased.

Analytical result: factors that caused error in discriminating between units

1   The operating environments of Units 1 and 2 were similar.

2   Operator B gave instructions to Operator A for the operation of both units.

3   Much work had to be done at one time.

4   The valve numbers of the two units were identical.

5   Operator A did not pay any special attention when moving to the place where the intended valve was located.

22.2.3 Investigation and analysis of the incident at Takahama Unit No. 2

Incident description

1   Maintenance Personnel A and B informed the shift operator in the main control room that they would inspect the transformer protective relay of Unit No. 1.

2   Maintenance Personnel A and B went to the relay room, where B completed work for the x-system relay circuits of Unit No. 1 (trip circuit and current transformer circuit) under the direction of Maintenance Personnel A.

3   Maintenance Personnel A went to the panel containing the trip circuit in the Y-system relay circuits of Unit No. 1, and successfully completed work for the trip circuit..

4   Maintenance Personnel A intended to go to the panel containing the current transformer circuit in the Y-system relay circuits of Unit No. 1, but he went to that of Unit No. 2 by mistake.

5   Maintenance Personnel A intended to short-circuit the current transformer circuit of Unit No. 1, but he short-circuited that of Unit No. 2.

6   The protective relay of Unit No. 2 was actuated, and the reactor was automatically shut down.

Factors that caused error in discriminating between units

1   The panels of Unit No. 1 and Unit No. 2 are adjacent to each other.

2   Panels of the two units are identical in shape and colour.

3   The terminal nameplates of the two units are identical in shape, colour, and name.

4   The maintenance personnel must move from the front to the back of the same panel many times.

5   The procedure requires frequent repetition of movements.

6   The unit number is located high on the panel, where it is difficult to see.

22.3   HUMAN FACTORS IMPROVEMENT PROMOTION COMMITTEE

Following the incident at the Ohi nuclear power station, the Kansai management took added measures to prevent operators from confusing the units. For example, the number of each unit is now posted in the passage, and the number of locked valves to which only the supervisor has a key has been increased. Nevertheless, units were again confused at the Takahama nuclear power station within less than five months. It was pointed out that the measures taken after the Ohi incident had been implemented in only a limited area, and that that was the reason why the same kind of incident recurred within such a short period. We managers had a feeling of crisis that the awareness of the need to avoid repeating human errors might not have spread sufficiently among all the employees in the nuclear power stations and that the same problems might occur in the future. To break through this difficulty, we determined that the people involved in nuclear power generation in the head office and the district office should also discuss the creation of activities designed to prevent human errors through mutual discussion in each nuclear power station. Thus, we established the Human Factors Improvement Promotion Committee with the general manager of the Wakasa District Office as the chairman. The main discussion points of this committee have been as follows:

1   What should be done to prevent units being confused?

2   Are the human-error prevention measures that have been implemented so far appropriate?

3   Are there any other human error factors that we managers are not aware of?

4   Are solutions to the problems in other sections being applied?

Specific activities of the committee have been as follows:

1   Investigation of human error prevention measures taken in other sections, such as thermal power stations and substations in Kansai.

2   Investigation of human error prevention measures taken in other utilities' nuclear power stations. ,

3   Confirmation of existing human error prevention measures.

4   Understanding of potential human error factors by all employees in each power station.

5   Discussion of countermeasures based on investigation results.

22.3.1 Additional preventive measures

As a result of the investigation performed by the Human Factors Improvement Promotion Committee, several measures to prevent human errors have been identified and implemented:

1   Increase the use of colour coding to make it easier to distinguish between units. Assign each unit a separate colour and change all work notices and work slips to the colour of the respective units.
Paint doors, walls, and columns in the unit colour.
Colour the panel nameplate in the unit colour everywhere except in the main control room.
(Unit colour: No. 1, white; No. 2, yellow; No. 3, blue; No. 4, pink)

2   Install large unit number signs in conspicuous places on the control panels, machines, paths, and other clearly visible locations.

3   Install physical partitions between adjacent panels of different units.

4   Install plastic covers on the control panel test terminals.

5   Utilise the bar codes for identification of the valves concerned. Put a bar code on the valve number sign and check the valve number with the portable terminal before opening or closing it. Valve opening or closing procedures are stored in the portable terminal in advance, so an error in selecting the valve concerned will activate an alarm.

6   Document any narrow escape from dangerous situations. To heighten awareness, prepare a pamphlet on narrow escapes from dangerous situations experienced by plant personnel.

7 Distribute notification of past incidents.
   Panels describing major incidents in Kansai are exhibited in the Nuclear
   Maintenance Training Center, so that Kansai employees will remember important experiences. However, some personnel who joined Kansai well after those
   incidents are not aware of them, so copies of these panels have been distributed
   to the employees in the power stations.

8 Have lecturers from other fields, such as railroads and aviation, speak about
   human factors affecting safety.

9 Hold discussions between Kansai's executive directors in charge of nuclear
   power generation and plant managers to enhance awareness of human factors
   affecting safety.

In conclusion, the Kansai management has implemented various measures to
enhance safety at Kansai nuclear power stations by reducing human error. None of
them are perfect, however, so continuous effort must be made to eradicate factors
that can cause errors.

Index

349